AF719840

MANUAL ON THE USE OF THERMOCOUPLES IN TEMPERATURE MEASUREMENT

Sponsored by ASTM
Committee E-20 on
Temperature Measurement
and Subcommittee E20.04 on
Thermocouples
AMERICAN SOCIETY FOR
TESTING AND MATERIALS

ASTM SPECIAL TECHNICAL PUBLICATION 470B

ASTM Publication Code Number (PCN)
04-470020-40

1916 Race Street, Philadelphia, Pa. 19103

Copyright © by AMERICAN SOCIETY FOR TESTING AND MATERIALS 1981
Library of Congress Catalog Card Number: 80-69066

NOTE

The Society is not responsible, as a body, for the statements and opinions advanced in this publication.

Printed in Baltimore, Md.
July 1981

Second Printing, Baltimore, Md. (b)
July 1982

Third Printing, Baltimore, Md. (b)
February 1983

Foreword

The Manual on the Use of Thermocouples in Temperature Measurement was sponsored and compiled by Committee E-20 on Temperature Measurement and Subcommittee E20.04 on Thermocouples of the American Society for Testing and Materials. The editorial work was co-ordinated by R. P. Benedict, Westinghouse Electric Corp. Helen Hoersch was the ASTM editor.

Related ASTM Publications

Evolution of the International Practical Temperature Scale of 1968, STP 565 (1974), $7.00, 04-565000-40

Contents

Acknowledgments

Editors for this Edition of the Manual

R. P. Benedict (chairman), Westinghouse Electric Corp.
E. L. Lewis (secretary), Consultant
J. A. Bard, Johnson Matthey, Inc.
P. Bliss, Pratt and Whitney Aircraft
G. W. Burns, National Bureau of Standards
G. J. Champagne, The Foxboro Co.
R. S. Flemons, Canadian General Electric Co., Ltd.
H. L. Kurtz, Driver-Harris Co.
R. M. Park, Marlin Mfg. Corp.
L. J. Pickering, Claud S. Gordon Co.
F. S. Sibley, Hoskins Manufacturing Co.

Officers of Committee E-20

E. D. Zysk (chairman), Engelhard Minerals and Chemical Corp.
R. P. Benedict (1st vice chairman), Westinghouse Electric Corp.
N. R. Corallo (2nd vice chairman), Becton Dickinson
R. L. Shepard (recording secretary), Oak Ridge National Lab.
A. E. Gealt (membership secretary), Honeywell, Inc.

Officers of Subcommittee E-20.04

G. J. Champagne (chairman), The Foxboro Corp.
F. S. Sibley (secretary), Hoskins Mfg. Co.

Those Primarily Responsible for Individual Sections of this Edition

Principles—R. P. Benedict, Westinghouse Electric Corp.
Common Thermocouples—G. J. Champagne, The Foxboro Co.
Extension Wires—F. S. Sibley, Hoskins Manufacturing Co.
Nonstandard Thermocouples—J. A. Bard, Johnson Matthey, Inc.
Typical Thermocouples Designs—L. J. Pickering, Claud S. Gordon Co.
Sheathed Thermocouples—P. Bliss, Pratt and Whitney Aircraft
EMF Measurements—A. S. Tenney, Leeds and Northrup Co.
Reference Junctions—R. S. Flemons, Canadian General Electric Co., Ltd.
Calibration—G. W. Burns, National Bureau of Standards
Single Element Calibration—H. L. Kurtz, Driver-Harris Co.

Installations—R. P. Benedict, Westinghouse Electric Corp.
Cryogenics—G. W. Burns, National Bureau of Standards
Uncertainty—P. Bliss, Pratt and Whitney Aircraft
Terminology—E. L. Lewis, Consultant
Index—R. M. Park, Marlin Mfg. Corp.

In addition to those listed, many other members of Committee E-20 have made substantial contributions to this manual as authors and reviewers. Their help is gratefully acknowledged.

List of Figures

List of Tables

Chapter 1—Introduction

First Edition, 1970

This manual was prepared by Subcommittee IV of ASTM Committee E-20 on Temperature Measurement. The responsibilities of ASTM Committee E-20 include "Assembling a consolidated source book covering all aspects relating to accuracy, application, and usefulness of thermometric methods." This manual was addressed to the thermocouple portion of this responsibility.

The contents include principles, circuits, standard electromotive force (emf) tables, stability and compatibility data, installation techniques, and other information required to aid both the beginner and the experienced user of thermocouples. While the manual is intended to be comprehensive, the material, however, will not be adequate to solve all the individual problems associated with many applications. To further aid the user in such instances, there are numerous references and an extensive bibliography. In addition to presenting technical information, an attempt is made to properly *orient* a potential user of thermocouples. Thus, it is hoped that the reader of this manual will make fewer mistakes than the nonreader.

Regardless of how many facts are presented herein and regardless of the percentage retained, all will be for naught unless one simple important fact is kept firmly in mind. The thermocouple reports only what it "feels." This may or may not be the temperature of interest. The thermocouple is influenced by its entire environment, and it will tend to attain thermal equilibrium with this environment, not merely part of it. Thus, the environment of each thermocouple installation should be considered unique until proven otherwise. Unless this is done, the designer will likely overlook some unusual, unexpected, influence.

Of all the available temperature transducers, why use a thermocouple in a particular application? There are numerous advantages to consider. Physically, the thermocouple is inherently simple, being only two wires joined together at the measuring end. The thermocouple can be made large or small depending on the life expectancy, drift, and response-time requirements. It may be flexible, rugged, and generally is easy to handle and install. A thermocouple normally covers a wide range of temperatures, and its output is reasonably linear over portions of that range. Unlike many temperature transducers, the thermocouple is not subject to selfheating problems. In practice, thermocouples of the same type are interchangeable within specified limits of error. Also, thermocouple materials are readily available at reasonable cost, the expense in most cases being nominal.

The bulk of the manual is devoted to identifying material characteristics and discussing application techniques. Every section of the manual is essential to an understanding of thermocouple applications. Each section should be studied carefully. Information should not be used out of context. The general philosophy should be—let the user beware.

Second Edition, 1974

In preparing this edition of the manual, the committee endeavored to include four major changes which greatly affect temperature measurement by means of thermocouples. In 1968, at the same time the First Edition was being prepared, the International Practical Temperature Scale was changed. This new scale (IPTS-68) is now the law of the land, and Chapter 8 has been completely rewritten to so reflect this. In 1972–1973, new Thermocouple Reference Tables were issued by the National Bureau of Standards. Accordingly, Chapter 10 has been revised to include the latest tables of temperature versus electromotive force for the thermocouple types most commonly used in industry. Also, along these same lines, the National Bureau of Standards has issued new methods for generating the new Reference Table values for computer applications. These power series relationships, giving emf as a function of a temperature, are now included in Chapter 10.3. Finally, there have been several important changes in thermocouple material compositions, and such changes have been noted in the appropriate places throughout the text. The committee has further attempted to correct any gross errors in the First Edition and has provided a more complete bibliography in Chapter 12.

Third Edition, 1980

This edition of the manual has been prepared by ASTM E-20.10, the publications subcommittee. The main impetus for this edition was the need for a reprinting. Taking advantage of this opportunity, the editors have carefully reviewed each chapter as to additions and corrections called for by developments in the field of temperature measurement by thermocouples since 1974. Chapters 3, 4, 5, 6, 7, and 8 have been completely revised and strengthened by the appropriate experts. An important addition is Chapter 12 on Measurement Uncertainty. This reflects the trend toward a more statistical approach to all measurements. A selected bibliography is still included at the end of each chapter. A final innovation of this edition is the index to help the users of this manual.

Chapter 2—Principles of Thermoelectric Thermometry

The principles, or theory, underlying thermoelectric effects were not established by one man at one time, but by several scientists working over a span of many years beginning with Alessandro Volta, who concluded in 1800 that the electricity which caused Galvani's frog to twitch was due to a contact of two dissimilar metals. This conclusion was the forerunner of the principle of the thermocouple. Others built on this base; for example, Thomas Johann Seebeck (1821), Jean Charles Althanase Peltier (1834), and William Thomson—later Lord Kelvin—(1848-1854). During this same period, Jean Baptiste Joseph Fourier published his basic heat-conduction equation (1821), Georg Simon Ohm discovered his celebrated equation for electrical conduction (1826), James Prescott Joule found the principle of the first law of thermodynamics and the important I^2R heating effect (1840-1848), and Rudolf Julius Emanuel Clausius announced the principle of the second law of thermodynamics and introduced the concept of entropy (1850).

2.1 Historical Development of Basic Relations

2.1.1 Seebeck

Seebeck discovered the existence of thermoelectric currents while observing electromagnetic effects associated with bismuth-copper and bismuth-antimony circuits. His experiments showed that, when the junctions of two dissimilar metals forming a closed circuit are exposed to different temperatures, a net thermal electromotive force is generated which induces a continuous electric current.

The Seebeck effect concerns the net conversion of thermal energy into electrical energy with the appearance of an electric current. The Seebeck voltage refers to the net thermal electromotive force set up in a thermocouple under zero-current conditions. The direction and magnitude of the Seebeck voltage, E_S, depend upon the temperature of the junctions and upon the materials making up the thermocouple. For a particular combination of materials, A and B, for a small temperature difference

$$dE_S = \alpha_{A,B}dT \tag{1}$$ [1]

[1]Nomenclature not defined in the text is given at the end of this chapter.

where $\alpha_{A,B}$ is a coefficient of proportionality called the Seebeck coefficient. (This commonly is called the thermoelectric power.) The Seebeck coefficient is obtained usually in one of two ways: (1) as an algebraic sum, $\alpha_{A,B}$, of relative Seebeck coefficients, α_{AR} and α_{BR}, where, for a given temperature difference and at given temperature levels, emf's of each of the substances, A and B, making up the thermocouple are obtained with respect to an arbitrary reference material, R; and (2) by numerically differentiating tabulated values of E_S versus T for a given reference temperature, T_R, according to the relation

$$E_S = \int_{T_R}^{T} \alpha_{A,B} dT \tag{2}$$

In either case, the Seebeck coefficient represents, for a given material combination, the net change in thermal emf caused by a unit temperature difference; that is

$$\alpha_{A,B} = \operatorname*{limit}_{\Delta T \to 0} \frac{\Delta E_S}{\Delta T} = \frac{dE_S}{dT} \tag{3}$$

Thus, if $E = aT + 0.5bT^2$ is determined by calibration, then $\alpha = a + bT$. Note that, based on the validity of the experimental relation

$$E_S = \int_{T_2}^{T} \alpha dT = \int_{T_1}^{T} \alpha dT - \int_{T_1}^{T_2} \alpha dT \tag{4}$$

where $T_1 < T_2 < T$, it follows that α is entirely independent of the reference temperature employed. In other words, for a given combination of materials, the Seebeck coefficient is a function of temperature level only.

2.1.2 Peltier

Peltier discovered peculiar thermal effects when he introduced small, external electric currents in Seebeck's bismuth-antimony thermocouple. His experiments show that, when a small electric current is passed across the junction of two dissimilar metals in one direction, the junction is cooled (that is, it acts as a heat sink) and thus absorbs heat from its surroundings. When the direction of the current is reversed, the junction is heated (that is, it acts as a heat source) and thus releases heat to its surroundings.

The Peltier effect concerns the reversible evolution, or absorption, of heat which usually takes place when an electric current crosses a junction between two dissimilar metals. (In certain combinations of metals, at certain temperatures, there are thermoelectric neutral points where no Peltier effect is apparent.) This Peltier effect takes place whether the current is introduced externally or is induced by the thermocouple itself. The Peltier heat was found early to be proportional to the current, and may be written

$$dQ_P = \pi I dt \tag{5}$$

where π is a coefficient of proportionality known as the Peltier coefficient or the Peltier voltage. Note that π represents the reversible heat which is absorbed, or evolved, at the junction when unit current passes across the junction in unit time, and that it has the dimensions of voltage. The direction and magnitude of the Peltier voltage depend upon the temperature of the junction and upon the materials making up the junction; however, π at one junction is independent of the temperature of the other junction.

External heating, or cooling, of the junctions results in the converse of the Peltier effect. Even in the absence of all other thermoelectric effects, when the temperature of one junction (the reference junction) is held constant and when the temperature of the other junction is increased by external heating, a net electric current will be induced in one direction. If the temperature of the latter junction is reduced below the reference-junction temperature by external cooling, the direction of the electric current will be reversed. Thus, the Peltier effect is seen to be related closely to the Seebeck effect. Peltier himself observed that, for a given electric current, the rate of absorption, or liberation, of heat at a thermoelectric junction depends upon the Seebeck coefficient, α, of the two materials.

2.1.3 Thomson

It remained for Thomson (see the Kelvin relations discussed next) to show that α and π are related by the absolute temperature. (We might appropriately mention at this time that the Peltier thermal effects build up a potential difference opposing the thermoelectric current, thus negating the perpetual-motion question.) Thomson came to the remarkable conclusion that an electric current produces different thermal effects, depending upon the direction of its passage from hot to cold or from cold to hot, in the same metal. By applying the (then) new principles of thermodynamics to the thermocouple, and by disregarding (with tongue in cheek) the irreversible I^2R and conduction-heating processes, Thomson reasoned that, if an electric current produces only the reversible Peltier heating effects, then the net Peltier voltage will equal the Seebeck voltage and will be linearly proportional to the temperature difference at the junctions of the thermocouple.

This reasoning led to requirements at variance with observed characteristics (that is, $dE_S/dT \neq$ constant). Therefore, Thomson concluded that the net Peltier voltage is not the only source of emf in a thermocouple circuit, but that the single conductor itself, whenever it is exposed to a longitudinal temperature gradient, must be also a seat of emf. (Becquerel had at that time already discovered a thermoelectric neutral point, that is, $E_S = 0$, for an iron-copper couple at about 280°C. Thomson agreed with Becquerel's conclusion and started his thermodynamic reasoning from there.)

The Thomson effect concerns the reversible evolution, or absorption, of heat occurring whenever an electric current traverses a single homogeneous conductor, across which a temperature gradient is maintained, regardless of external introduction of the current or its induction by the thermocouple itself. The Thomson heat absorbed, or generated, in a unit volume of a conductor is proportional to the temperature difference and to the current, that is

$$dQ_T = \left[\int \sigma dT\right] I dt \tag{6}$$

where σ is a coefficient of proportionality called the Thomson coefficient. Thomson refers to this as the specific heat of electricity because of an apparent analogy between σ and the usual specific heat, c, of thermodynamics. Note that σ represents the rate at which heat is absorbed, or evolved, per unit temperature difference per unit current, whereas c represents the heat transfer per unit temperature difference per unit mass. The Thomson coefficient is seen also to represent an emf-per-unit difference in temperature. Thus, the total Thomson voltage set up in a single conductor may be expressed as

$$E_T = \int_{T_1}^{T_2} \sigma dT \tag{7}$$

where its direction and magnitude depend upon temperature level, temperature difference, and material considered. Note that the Thomson voltage alone cannot sustain a current in a single homogeneous conductor forming a closed circuit, since equal and opposite emf's will be set up in the two paths from heated to cooled parts.

Soon after his heuristic reasoning, Thomson succeeded in demonstrating indirectly the existence of the predicted Thomson emf's. He sent an external electric current through a closed circuit, formed of a single homogeneous conductor which was subjected to a temperature gradient, and found the I^2R heat to be augmented slightly, or diminished, by the reversible Thomson heat in the paths from cold to hot or from hot to cold, depending upon the direction of the current and the material under test.

2.1.4 Interim Summary

In summary, thermoelectric currents may exist whenever the junctions of a circuit formed of at least two dissimilar metals are exposed to different temperatures. This temperature difference always is accompanied by irreversible Fourier heat conduction, while the passage of electric currents always is accompanied by irreversible Joule heating effects. At the same time, the passage of electric currents always is accompanied by reversible Peltier heating or cooling effects at the junctions of the dissimilar metals, while the combined temperature difference and passage of electric current always is accompanied by reversible Thomson heating or cooling effects along the conductors. The two reversible heating-cooling effects are manifestations of four distinct emf's which make up the net Seebeck emf

$$E_S = \pi_{A,B}|_{T_2} - \pi_{A,B}|_{T_1} + \int_{T_1}^{T_2} \sigma_A dT - \int_{T_1}^{T_2} \sigma_B dT = \int_{T_1}^{T_2} \alpha_{A,B} dT \tag{8}$$

where the three coefficients, α, π, σ, are related by the Kelvin relations.

2.1.5 Kelvin Relations

Assuming that the irreversible I^2R and heat-conduction effects can be disregarded completely (actually, they can be only minimized since, if thermal conductivity is decreased, electrical resistivity usually is increased, and vice versa), then the net rate of absorption of heat required by the thermocouple to maintain equilibrium in the presence of an electric current is

$$q = \frac{Q_{\text{net}}}{\Delta t} = \left[\pi_2 - \pi_1 + \int_1^2 (\sigma_A - \sigma_B) dT\right] I = E_S I \tag{9}$$

This is in accord with the first law of thermodynamics, according to which heat and work are mutually convertible. Thus, the net heat absorbed must equal the electric work accomplished or, in terms of a unit charge of electricity, the Seebeck emf, E_S, which may be expressed in the differential form

$$dE_S = d\pi + (\sigma_A - \sigma_B) dT \tag{10}$$

The second law of thermodynamics may be applied also to the thermocouple cycle, the unit charge of electricity again being considered, as

$$\Delta S_{\text{rev}} = \Sigma \frac{\Delta Q}{T_{\text{abs}}} = 0 \tag{11}$$

where ΔQ implies the various components of the net heat absorbed (that is, the components of E_S), and T_{abs} implies the temperature at which the heat is transferred across the system boundaries. Equation 11 can be expressed in the differential form

$$dS_{\text{rev}} = d\left(\frac{\pi}{T}\right) + \frac{(\sigma_A - \sigma_B)}{T} dT = 0 \tag{12}$$

Combining the differential expressions for the first and second laws of thermodynamics, we obtain the Kelvin relations

$$\pi_{A,B} = T_{\text{abs}}\left(\frac{dE_S}{dT}\right) = T_{\text{abs}}\alpha_{A,B} \tag{13}$$

$$(\sigma_A - \sigma_B) = -T_{\text{abs}}\left(\frac{d^2E_S}{dT^2}\right) \tag{14}$$

from which we can determine α, π, and $\Delta\sigma$, when E_S is obtained as a function of T. Thus, if

$$E_S = aT + \frac{1}{2}bT^2 + \cdots \tag{15}$$

is taken to represent the thermoelectric characteristics of a thermocouple whose reference junction is maintained at 0°C, and where the coefficients, a and b, are obtained (for example) by the curve fitting of calibration data, then

$$\alpha = (a + bT + \cdots) \tag{16}$$

$$\pi = T_{\text{abs}}(a + bT + \cdots) \tag{17}$$

$$\Delta\sigma = -T_{\text{abs}}(b + \cdots) \tag{18}$$

Examples of the use of these coefficients are given in Table 2.1.

2.1.6 Onsager Relations

The historical viewpoint presented thus far has avoided the very real irreversible I^2R and heat conduction in order to arrive at the useful and experimentally confirmed Kelvin relations. We shall now discuss how the present-day, irreversible thermodynamic viewpoint removes this flaw in our reasoning.

Basically, we judge whether a given process is reversible or irreversible by noting the change in entropy accompanying a given change in the thermodynamic state. Thus, if $dS > \delta Q_q/T_{\text{abs}}$, we say the process is irreversible; or, stated in a more useful manner

$$dS_{\text{system}} = dS_{\text{across boundary}} + dS_{\text{produced inside}} \tag{19}$$

or

$$dS_s = dS_0 + dS_i = \frac{\delta Q_q}{T_{\text{abs}}} + \frac{\delta F}{T_{\text{abs}}} \tag{20}$$

TABLE 2.1—*Determination of various thermoelectric quantities applied to thermocouples.*

Given, the two constants, a and b, as determined with respect to platinum via Eq 15:

Metal	a, μV/°C	b, μV/(°C)^2
Iron(Fe)	+16.7	−0.0297
Copper(Cu)	+2.7	+0.0079
Constantan(Cu-Ni)	−34.6	−0.0558

By way of illustration, consider the following combinations of materials: iron/copper and iron/constantan, with their measuring junctions at 200°C and their reference junctions at 0°C:

Iron/copper

$$a_{Fe/Cu} = a_{Fe} - a_{Cu} = 16.7 - 2.7 = 14\ \mu V/°C$$

$$b_{Fe/Cu} = b_{Fe} - b_{Cu} = -0.0297\ -0.0079$$

$$b_{Fe/Cu} = -0.0376\ \mu V/(°C)^2$$

Iron/constantan

$$a_{Fe/Cu\text{-}Ni} = a_{Fe} - a_{Con} = 16.7 - (-34.6) = 51.3\ \mu V/°C$$

$$b_{Fe/Cu\text{-}Ni} = b_{Fe} - b_{Con} = -0.0297 - (-0.0558)$$

$$b_{Fe/Cu\text{-}Ni} = 0.0261\ \mu V/(°C)^2$$

Since Seebeck voltage $E_s = aT + 1/2bT^2$,

Iron/copper

$$E_s = 14(200) + ½(-0.0376)(200)^2$$

$$E_s = 2048\ \mu V$$

Iron/constantan

$$E_s = 51.3(200) + ½(0.0261)(200)^2$$

$$E_s = 10\ 782\ \mu V$$

Note how different combinations of materials give widely different thermal emf's.

Now we proceed to write expressions for α, π, and $\Delta\sigma$, to note how the separate emf's combine to give the (net) Seebeck emf: Since $\alpha_{A,B} = a_{A,B} + b_{A,B}T$ = Seebeck coefficient

Iron/copper

$$\alpha_0 = 14 + (-0.0376)(0) = 14\ \mu V/°C$$

$$\alpha_{200} = 14 + (-0.0376)(200) = 6.48\ \mu V/°C$$

Iron/constantan

$$\alpha_0 = 51.3 + 0.0261(0) = 51.3\ \mu V/°C$$

$$\alpha_{200} = 51.3 + 0.0261(200) = 56.52\ \mu V/°C$$

Note that it is the great difference in Seebeck coefficients (thermoelectric powers) for the two combinations which accounts for the difference in thermal emf's:

$$E_s = \int_{TR}^{T} \alpha_{A,B}\,dT$$

Since $\pi_{A,B} = T_{abs}\alpha_{A,B}$ = Peltier coefficient = Peltier voltage

Iron/copper

$$\pi_0 = 273(14) = 3822\ \mu V$$

$$\pi_{200} = 473(6.48) = 3065\ \mu V$$

Iron/constantan

$$\pi_0 = 273(51.3) = 14\ 005\ \mu V$$

$$\pi_{200} = 473(56.52) = 26\ 734\ \mu V$$

TABLE 2.1—(*Continued*).

Note that, in the case of the iron/copper (Fe-Cu) thermocouple, $\pi_{cold} > \pi_{hot}$, whereas in the more usual Fe/Cu-Ni thermocouple, $\pi_{hot} > \pi_{cold}$.

Since $\Delta\sigma_{A,B} = -b_{A,B}T_{abs}$ = Thomson coefficient, and

$$E_T = \int_{T_{Rabs}}^{T_{abs}} \Delta\sigma \, dT = \tfrac{1}{2} b_{A,B}(T^2_{Rabs} - T^2_{abs}) = \text{Thomson voltage}$$

Iron/copper

$$E_T = -\frac{0.0376}{2}(273^2 - 473^2)$$

$$E_T = 2805 \ \mu V$$

Iron/constantan

$$E_T = \frac{0.0261}{2}(273^2 - 473^2)$$

$$E_T = -1947 \ \mu V$$

We sum the various components

$$E_S = \pi_2 - \pi_1 + \int_1^2 \Delta\sigma \, dT = \text{Seebeck voltage}$$

Iron/copper

$$E_s = 3065 - 3822 + 2805$$

$$E_s = 2048 \ \mu V$$

Iron/constantan

$$E_s = 26\,734 - 14\,005 - 1947$$

$$E_s = 10\,782 \ \mu V$$

These figures of course, check with the original calculations. Note that, in the Fe/Cu case, the net Thomson emf far outweighs in importance the net Peltier emf, whereas in the Fe/Cu-Ni case, the converse is true.

Hence, only in the absence of entropy within the system boundaries do we have the reversible case, $dS_{rev} = \delta Q_q/T_{abs}$, which may be handled adequately by classical thermodynamics in the steady and quasi-steady states. Evidently, the rate of production of entropy per unit volume, ξ, is an important quantity in irreversible thermodynamics, which may be expressed as

$$\xi = \left(\frac{1}{Adx}\right)\frac{dS_i}{dt} = \left(\frac{1}{Adx}\right)\frac{\delta F}{T_{abs}dt} \tag{21}$$

where Adx is the area times the differential length.

Another significant quantity, the product $T_{abs}\xi$ (called the dissipation), always can be split into two terms or a sum of two terms; one associated with a flow, J, and the other associated with a force, X. Furthermore, in many simple cases a linear relation is found (by experiment) to exist between the flow and force terms so defined. For example, in the one-dimensional, isothermal, steady flow of electric charges, $\delta Q_e/dt$ across a potential gradient, $-dE/dx$, it may be shown that

$$T_{abs}\xi = \left(\frac{I}{A}\right)\left(-\frac{dE}{dx}\right) = J_e X_e \tag{22}$$

where J_e and X_e represent the electric flow and force terms, respectively, as defined by the entropy production method. The term J_e represents the electric-current density and the term X_e the electric-field strength or the electromotive force, which of course are related by the linear Ohm's law (that is, $J_e = L_e X_e$, where L_e represents the electrical conductivity). Again, in the one-dimensional, steady flow of thermal charges, dQ_q/dt, across a temperature gradient, $-dT/dx$, it may be shown that

$$T_{abs}\xi = \left(\frac{Q}{A}\right)\left(\frac{1}{T_{abs}}\frac{dT}{dx}\right) = J_q X_q \tag{23}$$

where J_q and X_q represent the thermal flow and force, respectively, as defined by the entropy production method. The term J_q represents the thermal current density, and the term X_q represents the thermomotive force, which are, of course, related by the linear Fourier's law (that is, $J_q = L_q X_q$ where L_q represents the product of the thermal conductivity and the absolute temperature). It has been found that, even in complex situations, it always may be stated that

$$T_{abs}\xi = \Sigma J_K X_K \tag{24}$$

When several irreversible transport processes occur simultaneously (as, for example, the electric and thermal conduction in a thermocouple), they usually will interfere with each other; therefore, the linear relations must be generalized to include the various possible interaction terms. Thus, for the combined electric and thermal effects we would write

$$J_e = L_{ee} X_e + L_{eq} X_q \tag{25}$$

$$J_q = L_{qe} X_e + L_{qq} X_q \tag{26}$$

or, in general

$$J_i = \Sigma L_{ij} X_j \tag{27}$$

We have just seen that an entropy production necessarily accompanies both the I^2R and heat conduction effects (that is, they are irreversible); therefore, the Kelvin relations could not follow from reversible thermodynamic theory without certain intuitive assumptions. By reasoning that the electrical and thermal currents were independent, Thomson tacitly assumed that $L_{eq} = L_{qe}$ as we shall subsequently show. Experimentally, this reciprocal relationship often was found to be true. The American chemist, Lars Onsager, proved in 1931 from a statistical-mechanics viewpoint that the assumption

$$L_{ij} = L_{ji} \tag{28}$$

is always true when the linear relations between flows, J_k, and forces, X_k, are valid. The Onsager reciprocal relation forms the basis of irreversible thermodynamics. By applying these concepts to the processes involved in the thermocouple, we are led rationally and unambiguously to the Kelvin relations. Thus, whenever the junctions of a thermocouple are maintained at different temperatures, we expect that an electric potential difference, an electric current, and a thermal current will be present. The dissipation for this thermoelectric process is simply the sum of the electric and thermal terms previously given. That is

$$T_{\text{abs}}\xi = \frac{I}{A}\left(-\frac{dE}{dT}\right) + \frac{Q}{A}\left(\frac{1}{T_{\text{abs}}}\frac{dT}{dx}\right) \tag{29}$$

The generalized linear laws for this case also have been given as

$$J_e = L_{ee}\left(-\frac{dE}{dT}\right) + \frac{L_{eq}}{T_{\text{abs}}}\left(\frac{dT}{dx}\right) \tag{30}$$

$$J_q = L_{qe}\left(-\frac{dE}{dx}\right) + \frac{L_{qq}}{T_{\text{abs}}}\left(\frac{dT}{dx}\right) \tag{31}$$

Recalling that the Seebeck emf is determined under conditions of zero electric current, the Seebeck coefficient, α, may be expressed in terms of the Onsager coefficients as

$$\alpha = \left(\frac{dE_S}{dT}\right)_{I=0} = \frac{L_{eq}}{L_{ee}T_{\text{abs}}} \tag{32}$$

Recalling that the Peltier coefficient, π, represents the heat absorbed, or evolved, with the passage of an electric current across an isothermal junction, this too may be expressed in terms of the Onsager coefficients as

$$\pi = \left(\frac{J_q}{J_e}\right)_{dT=0} = \frac{L_{qe}}{L_{ee}} \tag{33}$$

Finally, we recall that Thomson found experimentally (and expressed in the Kelvin relations) that the Seebeck and Peltier coefficients are related, as shown in Eq 13.

$$\pi = T_{\text{abs}}\left(\frac{dE_S}{dT}\right) \tag{34}$$

In terms of the Onsager coefficients, this requires that

$$\frac{L_{qe}}{L_{ee}} = T_{\text{abs}}\left(\frac{L_{eq}}{L_{ee}T_{\text{abs}}}\right) \tag{35}$$

$$L_{qe} = L_{eq} \tag{36}$$

which indicates that the experimental results agree with those which are predicted by the entropy production-linear law-Onsager reciprocal relation approach; in other words, by irreversible thermodynamics, without using any intuitive assumption. The Kelvin relations, also in accord with experiment, must follow.

2.2 Laws of Thermoelectric Circuit

Numerous investigations of thermoelectric circuits in which accurate measurements were made of the current, resistance, and electromotive force have resulted in the establishment of several basic laws. These laws have been established experimentally beyond a reasonable doubt and may be accepted in spite of any lack of a theoretical development.

2.2.1 Law of Homogeneous Metals

A thermoelectric current cannot be sustained in a circuit of a single homogeneous material, however varying in cross section, by the application of heat alone.

A consequence of this law is that two different materials are required for any thermocouple circuit. Experiments have been reported suggesting that a nonsymmetrical temperature gradient in a homogenous wire gives rise to a measurable thermoelectric emf. A preponderance of evidence indicates, however, that any emf observed in such a circuit arises from the effects of local inhomogeneities. Furthermore, any current detected in such a circuit when the wire is heated in any way whatever is taken as evidence that the wire is inhomogenous.

2.2.2 *Law of Intermediate Metals*

The algebraic sum of the thermoelectromotive forces in a circuit composed of any number of dissimilar materials is zero if all of the circuit is at a uniform temperature.

A consequence of this law is that a third homogeneous material always can be added in a circuit with no effect on the net emf of the circuit so long as its extremities are at the same temperature. Therefore, it is evident that a device for measuring the thermoelectromotive force may be introduced into a circuit at any point without affecting the resultant emf, provided all of the junctions which are added to the circuit by introducing the device are all at the same temperature. It also follows that any junction whose temperature is uniform and which makes a good electrical contact does not affect the emf of the thermoelectric circuit regardless of the method employed in forming the junction (Fig. 2.1).

Another consequence of this law may be stated as follows. If the thermal emfs of any two metals with respect to a reference metal (such as C) are known, then the emf of the combination of the two metals is the algebraic sum of their emfs against the reference metal (Fig. 2.2).

2.2.3 *Law of Successive or Intermediate Temperatures*

If two dissimilar homogeneous metals produce a thermal emf of E_1, *when the junctions are at temperatures* T_1 *and* T_2, *and a thermal emf of* E_2, *when the junctions are at* T_2 *and* T_3, *the emf generated when the junctions are at* T_1 *and* T_3, *will be* $E_1 + E_2$.

One consequence of this law permits a thermocouple, calibrated for a given reference temperature, to be used with any other reference temperature through the use of a suitable correction (see Fig. 2.3 for a schematic example).

Another consequence of this law is that extension wires, having the same thermoelectric characteristics as those of the thermocouple wires, can be introduced in the thermocouple circuit (say from region T_2 to T_3 in Fig. 2.3) without affecting the net emf of the thermocouple.

2.3 Elementary Thermoelectric Circuits

Two continuous, dissimilar thermocouple wires extending from the measuring junction to the reference junction, when used together with cop-

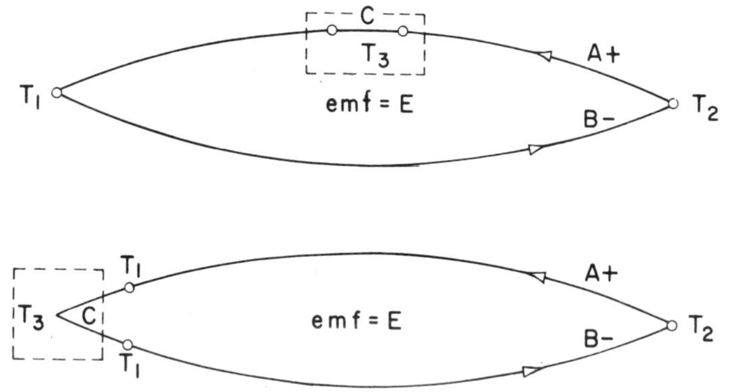

FIG. 2.1—E *unaffected by third material.* C.

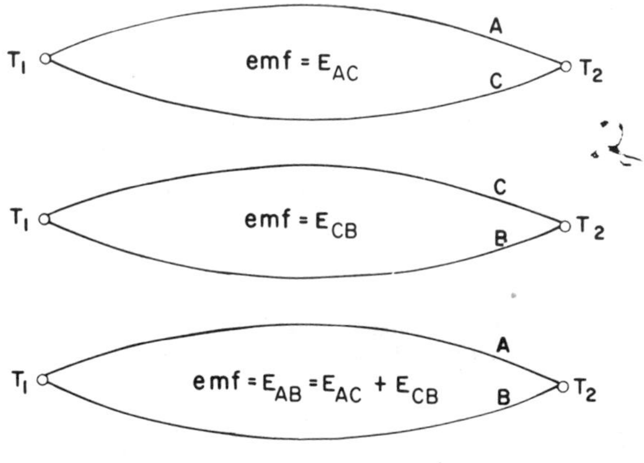

FIG. 2.2—*Emfs are additive for materials.*

per connecting wires and a potentiometer, connected as shown in Fig. 2.4, make up the basic thermocouple circuit.

An ideal circuit is given in Fig. 2.5 for use when more than one thermocouple is involved. The usual thermocouple circuit, however, includes: measuring junctions, thermocouple extension wires, reference junctions, copper connecting wires, a selector switch, and potentiometer, as indicated in Fig. 2.6. Many different circuit arrangements of the above components are also acceptable, depending on given circumstances, and these are discussed in the appropriate sections which follow.

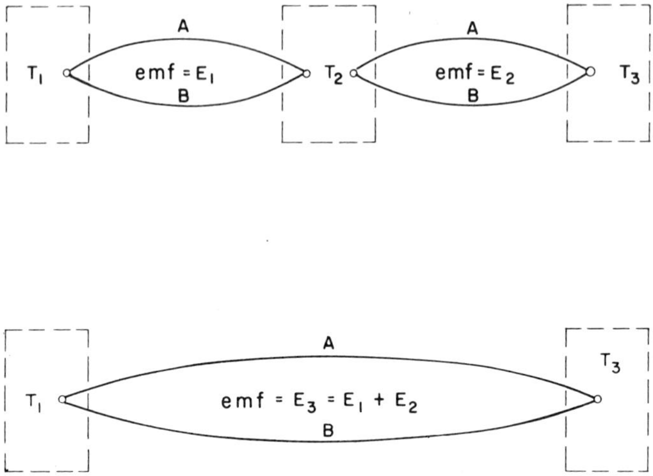

FIG. 2.3—*Emfs are additive for temperature intervals.*

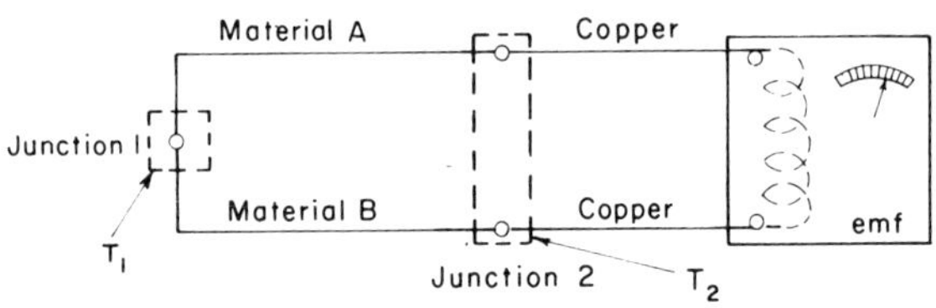

(a) For temperature level (Junction 2 is held at a constant, known reference temperature)

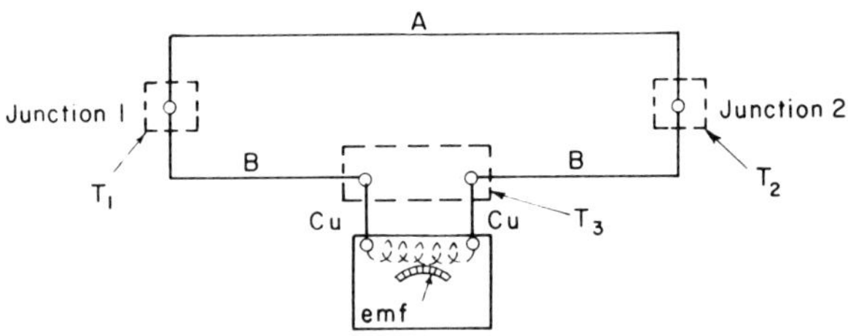

(b) For temperature difference (Junctions 1 and 2 are each exposed to unknown environment temperatures)

FIG. 2.4—*Several methods for introducing copper extension wires in elementary thermocouple circuits.*

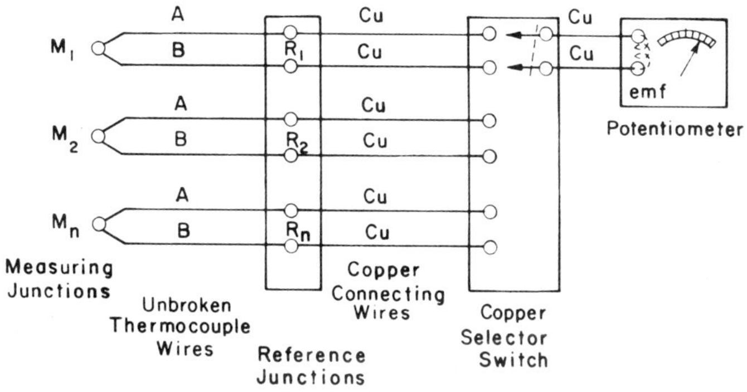

FIG. 2.5—*Basic thermocouple circuit.*

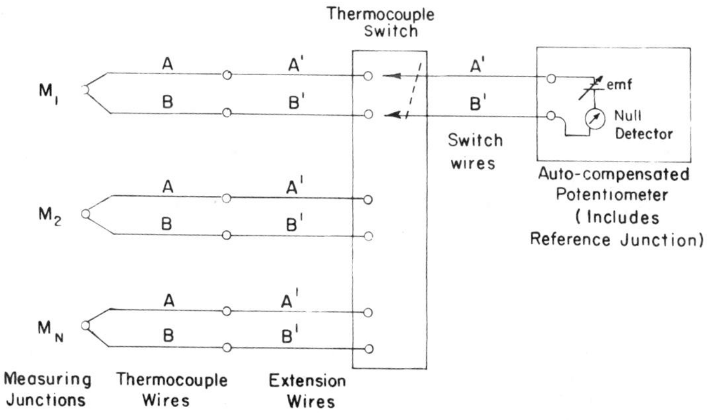

FIG. 2.6—*Typical industrial thermocouple circuits.*

2.4 Bibliography

2.4.1 Early Historical References

Volta, A., "On the Electricity Excited by Mere Contact of Conducting Substances of Different Kinds," *Philosophical Transactions,* 1800, p. 403.

Seebeck, T. J., "Evidence of the Thermal Current of the Combination Bi-Cu by Its Action on Magnetic Needle, *Royal Academy of Science*, Berlin, 1822-1823, p. 265.

Fourier, J. B. J., *Analytical Theory of Heat*, Gauthier-Villars et Cie., Paris, 1822; English translation by Freeman, A., Cambridge University Press, Cambridge, 1878.

Ohm, G. S., "Determination of the Laws by Which Metals Conduct the Contact Electricity, and also a Draft for a Theory of the Voltage Apparatus," *Journal for Chemie and Physik* (*Schweigger's Journal*), Vol. 46, 1826, p. 137.

Peltier, J. C. A., "Investigation of the Heat Developed by Electric Currents in Homogeneous Materials and at the Junction of Two Different Conductors," *Annalis de Chemie et de Physique*, Vol. 56 (2nd Series), 1834, p. 371.

Joule, J. P., "On the Production of Heat by Voltaic Electricity," *Proceedings of the Royal Society*, Dec. 1840.

Thomson, W., "On an Absolute Thermometric Scale Founded on Carnot's Theory of the Motive Power of Heat, and Calculated from Regnault's Observations," *Proceedings of the Cambridge Philosophical Society*, June 1848.

Clausius, R. J. E., "About the Motive Force of Heat," *Annalen der Physik und Chemie*, Vol. 79, 1850, pp. 368 and 500.

Thomson, W., "On a Mechanical Theory of Thermo-Electric Current," *Proceedings of the Royal Society of Edinburgh*, Dec. 1851.

Thomson, W., "On the Thermal Effects of Electric Currents in Unequal Heated Conductors," *Proceedings of the Royal Society*, Vol. VII, May 1854.

2.4.2 Recent References

Benedict, R. P., "Thermoelectric Effects," *Electrical Manufacturing*, Feb. 1960, p. 103.

Finch, D. I., "General Principles of Thermoelectric Thermometry," *Temperature*, Vol. 3, Part 2, Reinhold, New York, 1962.

Roeser, W. F., "Thermoelectric Circuitry," *Journal of Applied Physics*, Vol. 11, 1940, p. 388.

Dike, P. H., "Thermoelectric Thermometry," *Leeds and Northrup Technical Publication EN-33A (1a)*, 3rd ed., April 1955.

Stratton, R., "On the Elementary Theory of Thermoelectric Phenomena," *British Journal of Applied Physics*, Vol. 8, Aug. 1949, p. 33.

Miller, D. G., "Thermodynamic Theory of Irreversible Processes," *American Journal of Physics*, May 1956, p. 433.

Onsager, L., "Reciprocal Relations in Irreversible Processes," *Physical Review*, Vol. 37, Feb. 1931, p. 405.

Callen, H. B., "The Application of Onsager's Reciprocal Relations to Thermoelectric, Thermomagnetic, and Galvanomagnetic Effects," *Physical Review*, Vol. 73, June 1948, p. 3149.

Hatsopoulos, G. N. and Keenan, J. H., "Analyses of the Thermoelectric Effects of Methods of Irreversible Thermodynamics," *Journal of Applied Mechanics*, Vol. 25, Dec. 1958, p. 428.

Benedict, R. P., *Fundamentals of Temperature, Pressure, and Flow Measurements*, 2nd ed., Wiley, New York, 1977.

Pollock, D. D., *The Theory and Properties of Thermocouple Elements, ASTM STP 492*, American Society for Testing and Materials, 1971.

Benedict, R. P. and Russo, R. J., "A Note on Grounded Thermocouple Circuits," *Transactions*, American Society of Mechanical Engineers, *Journal of Basic Engineering*, June 1972, p. 377.

Fenton, A. W., "The Theory of Thermoelectric Thermometers," United Kingdom Atomic Energy Authority, 1969, unpublished.

Fenton, A. W., "Errors in Thermoelectric Thermometers," *Proceedings*, Institute of Electrical Engineers, Vol. 116, No. 7, July 1969.

Fenton, A. W. et al, "Thermocouples: Instabilities of Seebeck Coefficients," United Kingdom Atomic Energy Authority, Report No. TRG 1447(R), Jan. 1967.

Freeman, R. J., "Distributed Seebeck Effect at High Temperature," High Temperature Thermometry Seminar, Oct. 1959, see TID 7586, Part 1.

Fenton, A. W., "A Note on How Thermocouples Work," Report No. ND-R-327(R), United Kingdom Atomic Energy Authority, April 1974.

Broomfield, G. H., "Signals from Temperature Measuring Thermocouples," *The Metallurgist and Materials Technologist*, May 1979, p. 281.

Moffat, R. J., "Thermocouple Theory and Practice," in *Fundamentals of Aerospace Instrumentation*, Vol. 6, Instrument Society of America, 1974, pp. 111–124.

2.5 Nomenclature

Roman

a, b	Coefficients
A	Area
E	Electric potential
F	Frictional loss
I	Electric current
J	Flow
k	Thermal conductivity
L	Constant of proportionality
Q	Heat
S	Entropy
t	Time
T	Temperature
U	Internal energy
W	Work
x	Length or thickness
X	Force

Subscripts

$A, B, C,$	Thermocouple materials
e	Electrical
i, j, k	General subscripts
i	Internal
P	Peltier
q	Thermal
R	Reference
S	Seebeck
T	Thomson
abs	Absolute
rev	Reversible
1,2	States

Greek

α	Seebeck coefficient
Δ	Finite difference
π	Peltier coefficient
σ	Thomson coefficient
Σ	Sum
ξ	Entropy production/volume

Chapter 3—Thermocouple Materials

3.1 Common Thermocouple Types

The commonly used thermocouple types are identified by letter designations originally assigned by the Instrument Society of America (ISA) and adopted as an American Standard in ANSI MC 96.1. This chapter covers general application data on the atmospheres in which each thermocouple type can be used, recommended temperature ranges, limitations, etc. Physical and thermoelectric properties of the thermoelement materials used in each of these thermocouple types are also presented in this section.

The following thermocouple types are included (these are defined as having the emf-temperature relationship given in the corresponding letter-designated Table in Chapter 10 within the limits of error specified in Table 10.1 of that chapter):

Type T—Copper (+) versus constantan (−).

Type J—Iron (+) versus constantan (−).

Type E—Nickel-10 percent chromium (+) versus constantan (−).

Type K—Nickel-10 percent chromium (+) versus nickel-5 percent aluminum and silicon (−) (see note).

Type R—Platinum-13 percent rhodium (+) versus platinum (−).

Type S—Platinum-10 percent rhodium (+) versus platinum (−).

Type B—Platinum-30 percent rhodium (+) versus platinum-6 percent rhodium (−).

Temperature limits stated in the text are maximum values. Table 3.1 gives recommended maximum temperature limits for various gage sizes of wire. Figure 3.1 is a graphical presentation of maximum temperature limits from Table 3.1 and permits interpolation based on wire size. Table 3.2 gives nominal Seebeck coefficients for the various types. Temperature-emf equivalents and commercial limits of error for these common thermocouple types are given in Chapter 10.

3.1.1 General Application Data

Type T—These thermocouples are resistant to corrosion in moist atmospheres and are suitable for subzero temperature measurements (see Table 10.1 for limits of error in the subzero region.) Their use in air or in oxidizing environments is restricted to 370°C (700°F) due to oxidation of the

TABLE 3.1—*Recommended upper temperature limits for protected thermocouples.*

	Upper Temperature Limit for Various Wire Sizes (awg), °C (°F)					
Thermocouple Type	No. 8 Gage, 3.25 mm (0.128 in.)	No. 14 Gage, 1.63 mm (0.064 in.)	No. 20 Gage, 0.81 mm (0.032 in.)	No. 24 Gage, 0.51 mm (0.020 in.)	No. 28 Gage, 0.33 mm (0.013 in.)	No. 30 Gage, 0.25 mm (0.010 in.)
T		370 (700)	260 (500)	200 (400)	200 (400)	150 (300)
J	760 (1400)	590 (1100)	480 (900)	370 (700)	370 (700)	320 (600)
E	870 (1600)	650 (1200)	540 (1000)	430 (800)	430 (800)	370 (700)
K	1260 (2300)	1090 (2000)	980 (1800)	870 (1600)	870 (1600)	760 (1400)
R and S				1480 (2700)		
B				1700 (3100)		

NOTE—This table gives the recommended upper temperature limits for the various thermocouples and wire sizes. These limits apply to protected thermocouples, that is, thermocouples in conventional closed-end protecting tubes. They do not apply to sheathed thermocouples having compacted mineral oxide insulation. Properly designed and applied sheathed thermocouples may be used at temperatures above those shown in the tables. Other literature sources should be consulted.

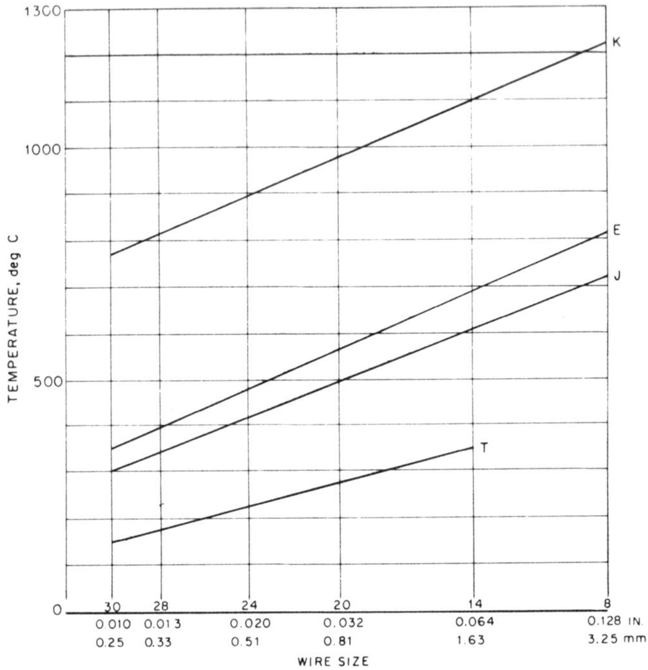

NOTE—This graph gives the recommended upper temperature limits for the various thermocouples and wire sizes. These limits apply to thermocouples used under the atmospheric limitations outlined in the text. They do not apply to sheathed thermocouples having compacted mineral oxide insulation. In any general recommendation of thermocouple temperature limits, it is not practicable to take into account special cases. In actual operation, there may be instances where the temperature limits recommended can be exceeded. Likewise, there may be applications where satisfactory life will not be obtained at the recommended temperature limits. However, in general, the temperature limits listed are such as to provide satisfactory thermocouple life when the wires are operated continuously at these temperatures.

FIG 3.1—*Recommended upper temperature limits for types K, E, J, T thermocouples.*

copper thermoelement. They may be used to higher temperatures in some other atmospheres.

They can be used in a vacuum and in oxidizing, reducing or inert atmospheres over the temperature range of −200 to 370°C (−330 to 700°F). The upper temperature limit is due primarily to oxidation of the copper element.

This is the only thermocouple type for which limits of error are established in the subzero temperature range (see rates under Table 10.1, Chapter 10).

Type J—These thermocouples are suitable for use in vacuum and in oxidizing, reducing, or inert atmospheres, over the temperature range of 0 to 760°C (32 to 1400°F). The rate of oxidation of the iron thermoelement is rapid above 540°C (1000°F), and the use of heavy-gage wires is recom-

TABLE 3.2—*Nominal Seebeck coefficients (thermoelectric power).*

Temperature	Thermocouple Type E	J	K	R	S	T	B
°C	Seebeck Coefficient-Microvolts/°C						
−190	27.3	24.2	17.1	...	...	17.1	
−100	44.8	41.4	30.6	...	...	28.4	
0	58.5	50.2	39.4	...	...	38.0	...
200	74.5	55.8	40.0	8.8	8.5	53.0	2.0
400	80.0	55.3	42.3	10.5	9.5	...	4.0
600	81.0	58.5	42.6	11.5	10.3	...	6.0
800	78.5	64.3	41.0	12.3	11.0	...	7.7
1000	...	...	39.0	13.0	11.5	...	9.2
1200	...	...	36.5	13.8	12.0	...	10.3
1400	...	...	...	13.8	12.0	...	11.3
1600	...	...	...	...	11.8	...	11.6
°F	Seebeck Coefficient-Microvolts/°F						
−300	15.5	14.4	...	...	...	9.7	
−200	22.0	20.6	...	...	...	13.7	
−100	27.0	24.6	...	...	...	17.3	
32	32.5	28.0	21.7	3.0	3.0	21.3	...
200	37.5	30.1	23.2	4.1	4.0	25.7	0.5
400	41.5	30.9	22.3	4.9	4.8	29.8	1.1
600	43.5	30.7	23.1	5.5	5.1	32.7	1.8
800	45.0	30.6	23.5	5.9	5.3	...	2.4
1000	45.0	31.7	23.7	6.2	5.5	...	3.0
1500	44.0	35.7	22.8	6.9	6.1	...	4.4
2000	...	...	21.1	7.6	6.6	...	5.4
2500	...	...	...	7.6	6.6	...	6.2
3000	...	...	...	7.6	6.5	...	6.5

mended when long life is required at the higher temperatures. Type J may be used to higher temperatures in some atmospheres. However, they should not be used in sulfurous atmospheres above 540°C (1000°F).

This thermocouple is not recommended for use below the ice point because rusting and embrittlement of the iron thermoelement make its use less desirable than Type T. Limits of error have not been established for Type J thermocouples at subzero temperatures.

Type E—Type E thermocouples are recommended for use over the temperature range of −200 to 900°C (−330 to 1600°F) in oxidizing or inert atmospheres. In reducing atmospheres, alternately oxidizing and reducing atmospheres, marginally oxidizing atmospheres, and in vacuum, they are subject to the same limitations as Type K thermocouples.

These thermocouples are suitable for subzero temperature measurements since they are not subject to corrosion in atmospheres with high moisture content. Limits of error are shown in Table 10.1, Chapter 10.

Type E thermocouples develop the highest emf per degree of all the commonly used types and are often used primarily because of this feature.

Type K—Type K thermocouples are recommended for use in an oxidizing or completely inert atmosphere over the temperature range of −200 to 1260°C (−330 to 2300°F). Because their oxidation resistance characteristics are better than those of Types E, J, and T thermocouples, they find widest use at temperatures above 540°C (1000°F).

Type K thermocouples are suitable for temperature measurements as low as −250°C (−420°F), although limits of error have been established only for the temperature range given previously.

Type K thermocouples should not be used in:

1. Atmospheres that are reducing or alternately oxidizing and reducing.
2. Sulfurous atmospheres, since sulfur will attack both thermoelements and will cause rapid embrittlement and breakage of the negative thermoelement wire through integranular corrosion.
3. Vacuum, except for short time periods, since preferential vaporization of chromium from the positive element may alter calibration.
4. Atmospheres that promote "green-rot" corrosion of the positive thermoelement. Such corrosion results from preferential oxidation of chromium when the oxygen content of the atmosphere surrounding the thermocouple is low. Green-rot corrosion can cause large negative errors in calibration and is most serious at temperatue levels of 815 to 1040°C (1500 to 1900°F).

Green-rot corrosion frequently occurs when thermocouples are used in long unventilated protecting tubes of small diameter. It can be minimized by increasing the oxygen supply through the use of large diameter protecting tubes or ventilated protecting tubes. Another approach is to decrease the oxygen content below that which will promote preferential oxidation by inserting a "getter" to absorb the oxygen in a sealed protecting tube.

Types R and S—Types R and S thermocouples are recommended for continuous use in oxidizing or inert atmospheres over the temperature range of 0 to 1480°C (32 to 2700°F).

They should not be used in reducing atmospheres, nor those containing metallic or nonmetallic vapors. They never should be inserted directly into a metallic primary protecting tube.

Types R and S—Types R and S thermocouples are recommended for continuous use in oxidizing or inert atmospheres over the temperature range of 0 to 1480°C (32 to 2700°F).

Continued use of Types R and S thermocouples at high temperatures causes excessive grain growth which can result in mechanical failure of the platinum element. It also renders the platinum susceptible to contamination which causes a reduction in the emf output of the thermocouple.

Calibration changes also are caused by diffusion of rhodium from the alloy wire into the platinum, or by volatilization of rhodium from the alloy. All of these effects tend to produce inhomogeneity.

Type B—Type B thermocouples are recommended for continuous use in oxidizing or inert atmospheres over the temperature range of 870 to 1700°C (1000 to 3100°F). They are also suitable for short term use in vacuum. They should not be used in reducing atmospheres, nor those containing metallic or nonmetallic vapors. They should never be inserted directly into a metallic primary protecting tube or well.

Under corresponding conditions of temperature and environment, Type B thermocouples will show less grain growth and less drift in calibration that Types R and S thermocouples.

3.1.2 Properties of Thermoelement Materials

This section indicates in Tables 3.3 to 3.9 and in Figure 3.2 the physical and electrical properties of thermoelement materials as used for the common letter-designated thermocouple types (Types T, J, E, K, R, S, and B). These are typical data and are listed for information only. They are not intended for use as specifications for ordering thermocouple materials.

Thermoelement materials are designated in the tables by the established American Standard letter symbols JP, JN, etc. The first letter of the symbol designates the type of thermocouple. The second letter, P or N, denotes the positive or negative thermoelement. Typical materials to which these letter designations apply are:

TP	Copper
JP	Iron, ThermoKanthal JP[1]
TN, JN, or EN	Constantan, Cupron[2], Advance[3], ThermoKanthal JN[1]
KP or EP	Nickel-chrome, Chromel[4], Tophel[2], T-1[3], ThermoKanthal KP [1]
KN	Nickel-silicon, Alumel[4], Nial[2], T-2[3], ThermoKanthal KN[1]
RP	Platinum-13 percent rhodium
SP	Platinum-10 percent rhodium
RN or SN	Platinum
BP	Platinum-30 percent rhodium
BN	Platinum-6 percent rhodium

Note that TN, JN, and EN thermoelements, as just listed, are composed of the same basic types of material. The typical data contained in the following pages are applicable to any of these thermoelements. The thermal emf of these thermoelements when referenced to platinum may differ significantly depending on the type of thermocouple of which each is intended.

It also should be noted that positive and negative thermoelements for a given type of thermocouple, as supplied by any one manufacturer, will con-

[1]Trademark of the Kanthal Corporation.
[2]Trademark of American Metal Climax Incorporated.
[3]Trademark of the Driver-Harris Company.
[4]Trademark of the Hoskins Manufacturing Company.

TABLE 3.3—*Nominal chemical composition of thermoelements.*

Element	JP	JN, TN, EN[a]	TP	KP, EP	KN	RP	SP	RN, SN	BP	BN
	Nominal Chemical Composition, %									
Iron	99.5	...	...	...	...	...	...	...	...	...
Carbon	...[b]	...	...	...	...	...	...	...	...	...
Manganese	...[b]	...	...	...	2	...	...	...	...	...
Sulfur	...[b]	...	...	...	...	...	...	...	...	...
Phosphorus	...[b]	...	...	...	...	...	...	...	...	...
Silicon	...[b]	...	...	...	1	...	...	...	...	...
Nickel	...[b]	45	...	90	95	...	...	...	...	...
Copper	...[b]	55	100	...	...	...	...	...	...	...
Chromium	...[b]	...	...	10	...	...	...	...	...	...
Aluminum	...	...	...	...	2	...	...	...	...	...
Platinum	...	...	...	...	...	87	90	100	70.4	93.9
Rhodium	...	...	...	...	...	13	10	...	29.6	6.1

[a]Types JN, TN, and EN thermoelements usually contain small amounts of various elements for control of thermal emf, with corresponding reductions in the nickel or copper content, or both.

[b]Thermoelectric iron (JP) contains small but varying amounts of these elements.

TABLE 3.4—*Environmental limitations of thermoelements.*

Thermoelement	Environmental Recommendations and Limitations (see notes)
JP	For use in oxidizing, reducing, or inert atmospheres or in vacuum. Oxidizes rapidly above 540°C (1000°F). Will rust in moist atmospheres as in subzero applications. Stable to neutron radiation transmutation. Change in composition is only 0.5 percent (increase in manganese) in 20-year period.
JN, TN, EN	Suitable for use in oxidizing, reducing, and inert atmospheres or in vacuum. Should not be used unprotected in sulfurous atmospheres above 540°C (1000°F). Composition changes under neutron radiation since copper content is converted to nickel and zinc. Nickel content increases 5 percent in 20-year period.
TP	Can be used in vacuum or in oxidizing, reducing or inert atmospheres. Oxidizes rapidly above 370°C (700°F). Preferred to Type JP element for subzero use because of its superior corrosion resistance in moist atmospheres. Radiation transmutation causes significant changes in composition. Nickel and zinc grow into the material in amounts of 10 percent each in a 20-year period.
KP, EP	For use in oxidizing or inert atmospheres. Can be used in hydrogen or cracked ammonia atmospheres if dew point is below −40°C (−40°F). Do not use unprotected in sulfurous atmospheres above 540°C (1000°F).

TABLE 3.4—*(Continued)*.

Thermoelement	Environmental Recommendations and Limitations (see notes)
	Not recommended for service in vacuum at high temperatures except for short time periods because preferential vaporization of chromium will alter calibration. Large negative calibration shifts will occur if exposed to marginally oxidizing atmospheres in temperature range 815 to 1040°C (1500 to 1900°F).
	Quite stable to radiation transmutation. Composition change is less than 1 percent in 20-year period.
KN	Can be used in oxidizing or inert atmospheres. Do not use unprotected in sulfurous atmospheres as intergranular corrosion will cause severe embrittlement.
	Relatively stable to radiation transmutation. In 20-year period, iron content will increase approximately 2 percent. The manganese and cobalt contents will decrease slightly.
RP, SP, SN, RN, BP, BN	For use in oxidizing or inert atmospheres. Do not use unprotected in reducing atmospheres in the presence of easily reduced oxides, atmospheres containing metallic vapors such as lead or zinc, or those containing nonmetallic vapors such as arsenic, phosphorus, or sulfur. Do not insert directly into metallic protecting tubes. Not recommended for service in vacuum at high temperatures except for short time periods.
	Types RN and SN elements are relatively stable to radiation transmutation. Types BP, BN, RP, and SP elements are unstable because of the rapid depletion of rhodium. Essentially, all the rhodium will be converted to palladium in a 10-year period.

NOTE 1—Refer to Table 3.5 for recommended upper temperature limits.

NOTE 2—Stability under neutron radiation refers to chemical composition of thermoelement, not to stability of thermal emf.

NOTE 3—Radiation transmutation rates[a] are based on exposure to a thermal neutron flux of 1×10^{14} neutrons/cm^2·s.

[a] Browning, W. E., Jr., and Miller, C. E., Jr., "Calculated Radiation Induced Changes in Thermocouple Composition," *Temperature, Its Measurement and Control in Science and Industry*, Part 2, Rheinhold, New York, Vol. C, 1962, p. 271.

form to the calibration curve for that thermocouple within specified limits of error. However, because materials used for a given thermoelement by various manufacturers may differ slightly in thermal emf, larger errors may occur if positive and negative thermoelements from different sources are combined.

3.2 Extension Wires

3.2.1 General Information

Extension wires are inserted between the measuring junction and the reference junction and have approximately the same thermoelectric proper-

TABLE 3.5—*Recommended upper temperature limits for protected thermoelements.*

	Upper Temperature Limits for Various Wire Sizes (awg), °C (°F)					
Thermoelement	No. 8 Gage, 3.25 mm (0.128 in.)	No. 14 Gage, 1.63 mm (0.064 in.)	No. 20 Gage, 0.81 mm (0.032 in.)	No. 24 Gage, 0.51 mm (0.020 in.)	No. 28 Gage, 0.33 mm (0.013 in.)	No. 30 Gage, 0.25 mm (0.010 in.)
JP	760°C (1400°F)	593°C (1100°F)	482°C (900°F)	371°C (700°F)	371°C (700°F)	320°C (600°F)
JN, TN, EN	871°C (1600°F)	649°C (1200°F)	538°C (1000°F)	427°C (800°F)	427°C (800°F)	427°C (800°F)
TP	...	371°C (700°F)	260°C (500°F)	204°C (400°F)	204°C (400°F)	150°C (300°F)
KP, EP, KN	1260°C (2300°F)	1093°C (2000°F)	982°C (1800°F)	871°C (1600°F)	871°C (1600°F)	760°C (1400°F)
RP, SP, RN, SN	...	...	...	1482°C (2700°F)	...	...
BP, BN	...	...	...	1705°C (3100°F)	...	...

NOTE—This table gives the recommended upper temperature limits for the various thermoelements and wire sizes. These limits apply to protected thermoelements, that is, thermoelements in conventional closed-end protecting tubes. They do not apply to sheathed thermoelements having compacted mineral oxide insulation. In any general recommendation of thermoelement temperature limits,it is not practicable to take into account special cases. In actual operation, there may be instances where the temperature limits recommended can be exceeded. Likewise, there may be applications where satisfactory life will not be obtained at the recommended temperature limits. However, in general, the temperature limits listed are such as to provide satisfactory thermoelement life when the wires are operated continuously at these temperatures.

TABLE 3.6—*Seebeck coefficient (thermoelectric power) of thermoelements with respect to Platinum 67 (typical values).*

Thermo-element	JP	JN, TN, EN	TP	KP, EP	KN	RP	SP	BP	BN
Tempera-ture, °C				Seeback Coefficient, μV/°C					
−190	+6.3	−20.9	−4.1	. . .	. . .	. . .	. . .	. . .	. . .
−100	14.4	27.0	+1.1	. . .	. . .	. . .	. . .	. . .	. . .
0	17.8	32.2	5.9	+25.7	−13.5	+5.5	+5.5	. . .	
200	14.6	41.0	12.0	32.7	7.4	8.5	8.5	+9.2	+7.2
400	9.7	45.5	16.2	34.6	7.7	10.5	9.5	11.7	7.6
600	11.7	46.8	. . .	33.8	8.8	11.5	10.0	13.8	7.9
800	17.8	46.4	. . .	32.2	8.8	12.5	11.0	15.8	8.2
1000	. . .	. . .	. . .	30.8	8.3	13.0	11.5	17.7	8.5
1200	. . .	. . .	. . .	29.1	7.4	14.0	12.0	19.1	8.7
1400	. . .	. . .	. . .	. . .	. . .	14.0	12.0	20.0	8.7
1600	. . .	. . .	. . .	. . .	. . .	13.5	12.0	20.4	8.7
Tempera-ture, °F				Seebeck Coefficient, μV/°F					
−300	+2.5	−11.9	−2.1	. . .	. . .	. . .	. . .	. . .	. . .
−200	6.7	14.0	+0.2	. . .	. . .	. . .	. . .	. . .	. . .
−100	8.8	15.8	1.5	. . .	. . .	. . .	. . .	. . .	. . .
32	9.9	17.9	3.3	+14.3	−7.5	+3.0	+3.0	. . .	. . .
200	9.6	20.5	5.0	16.7	6.5	4.1	4.0	+4.1	+3.6
400	8.0	22.9	6.7	18.3	4.0	4.9	4.7	5.1	4.0
600	6.2	24.5	8.2	19.0	4.1	5.5	5.2	5.8	4.2
800	5.3	25.3	. . .	19.1	4.4	5.8	5.4	6.5	4.2
1000	5.7	26.0	. . .	18.9	4.8	6.2	5.5	7.4	4.3
1500	9.9	25.8	. . .	17.8	4.9	6.8	6.1	8.8	4.6
2000	. . .	. . .	. . .	16.7	4.3	7.6	6.6	10.2	4.8
2500	. . .	. . .	. . .	14.9	4.0	7.7	6.7	11.0	4.9
3000	. . .	. . .	. .	. . .	. . .	7.6	6.5	11.3	4.9

ties as the thermocouple wires with which they are used. Table 3.10 gives comparative data on extension wires available for thermocouples in common use. Extension wires are normally available as single or duplex, solid or stranded, insulated wires in sizes ranging from 14 to 20 B&S gage. A variety of insulations and protective coverings is available in several combinations to suit the many types of environments encountered in industrial service (see Chapter 4).

Some advantages of using extension wires are:

1. Improvement in mechanical or physical properties of the thermoelectric circuit. For example, the use of stranded construction or smaller diameter solid wire may increase the flexibility of a portion of the circuit. Extension wires also may be selected to adjust the electrical resistance of the circuit.

TABLE 3.7—*Typical physical*

Property	JP	JN, EN, TN	TP	KP, EP
	Thermoelement			
Melting point (solidus temperatures):				
°C	1 490	1 220	1 083	1 427
°F	2 715	2 228	1 981	2 600
Resistivity:				
μΩ·cm:				
at 0°C	8.57	48.9	1.56	70
at 20°C	9.67	48.9	1.724	70.6
Ω cmil/ft:				
at 0°C	51.5	294.2	9.38	421
at 20°C	58.2	294	10.37	425
Temperature coefficient of resistance, Ω/Ω·°C (0 to 100°C)	65×10^{-4}	-0.1×10^{-4}	43×10^{-4}	4.1×10^{-4}
Coefficient of thermal expansion, in./in.·°C (20 to 100°C)	11.7×10^{-6}	14.9×10^{-6}	16.6×10^{-6}	13.1×10^{-6}
Thermal conductivity at 100°C:				
Cal·cm/s·cm^2·°C	0.162	0.0506	0.901	0.046
Btu·ft/h·ft^2·°F	39.2	12.2	218	11.1
Specific heat at 20°C, cal/g·°C	0.107	0.094	0.092	0.107
Density:				
g/cm^3	7.86	8.92	8.92	8.73
lb/in.^{3}	0.284	0.322	0.322	0.315
Tensile strength (annealed):				
kgf/cm^2	3 500	5 600	2 500	6 700
psi	50 000	80 000	35 000	95 000
Magnetic attraction	strong	none	none	none

2. Cost improvement in thermoelectric circuitry. For example, certain base metal extension wires may be substituted for noble metal wires when the reference junction is situated at a distance from a noble metal thermocouple.

Extension wires may be separated into two categories having the following characteristics:

Category 1—Alloys substantially the same as used in the thermocouple. This type of extension wire normally is used with base metal thermocouples.

Category 2—Alloys differing from those used in the thermocouple. This type of extension wire normally is used with noble metal thermocouples and with several of the nonstandardized thermocouples (see Section 3.3).

3.2.2 Sources of Error

Several possible sources of error in temperature measurement accompany the use of extension wires in thermocouple circuits. Most of the errors can be avoided, however, by exercising proper precautions.

One type of error arises from the disparity in thermal emf between thermocouples and nominally identical extension wire components of Category 1.

properties of thermoelement materials.

Material					
KN	RP	SP	RN, SN	BP	BN
1 399	1 860	1 850	1 769	1 927	1 826
2 550	3 380	3 362	3 216	3 501	3 319
28.1	19.0	18.4	9.83	...	...
29.4	19.6	18.9	10.4	19.0	17.5
169	114.3	110.7	59.1	...	...
177	117.7	114.0	62.4	114.5	106
23.9×10^{-4}	15.6×10^{-4}	16.6×10^{-4}	39.2×10^{-4}	13.3×10^{-4}	20.0×10^{-4}
12.0×10^{-6}	9.0×10^{-6}	9.0×10^{-6}	9.0×10^{-6}	...	...
0.071	0.088	0.090	0.171	...	...
17.2	21.3	21.8	41.4	...	...
0.125	...	...	0.032	...	...
8.60	19.61	19.97	21.45	17.60	20.55
0.311	0.708	0.721	0.775	0.636	0.743
6 000	3 200	3 200	1 400	4 900	2 800
85 000	46 000	45 000	20 000	70 000	40 000
moderate	none	none	none	none	none

The disparity results from the variations occurring among thermoelements lying within the standard limits of error for each type of thermocouple and extension wire. Thus, for example, it is possible that an error as great as ±4.4°C (±8°F) could occur in the Type K/KX and J/JX thermocouple-extension wire combinations, where the standard limits of error are ±2.2°C (±4°F) for the thermocouple and the extension wires treated as separate combinations. Such errors can be eliminated substantially by selecting extension wires whose emf closely matches that of the specific thermocouple, up to the maximum temperature of the thermocouple-extension wire junction.

A second source of error can arise if a temperature difference exists between the two thermoelement-extension wire junctions. Errors of this type are potentially greater in circuits employing extension wires of Category 2, where each extension element may differ significantly in emf from the corresponding thermoelement. Such errors may occur even though the extension pair emf exactly matches the thermocouple emf at each temperature. Referring to Fig. 3.3, schematically representing emf versus temperature curves for positive and negative thermoelements P and N, and corresponding extension wire elements PX and NX, the following relationships apply at any temperature T within the operating range of the extension wires:

TABLE 3.8—*Thermoelements—resistance to change with increasing temperature.*

Thermo-elements	Ratio of Resistance at Temperature Indicated to Resistance at 0°C (32°F)									
	0°C (32°F)	20°C (68°F)	200°C (392°F)	400°C (752°F)	600°C (1112°F)	800°C (1472°F)	1000°C (1832°F)	1200°C (2192°F)	1400°C (2552°F)	1500°C (2732°F)
JP	1.00	1.13	2.46	4.72	7.84	12.00	13.07	...	...	...
JN, TN, EN	1.00	0.999	0.996	0.994	1.024	1.056	1.092	...	...	...
TP	1.00	1.11	1.86	2.75	3.70	4.75	5.96	...	...	...
KP, EP	1.00	1.01	1.09	1.19	1.25	1.30	1.37	1.43	...	...
KN	1.00	1.05	1.43	1.64	1.82	1.98	2.15	2.32	...	...
RP	1.00	1.03	1.31	1.60	1.89	2.16	2.41	2.66	2.90	3.01
SP	1.00	1.03	1.33	1.65	1.95	2.23	2.50	2.76	3.01	3.13
RN, SN	1.00	1.06	1.77	2.50	3.18	3.81	4.40	4.94	5.42	5.66
BP	1.00	1.03	1.26	1.51	1.76	1.98	2.20	2.41	2.62	2.73
BN	1.00	1.03	1.40	1.78	2.14	2.47	2.78	3.08	3.37	3.51

TABLE 3.9—*Nominal resistance of thermoelements.*

Awg No.	Diameter, in.[b]	KN	KP, EP	TN, JN, EN	TP	JP	RN, SN	RP	SP	BP	BN
		Nominal Resistance, ohms per foot at 20°C (68°F)									
6	0.1620	0.0067	0.0162	0.0112	0.000395	0.0027	0.00243	0.00448	0.00438	0.00440	0.00407
9[a]	0.148	...	...	...	...	0.0033	...	0.00537	...	...	...
8	0.1285	0.0107	0.0257	0.0179	0.000628	0.0043	0.00386	0.00713	0.00697	0.00700	0.00648
10	0.1019	0.0170	0.041	0.0283	0.000999	0.0069	0.00614	0.00113	0.01108	0.01113	0.01030
12	0.0808	0.0270	0.065	0.0448	0.00159	0.0109	0.00976	0.01803	0.01761	0.01769	0.01637
14	0.0641	0.0432	0.104	0.0718	0.00253	0.0174	0.0155	0.0286	0.0280	0.0281	0.0260
16	0.0508	0.0683	0.164	0.113	0.00402	0.0276	0.0247	0.0456	0.0445	0.0447	0.0414
17	0.0453	0.0874	0.209	0.145	0.00506	0.0349	0.0311	0.0574	0.0562	0.0564	0.0523
18	0.0403	0.111	0.266	0.184	0.00648	0.0446	0.0399	0.0725	0.0719	0.0722	0.0669
20	0.0320	0.173	0.415	0.287	0.0102	0.0699	0.0624	0.1149	0.1125	0.1130	0.1046
22	0.0253	0.276	0.663	0.456	0.0161	0.1111	0.0993	0.1839	0.1790	0.1798	0.1664
23	0.0226	0.347	0.833	0.576	0.0204	0.1401	0.1251	0.2913	0.2257	0.2267	0.2099
24	0.0201	0.438	1.05	0.728	0.0257	0.1767	0.1578	0.4656	0.2847	0.2859	0.2647
25	0.0179	0.553	1.33	0.918	0.0324	0.2228	0.1990	0.7414	0.3589	0.3605	0.3337
26	0.0159	0.700	1.68	1.16	0.0408	0.281	0.2509	1.177	0.4526	0.4546	0.4208
28	0.0126	1.11	2.48	1.85	0.0649	0.447	0.3989	1.839	0.7197	0.7229	0.6692
30	0.0100	1.77	4.25	2.94	0.1032	0.710	0.6344	2.965	1.144	1.149	1.064
32	0.0080	2.76	6.65	4.59	0.1641	1.13	1.009	4.708	1.819	1.827	1.691
34	0.0063	4.45	10.7	7.41	0.2609	1.80	1.604	7.356	2.893	2.906	2.690
36	0.0050	7.08	17.0	11.8	0.4148	2.86	2.550	12.25	4.600	4.620	4.277
38	0.0040	11.1	26.6	18.4	0.6597	4.54	4.056		7.316	7.348	6.803
40	0.0031	18.4	44.2	30.6	1.049	7.22	6.448		11.63	11.68	10.81

[a] #9 Birmingham wire gage.
[b] 1 in. = 25.4 mm.

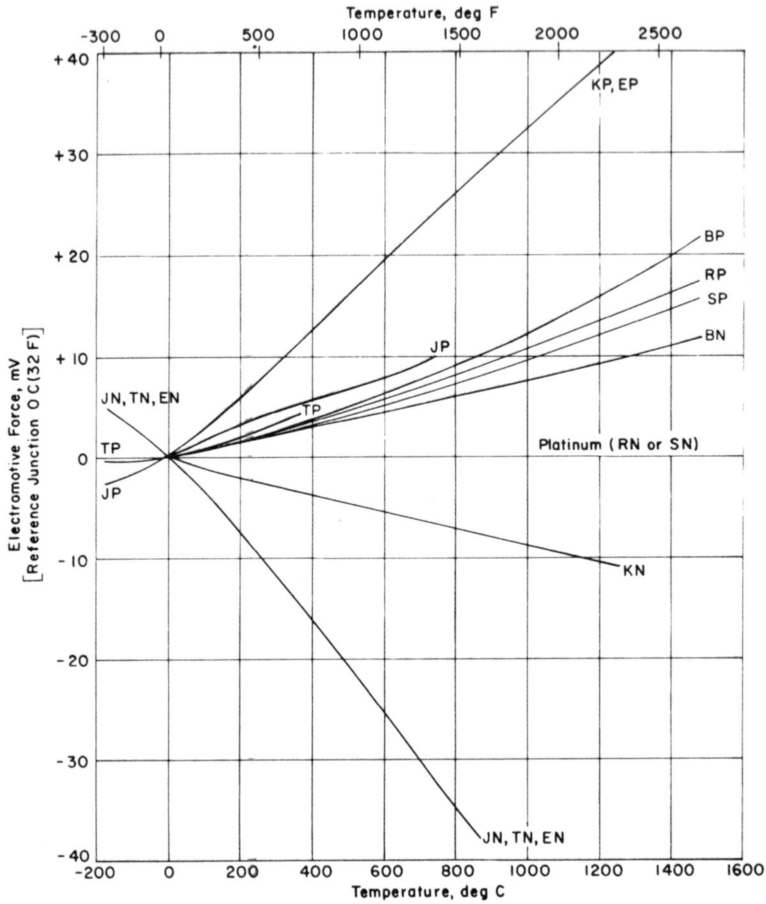

FIG. 3.2—*Thermal emf of thermoelements relative to platinum.*

Thermocouple output = extension pair output

That is

$$E_P - E_N = E_{PX} - E_{NX} \tag{1}$$

Rearranging to

$$E_P - E_{PX} = E_N - E_{NX} \tag{2}$$

If a temperature difference exists between the two junctions such that P joins PX at T_P, and N joins NX at T_N, an unwanted emf will exist across the two junctions, of magnitude

$$\Delta E = (E_P - E_{PX})_{TP} - (E_N - E_{NX})_{TN} \tag{3}$$

Rearranging Eq 39 according to Eq 38

$$\Delta E = (E_P - E_{PX})_{TP} - (E_P - E_{PX})_{TN} \quad (4)$$

The sign of ΔE will depend on the relationship of temperature T_P to T_N and the emf relationship of *PX* and *NX* to *P* and *N*.

This ΔE will be interpreted as an error in the output of the measuring thermocouple. Such errors do not exceed about one degree at the measuring junction, per degree of ΔT between the thermocouple-extension wire junctions. These errors can be eliminated by equalizing the temperatures of the two junctions.

A third source of error lies in the presence of reversed polarity at the thermocouple-extension wire junctions, or at the extension wire-instrument junctions. Although a single reversal of polarity in the assembly would be noticeable, an inadvertent double reversal likewise may produce measurement errors, but could escape immediate detection.

A fourth source of error concerns the use of connectors in the thermocouple assembly. If the connector material has thermal emf characteristics which differ appreciably from those of the thermocouple extension wires, then it is important that a negligible temperature difference be maintained across the connector. This follows directly from the Law of Intermediate Metals (see Section 2.2.2). Thus, in situations where a connector made of a third metal spans a substantial temperature gradient, unwanted emfs are generated between the thermoelectric materials and the extremities of the connector, and they appear as errors in the output of the thermocouple. The magnitude of errors of this type can vary over a wide range depending on the materials involved and the temperature difference spanned by the connector.

If the emf errors arising from the use of extension wires or from other sources are to be expressed as temperature errors, the Seebeck coefficient of the thermocouple at the measuring junction temperature must be used.

A useful graphical method of evaluating error sources in thermoelectric circuits is detailed in a paper by Moffat (see Ref *1*).[5]

3.3 Nonstandardized Thermocouple Types

Newer thermocouple materials are being evaluated constantly to find combinations which perform special functions more reliably than the common thermocouples. The specials functions for which these newer combinations are required frequently involve very high temperatures, but also may include unusual environments such as special atmospheres or areas susceptible to vibration.

Each of the combinations described in this section has been designed to measure temperatures under specific conditions and to perform with a degree of reliability superior to other combinations under these same condi-

[5]The italic numbers in brackets refer to the list of references appended to this chapter.

TABLE 3.10—*Extension wires for*

Thermocouple Type	Extension Wire Type	Alloy Type: Positive	Alloy Type: Negative
			STANDARD
Base Metal	Category 1		
E	EX	Ni-Cr	Constantan
J	JX	iron	Constantan
K[c]	KX	Ni-Cr	Ni-Al
T	TX	copper	Constantan
Noble Metal	Category 2		
B	BX	copper	copper
	proprietary	Cu-Mn[f]	copper
R	SX	copper	Cu-Ni[d,f]
S	SX	copper	Cu-Ni[d,f]
	proprietary	Ni-Cr[e]	Fe-Cr[e]
			OTHER
Refractory metal	Category 2		
W/W-26Re	proprietary	Ni-Cr[e]	Ni-Cr[e]
W-5Re/W-26Re	proprietary	Ni-Al[e]	Ni-Cu[e]
W-3Re/W-25Re	proprietary	Ni-Cr[e]	Ni-Cr[e]
Base metal			
Ni-14Cr/Ni-4Si	Category 1	Ni-Cr	Ni-Si
Ni-20Cr/Ni-3Si	Category 1	Ni-Cr	Ni-Si
Ni-18Mo/Ni-1Co	Category 1	Ni-Mo	Ni-Co
Noble metal			
Pt-20Rh/Pt-5Rh	Category 2	copper	copper
Pt-40Rh/Pt-20Rh	Category 2	copper	copper
Pt-13Rh/Pt-1Rh	Category 2	copper	copper
Pt-15Ir/Pd	Category 2	base metal	base metal
Pt-5Mo/Pt-0.1Mo	Category 2	copper	Cu-Ni
Ir-40,50,60Rh/Ir	Category 2	copper	AlSi 347 SS or aluminum
83Pd-14Pt-3Au/65Au-35Pd	Category 2	KP	KN
55Pd-31Pt-14Au/65Au-35Pd	Category 2	KP	KN
	Category 2	base metal	base metal

[a] See also ANSI MC96.1. Reference junction 0°C (32°F).

[b] M denotes strong magnetic response. O denotes little or no magnetic response at room temperature.

[c] Includes special Type K alloys discussed in 3.3.4.1.

[d] Driver Harris Co.

[e] Hoskins Manufacturing Co.

[f] AMAX Specialty Metals Corp.

thermocouples mentioned in Chapter 3.

Temperature Range		Limits of Error[a]				Magnetic[b] Response	
		Normal		Special			
°C	°F	±°C	±°F	±°C	±°F	P	N
THERMOCOUPLES							
0 to 200	32 to 400	1.7	3.0			O	O
0 to 200	32 to 400	2.2	4.0	1.1	2.0	M	O
0 to 200	32 to 400	2.2	4.0			O	M
−60 to 100	−75 to 200	1.0	1.8	0.5	0.9	O	O
0 to 100	32 to 200	+0.0 −3.7	+0.0 −6.7	measured junction >1000°C (1830°F)		O	O
0 to 320	32 to 600	...	...			O	O
0 to 200	32 to 400	5.0	9.0	measured junction >870°C (1600°F) or ±1% (whichever is greater)		O	O
0 to 200	32 to 400	5.0	9.0			O	O
0 to 540	32 to 1000	2.8	5.0			O	M
THERMOCOUPLES							
0 to 260	32 to 500	±0.14 MV		Equivalent to less than ±0.5% of measured temperature in range 1370 to 2200°C (2500 to 4000°F)		O	M
0 to 870	32 to 1600	±0.11 MV				M	M
0 to 260	32 to 500	±0.11 MV				O	M
0 to 200	32 to 400	2.2	4.0			O	M
0 to 200	32 to 400	2.2	4.0			O	M
0 to 200	32 to 400	2.2	4.0			O	M
0 to 175	32 to 350	not established				O	O
0 to 175	32 to 350	not established				O	O
0 to 175	32 to 350	not established				O	O
0 to 700	32 to 1300	not established				...	...
0 to 70	32 to 160	not established				O	O
not established		not established				O	O
at about 800	at about 1470	not established				O	M
at about 800	at about 1470	not established				O	M
0 to 160	32 to 320	...				...	...

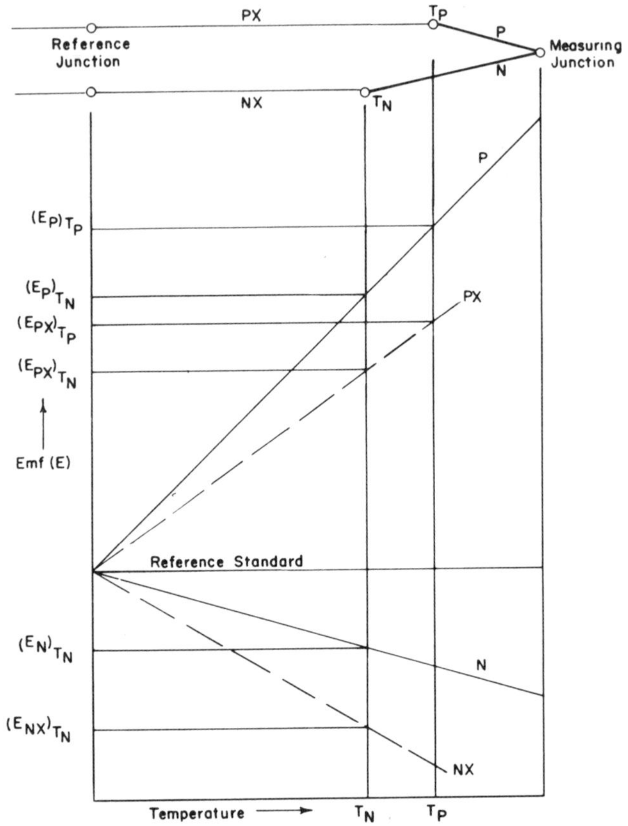

FIG. 3.3—*Error due to* ΔT *between the thermocouple-extension wire junctions.*

tions. The properties of each combination are detailed to allow a quick selection of a combination which is most likely to be suitable for a special condition. Thermocouple compositions are given in weight percent with the positive thermoelement of the thermocouple named first.

The information on newer thermocouple materials is presented using comments, tables, and curves. The comments made for the various thermocouple systems are intended to convey information not easily shown by tables or curves. The information contained in the tables is intended to help the reader to quickly decide if a certain thermocouple system, or a specific thermocouple, is suited to his particular needs. The information given is general and nominal, and cannot be used too literally. For example, the useful maximum temperature of a thermocouple depends in part on wire size, insulation used, method of installation, atmosphere conditions, vibration present, etc. The evaluation of certain properties as good, fair, or poor is subject to wide ranges of interpretation in terms of a particular application; hence, no attempt is made to define these terms. Approximate millivolt-versus-temperature relations for the various thermocouples are shown by curves. The curves are presented to show general temperature ranges for the various thermocouples but are not intended for use in converting emf to temperature. The reader should contact the wire manufacturer for temperature-emf tables.

The thermocouples described here are in use, and sufficient data and experience are available to warrant their inclusion. No attempt is made to include the many other thermocouple materials described in the literature which may have limited uses, or for which there are limited data, or for which there are serious problems of stability, emf reversibility, structural strength, etc.

The best source of information for a specific thermocouple in the "newer material" classifcation is considered to be the manufacturer of the particular thermocouple under consideration. Other useful information can be found in Ref *2*.

3.3.1 Platinum Types

3.3.1.1 Platinum-Rhodium Versus Platinum-Rhodium Thermocouples—The standard Types R and S thermocouples can be used for temperature measurement to the melting point of platinum, 1769°C (3216°F) on a short-term basis, but, for improved service life at temperatures over 1200°C (2192°F), special platinum-rhodium thermocouples are recommended.

The platinum-40 percent rhodium versus platinum-20 percent rhodium thermocouple, called the "Land-Jewell" thermocouple, is especially useful for continuous use to 1800°C (3272°F) or occasional use to 1850°C (3362°F). However, it is seldom used where the Type B thermocouple will suffice because of lower output and greater cost.

Other thermocouples suggested for high-temperature measurement have been a platinum-13 percent rhodium versus platinum-1 percent rhodium combination and a platinum-20 percent rhodium versus platinum-5 percent rhodium combination. The former shows slightly less tendency toward mechanical failure or contamination at high temperatures than the standard Types R and S thermocouples, while the latter has properties very similar to those of the Type B thermocouple.

Figure 3.4 and Table 3.11 show the characteristics of these alloys.

All special platinum-rhodium versus platinum-rhodium thermocouples, like the standard Types R, S, and B thermocouples, show improved life at high temperatures when protected by double-bore, full-length insulators of high-purity alumina.

3.3.1.2 Platinum-15 Percent Iridium Versus Palladium Thermocouples—The platinum-15 percent iridium versus palladium combination was developed as a high-output noble-metal thermocouple. It conbines the desirable attributes of noble metals with a high emf output as a lower cost than other noble-metal thermocouples.

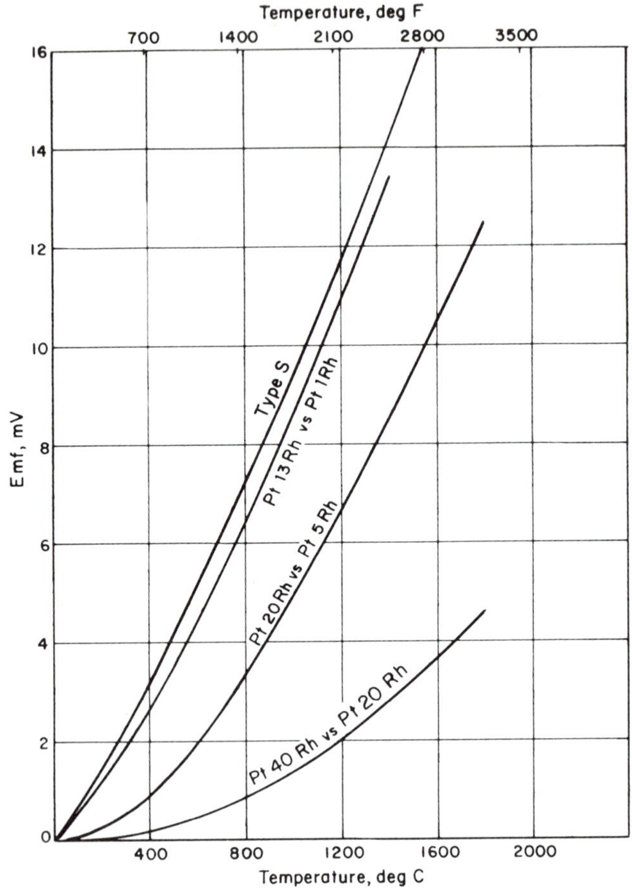

FIG. 3.4—*Thermal emf of platinum-rhodium versus platinum-rhodium thermocouples.*

The output becomes more linear and the Seebeck coefficient (thermoelectric power) increases with increasing temperature. In the absence of vibration, the useful range can probably be extended closer to the melting point of palladium, 1550°C (2826°F).

Figure 3.5 and Table 3.12 show the characteristics of these alloys.

Extension wires of base metals have been developed to provide a reasonable match with the thermocouple to about 700°C (1292°F).

Resistance to corrosion of the platinum-15 percent iridium alloy is better than that of the platinum-rhodium alloys in current use. Palladium is slightly less resistant to corrosion than the platinum alloy group. It will superficially oxidize at 700°C (1292°F). The oxide decomposes at about 875°C (1607°F) leaving a bright metal. When subjected to alternating oxidizing and reducing

TABLE 3.11—*Platinum-rhodium versus platinum-rhodium thermocouples.*

	Pt-20Rh Versus Pt-5Rh	Pt-40Rh Versus Pt-20Rh	Pt-13Rh Versus Pt-1Rh
Nominal operating temperature range, in:			
Reducing atmosphere (nonhydrogen)	NR[a]	NR	NR
Wet hydrogen	NR	NR	NR
Dry hydrogen	NR	NR	NR
Inert atmosphere	1700°C (3092°F)	1800°C (3272°F)	1600°C (2912°F)
Oxidizing atmosphere	1700°C (3092°F)	1800°C (3272°F)	1600°C (2912°F)
Vacuum (short-time use)	1700°C (3092°F)	1800°C (3272°F)	1600°C (2912°F)
Maximum short-time temperature	1770°C (3218°F)	1850°C (3362°F)	1770°C (3218°F)
Approximate microvolts per degree:			
Mean, over nominal operating range	6.8/°C (12.2/°F)	2.5/°C (4.5/°F)	9.9/°C (17.8/°F)
At top temperature of normal range	9.9/°C (17.8/°F)	4.7/°C (8.45/°F)	12.2/°C (22.0/°F)
Melting temperature, nominal:			
Positive thermoelement	1900°C (3452°F)	1930°C (3520°F)	1865°C (3389°F)
Negative thermoelement	1820°C (3308°F)	1900°C (3452°F)	1771°C (3220°F)
Stability with thermal cycling	good	good	good
High-temperature tensile properties	good	good	good
Stability under mechanical working	good	fair	good
Ductility (of most brittle thermoelement) after use	good	fair	good
Resistance to handling contamination	fair	fair	fair
Recommended extension wire, 175°C (347°F) max:			
Positive conductor	Cu	Cu	Cu
Negative conductor	Cu	Cu	Cu

[a]NR = not recommended.

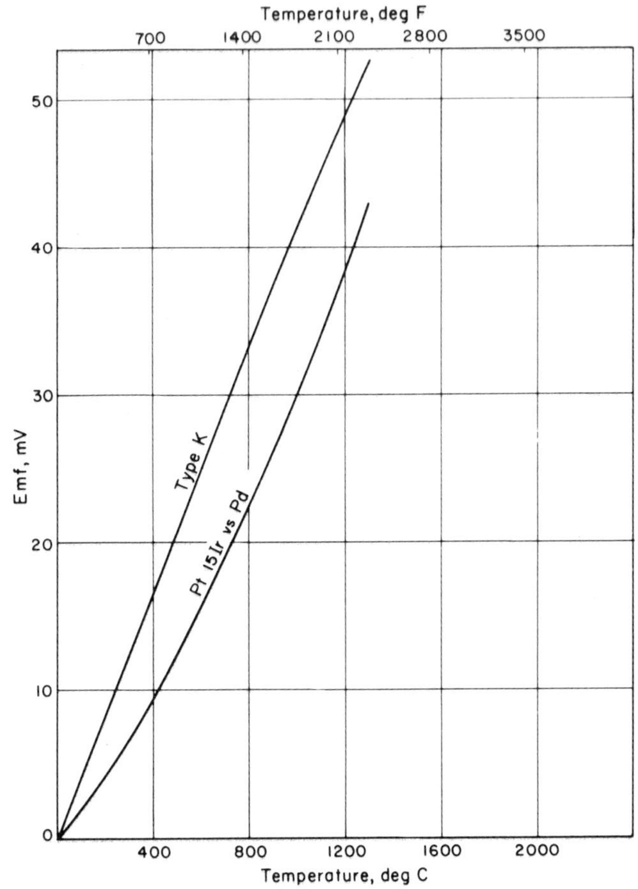

FIG. 3.5—*Thermal emf of platinum-iridium versus palladium thermocouples.*

atmospheres, surface blistering may result. As with all noble metals, the catalytic effect of the wires must be considered in combustible atmospheres. Its use may be preferred to base metals, however, for many applications. Both wires are ductile and may be reduced to very small sizes and still be handled with relative ease.

3.3.1.3 Platinum-5 Percent Molybdenum Versus Platinum-0.1 Percent Molybdenum Thermocouple—Platinum alloys containing rhodium are not suitable for use under neutron irradiation since the rhodium changes slowly to palladium. This causes a drift in the calibration of thermocouples containing rhodium. However, a thermocouple of platinum-5 percent molybdenum versus platinum-0.1 percent molybdenum is suitable for use in the helium atmosphere of a gas-cooled atomic reactor. Good stability at temperatures up

TABLE 3.12—*Platinum-iridium versus palladium thermocouple.*

	Pt-15Ir Versus Pd
Nominal operating temperature range, in:	
Reducing atmosphere (nonhydrogen)	NR[a]
Wet hydrogen	NR
Dry hydrogen	NR
Inert atmosphere	1370°C (2500°F)
Oxidizing atmosphere	1370°C (2500°F)
Vacuum	NR
Maximum short-time temperature	1550°C (2826°F)
Approximate microvolts per degree:	
Mean, over nominal operating range	12/°C (22/°F)
At top temperature of normal range	13.6/°C (24.6/°F)
Melting temperature, nominal:	
Positive thermoelement	1785°C (3245°F)
Negative thermoelement	1550°C (2826°F)
Stability with thermal cycling	good
High-temperature tensile properties	fair
Stability under mechanical working	good
Ductility (of most brittle thermoelement) after use	good
Resistance to handling contamination	fair
Recommended extension wire:	
Positive conductor	base metal alloys[b]
Negative conductor	base metal alloys[b]

[a] NR = not recommended.
[b] General Electric Company.

to 1400°C (2552°F) has been reported. The output of the thermocouple is high and increases in a fairly uniform manner with increasing temperature.

Figure 3.6 and Table 3.13 show the characteristics of these alloys.

The thermocouple usually is used in an insulated metallic sheath of platinum-5 percent molybdenum alloy. The sheath may be joined to a Type 321 stainless steel sheath beyond the area of the helium atmosphere. Both the platinum-molybdenum alloy and the Type 321 stainless steel behave well under neutron irradiation and are compatible with graphite which normally is used in the reactor.

Extension wires for this thermocouple can be copper for the positive conductor and copper-1.6 percent nickel for the negative conductor. Using these materials the junctions between the thermocouple and the extension wires should be maintained below 70°C (158°F).

3.3.2 Iridium-Rhodium Types

3.3.2.1 Iridium-Rhodium Versus Iridium Thermocouples—Iridium-rhodium versus iridium thermocouples are suitable for measuring temperature

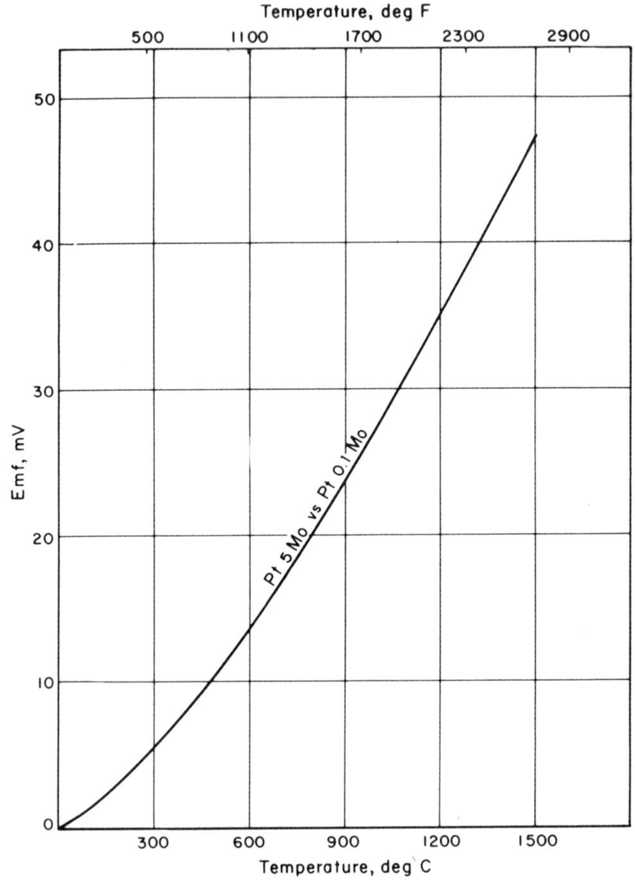

FIG. 3.6—*Thermal emf of platinum-molybdenum versus platinum-molybdenum thermocouples.*

to approximately 2000°C (3632°F), and generally are used above the range served by platinum-rhodium versus platinum thermocouples. They can be used in inert atmospheres and in vacuum, but not in reducing atmospheres, and they may be used in oxidizing atmospheres with shortened life.

The alloys of principal interest are those containing 40, 50, and 60 percent rhodium. They may be used for short times at maximum temperatures 2180, 2140, and 2090°C (3956, 3884, and 3794°F), these temperatures being 60°C (140°F) or more below the respective melting points.

Figure 3.7 and Table 3.14 show the characteristics of these alloys. The wires must be handled carefully. They are flexible in the fibrous (as drawn) state, but when annealed are broken easily by repeated bending.

TABLE 3.13—*Platinum-molybdenum versus platinum-molybdenum thermocouple.*

	Pt-5Mo Versus Pt-0.1Mo
Nominal operating temperature range, in:	
Reducing atmosphere (nonhydrogen)	NR[a]
Wet hydrogen	NR
Dry hydrogen	NR
Inert atmosphere (helium)	1400°C (2552°F)
Oxidizing atmosphere	NR
Vacuum	NR
Maximum short-time temperature	1550°C (2822°F)
Approximate microvolts per degree:	
Mean, over nominal operating range	29/°C (51.2/°F)
At top temperature of normal range	30/°C (54/°F)
Melting temperature, nominal:	
Positive thermoelement	1788°C (3250°F)
Negative thermoelement	1770°C (3218°F)
Stability with thermal cycling	good
High-temperature tensile properties	fair
Stability under mechanical working	good
Ductility (of most brittle thermoelement) after use	good
Resistance to handling contamination	fair
Recommended extension wire 70°C (158°F) max:	
Positive conductor	Cu
Negative conductor	Cu-1.6Ni

[a]NR = not recommended.

Metals said to be suitable for extension wires are 85 percent copper-15 percent nickel alloy for the positive conductor and 81 percent copper-19 percent nickel alloy for the negative conductor.[6]

3.3.2.2 Iridium-Rhodium Versus Platinum-Rhodium Thermocouples—Platinum-40 percent rhodium alloy[7] has been chosen by Lewis Research Center (NASA) as a substitute for an iridium thermoelement in combustor gas streams at pressures above 20 atmospheres and temperatures approaching 1600°C (2912°F). The thermocouple, consisting of a positive element of iridium-40 percent rhodium and a negative element of platinum-40 percent rhodium, showed reasonable oxidation resistance under these conditions.

Calibration to 1400°C (2552°F) showed the thermocouple output to be nearly linear and the absolute emf to be close to that of the iridium-40 percent rhodium versus iridium thermocouple.

[6]Johnson Matthey Incorporated.
[7]National Aeronautics and Space Administration Technical Brief, Nov. 1975.

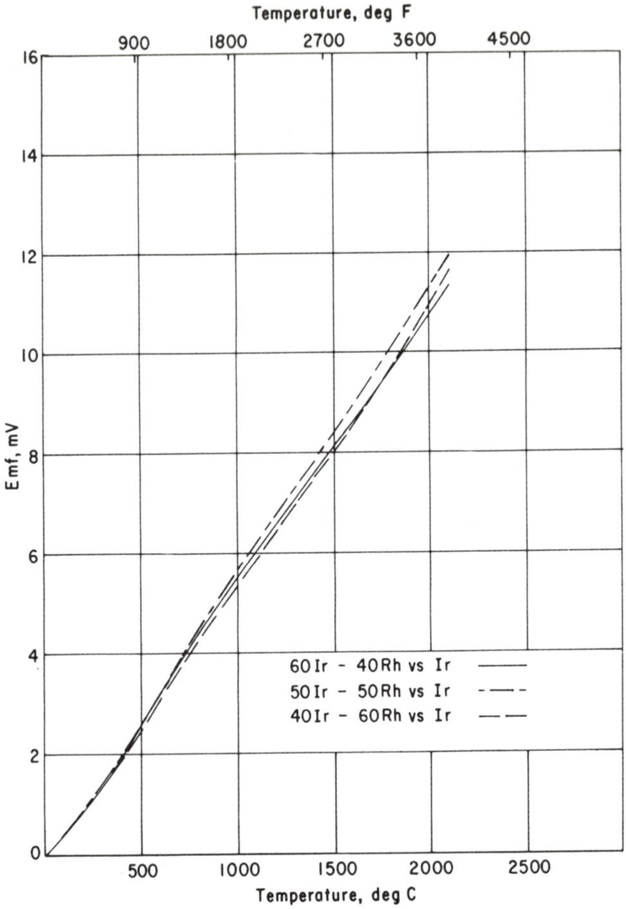

FIG. 3.7—*Thermal emf of iridium-rhodium versus iridium thermocouples.*

3.3.3 Platinel Types

3.3.3.1 Platinel Thermocouples—Platinel,[8] a noble-metal thermocouple combination, was metallurgically designed for high-temperature indication and control in turbo-prop engines. This combination approximates within reasonable tolerances the Type K thermocouple curve.

Actually, two combinations have been produced and are called Platinel I and Platinel II. The negative thermoelement in both thermocouples is a 65 percent gold-35 percent palladium alloy (Platinel 1503), but the positive one in Platinel I is composed of 83 percent palladium, 14 percent platinum, and

[8]Trademark of the Engelhard Industries, Inc.

TABLE 3.14—*Iridium-rhodium versus iridium thermocouples.*

	60Ir-40Rh Versus Ir	50Ir-50Rh Versus Ir	40Ir-60Rh Versus Ir
Nominal operating temperature range in:			
Reducing atmosphere (nonhydrogen)			
Wet hydrogen	NR[a]	NR	NR
Dry hydrogen	NR	NR	NR
Inert atmosphere	2100°C (3812°F)	2050°C (3722°F)	2000°C (3632°F)
Oxidizing atmosphere	NR	NR	NR
Vacuum	2100°C (3812°F)	2050°C (3722°F)	2000°C (3632°F)
Maximum short-time temperature	2190°C (3974°F)	2140°C (3884°F)	2090°C (3794°F)
Approximate microvolts per degree:			
Mean over nominal operating range	5.3°C (2.9°F)	5.7°C (3.2°F)	5.2°C (2.9°F)
At top temperature of normal range	5.6°C (3.1°F)	6.2°C (3.5°F)	5.0°C (2.8°F)
Melting temperature, nominal:			
Positive thermoelement	2250°C (4082°F)	2202°C (3996°F)	2153°C (3907°F)
Negative thermoelement	2443°C (4429°F)	2443°C (4429°F)	2443°C (4429°F)
Stability with thermal cycling	fair	fair	fair
High-temperature tensile properties	...	...	...
Stability under mechanical working	...	...	...
Ductility (of more brittle thermoelement) after use	poor	poor	poor
Resistance to handling contamination:			
Recommended extension wire	...	...	...
Positive conductor	...	...	...
Negative conductor	...	...	...

[a]NR = not recommended.

3 percent gold (Platinel 1786), while that used in Platinel II contains 55 percent palladium, 31 percent platinum, and 14 percent gold (Platinel 1813). Platinel II is the preferred type and has superior mechanical fatigue properties. The thermal emfs of these combinations differ little, as shown in Fig. 3.8. Other properties are given in Table 3.15.

From Fig. 3.8 it is apparent that the emf match with the Type K thermocouple is excellent at high temperatures, but some departure occurs at low temperatures. Generally, the user of Platinel makes the connection between the thermocouple and the extension wire (Type K thermocouple wire) at an elevated temperature (800°C) where the match is good. However, if this is done, care should be taken to ensure that the junctions of both conductors

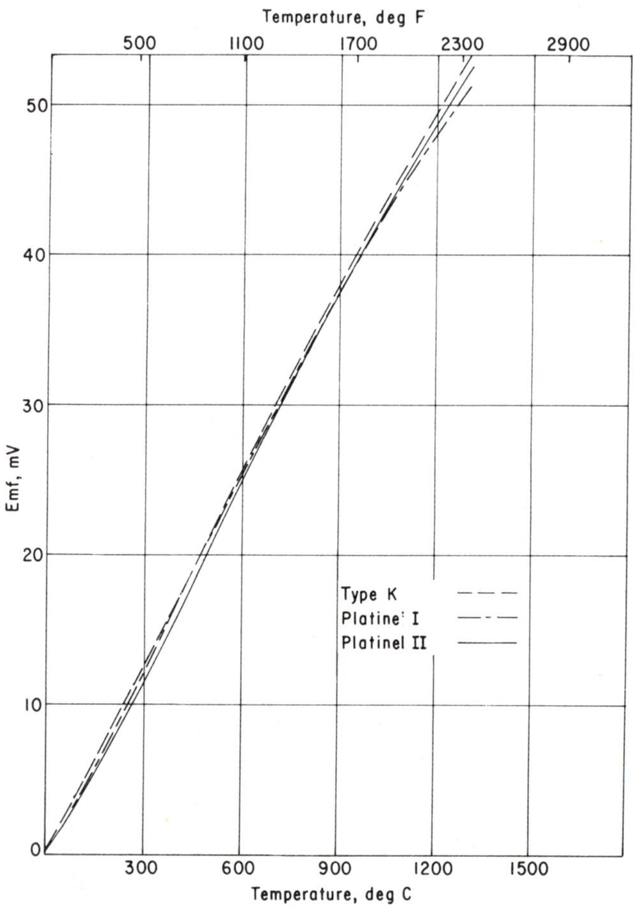

FIG. 3.8—*Thermal emf of platinel thermocouples.*

TABLE 3.15—*Platinel thermocouples.*

	Platinel II	Platinel I
Nominal operating temperature range, in:		
Reducing atmosphere (nonhydrogen)	NR[b]	NR
Wet hydrogen	NR	NR
Dry hydrogen[a]	1010°C (1850°F)	1010°C (1850°F)
Inert atmosphere	1260°C (2300°F)	1260°C (2300°F)
Oxidizing atmosphere	1260°C (2300°F)	1260°C (2300°F)
Vacuum	NR	NR
Maximum short-time temperature (<1 h)	1360°C (2480°F)	1360°C (2480°F)
Approximate microvolts per degree:		
Mean, over nominal operating range (100 to 1000°C)	42.5/°C (23.5/°F)	41.9/°C (23.3/°F)
At top temperature of normal range (1000 to 1300°C)	35.5/°C (19.6/°F)	33.1/°C (18.4/°F)
Melting temperature, nominal:		
Positive thermoelement—solidus	1500°C (2732°F)	1580°C (2876°F)
Negative thermoelement—solidus	1426°C (2599°F)	1426°C (2599°F)
Stability with thermal cycling	good	good
High-temperature tensile properties	fair	fair
Stability under mechanical working	?	?
Ductility (of most brittle thermoelement) after use	good	good
Resistance to handling contamination	?	?
Recommended extension wire at approximately 800°C (1472°F):		
Positive conductor	Type KP	Type KP
Negative conductor	Type KN	Type KN

[a] High-purity alumina insulators are recommended.
[b] NR = not recommended.

are at the same temperature. If the junction is made at a temperature where the extension wire/thermocouple emf match is not too close, then corrections should be made. Other base-metal extension wires capable of matching the emf of the Platinels very closely at low temperatures [to 160°C (320°F)] are also available.

It is recommended that precautions usually followed with the use of platinum-rhodium versus platinum thermocouples be observed when the Platinels are employed. Tests have shown that phosphorus, sulfur, and silicon have a deleterious effect on the life of the thermocouples.

3.3.4 Nickel-Chromium Types

3.3.4.1 Nickel-Chromium Alloy Thermocouples—Special nickel-chromium alloys are supplied by various manufacturers as detailed in the following paragraphs. Figure 3.9 and Table 3.16 give characteristics of these alloys.

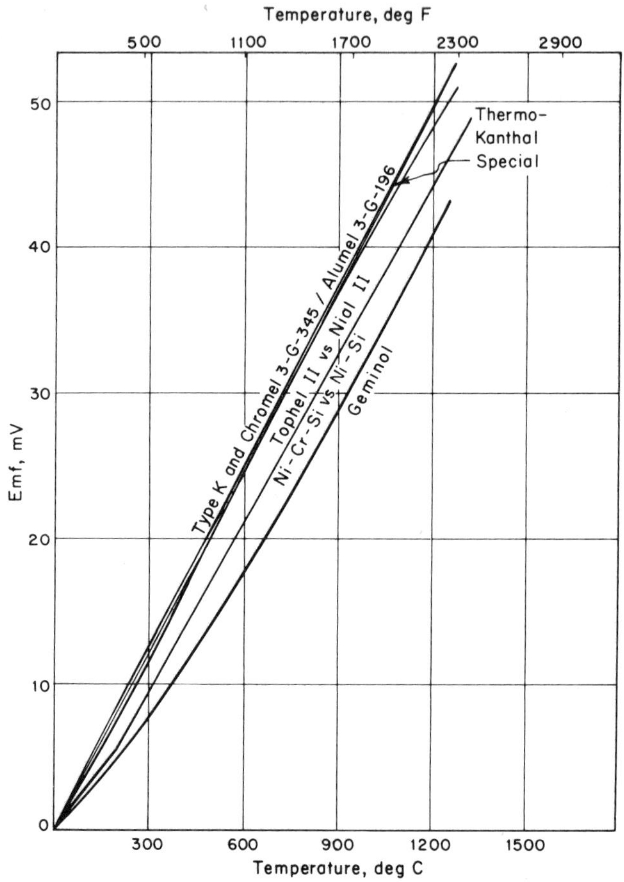

FIG. 3.9—*Thermal emf of nickel-chromium alloy thermocouples.*

3.3.4.1.1 Geminol—The Geminol[9] thermocouple was developed primarily for improved resistance to deterioration in reducing atmospheres.

The composition of the positive thermoelement has been adjusted specifically to combat in reducing atmospheres the destructive corrosion known as "green rot".

The substitution of an 80 percent nickel-20 percent chromium type alloy for conventional (Type KP) 90 percent nickel-10 percent chromium alloy positive thermoelement, and a 3 percent silicon in nickel alloy for the conventional (Type KN) manganese-aluminum-silicon in nickel alloy negative thermoelement, results in a more oxidation-resisting thermocouple.

[9]Trademark of the Driver-Harris Company.

TABLE 3.16—*Nickel-chromium alloy thermocouples.*

	Geminol	Thermo-Kanthal Special	Tophel II-Nial II	Chromel 3-G-345 Alumel 3-G-196	Ni-Cr-Si/ Ni-Si
Nominal operating temperature range, in:					
Reducing atmosphere (nonhydrogen)	1205°C (2200°F)		1205°C (2200°F)	1205°C (2200°F)	1205°C (2200°F)
Wet hydrogen	1205°C (2200°F)	1205°C (2200°F)	1205°C (2200°F)	1205°C (2200°F)	NR
Dry hydrogen	1205°C (2200°F)	1205°C (2200°F)	1205°C (2200°F)	1205°C (2200°F)	1205°C (2200°F)
Inert atmosphere	1205°C (2200°F)	1205°C (2200°F)	1205°C (2200°F)	1205°C (2200°F)	1250°C (2280°F)
Oxidizing atmosphere	1205°C (2200°F)	1205°C (2200°F)	1205°C (2200°F)	1205°C (2200°F)	1250°C (2280°F)
Vacuum	1040°C (1904°F)		1040°C (2000°F)	1040°C (2000°F)	NR[a]
Maximum short-time temperature	1260°C (2300°F)	1260°C (2300°F)	1260°C (2300°F)	1260°C (2300°F)	1300°C (2370°F)
Approximate microvolts per degree:					
Mean, over nominal operating range	18.7/°C (10.4/°F)	22.6/°C (12.6/°F)	40 μV/°C (22.5 μV/°F)	40.7 μV/°C (22.6 μV/°F)	37 μV/°C (21 μV/°F)
At top temperature of normal range	22.2/°C (12.3/°F)	20.0/°C (11.1/°F)	36 μV/°C (20μV/°F)	36 μV/°C (20 μV/°F)	36 μV/°C (20μV/°F)
Melting temperature, nominal:					
Positive thermoelement	1400°C (2550°F)	1432°C (2610°F)	1430°C (2600°F)	1430°C (2600°F)	1410°C (2570°F)
Negative thermoelement	1430°C (2600°F)	1410°C (2570°F)	1400°C (2550°F)	1400°C (2550°F)	1340°C (2445°F)
Stability with thermal cycling	good	good	good	good	good
High-temperature tensile properties	good	good	good	good	fair
Stability under mechanical working	intermediate	intermediate	fair	fair	
Ductility (of most brittle thermoelement) after use	good	good	good	good	good
Resistance to handling contamination	good	good	good	good	good
Recommended extension wire:					
Positive conductor	Geminol P	Thermo-Kanthal P	Tophel II or any Type K(+)	Chromel 3-G-345 or any Type K(+)	Ni-Cr-Si
Negatice conductor	Geminol N	Thermo-Kanthal N	Nial II	Alumel 3-G-196 and Type K(--)	Ni-Si

[a] NR = not recommended.

The temperature-emf curve is practically parallel to that of the conventional Type K thermocouple above 760°C (1400°F).

3.3.4.1.2 Thermo-Kanthal special—The Thermo-Kanthal special thermocouple was developed to give improved stability at temperatures between 982°C(1800°F) and 1260°C (2300°F) over that obtained with conventional base-metal thermocouple materials.

3.3.4.1.3 Tophel II-Nial II—

1. The Tophel II-Nial II thermocouple was developed for improved oxidation resistance and emf stability over the conventional Type K thermocouple alloys in both oxidizing and reducing atmospheres at elevated temperatures.
2. Tophel II, which is the positive thermoelement, is a nickel-10 percent chromium base alloy with additions to resist "green rot" attack in reducing atmospheres at elevated temperatures. The emf of Tophel II is within the standard tolerance of the conventional Type K positive thermoelement over the entire temperature range of 32 to 2300°F (0 to 1260°C). Tophel II can be matched with any acceptable Type K negative thermoelement to form a couple which is within the standard tolerance for the Type K thermocouple.
3. Nial II, which is the negative thermoelement, is a nickel-2.5 percent silicon base alloy with additions to improve the oxidation resistance and emf stability in an oxidizing atmosphere at elevated temperatures. The emf of Nial II is within the standard tolerance of the conventional Type K thermoelement between the range of 149 to 1260°C (300 to 2300°F). From 0 to 149°C (32 to 300°F), the Nial II is about 0.1 mV less negative than the conventional Type K negative thermoelement with reference to platinum.
4. The Tophel II-Nial II thermocouple meets the emf tolerances designated in ASTM Specification E 230 for the Type K thermocouple from 149 to 1093°C (300 to 2000°F). From 0 to 149°C (32 to 300°F), the Tophel II- Nial II thermocouple generates 0.1 mV (or 5 deg equivalent) less than the standard Type K thermocouple at the same temperature.
5. Tophel II-Nial II thermocouples can be used on existing instruments designed for the Type K thermocouples for temperatures sensing and control within the range of 149 to 1260°C (300 to 2300°F). If extension wire is needed, the negative extension wire should be Nial II, while the positive extension wire could either be Tophel II or any acceptable Type K (+) extension wire.
6. Through the improvements in both oxidation resistance and emf stability, Tophel II-Nial II thermocouples offers longer useful and total service life than conventional Type K thermocouples of the same size. As a corollary benefit, finer size Tophel II-Nial II thermocouples can be used to achieve equivalent or better stability than conventional Type K couples of larger sizes.

3.3.4.1.4 Chromel 3-G-345-Alumel 3-G-196—The Chromel 3-G-345-Alumel 3-G-196 thermocouple is designed to provide improved performance under extreme environmental conditions where the conventional Type K thermocouple is subject to accelerated loss of stability.

More specifically, Chromel 3-G-345 is a Type K positive thermoelement in which the basic 10 percent chromium-nickel alloy is modified to give improved resistance to preferential chromium oxidation ("green rot"). At temperatures from 871 to 1038°C (1600 to 1900°F), conventional Type K positive thermoelements operating in marginally oxidizing environments are subject to embrittlement and loss of output as a result of such attack.

Under those conditions, Type K thermocouples employing Chromel 3-G-345 positive thermoelements offer greater stability than conventional Type K thermocouples. The usual precautions regarding protection of Type K thermocouples in corrosive environments apply to the special thermocouple as well.

The modified Chromel thermoelement meets the accepted curve of emf versus platinum for Type K positive thermoelements, within standard tolerances. It can be combined with either Alumel 3-G-196 or conventional Alumel to form Type K thermocouples meeting standard emf tolerances.

Alumel 3-G-196 is a Type K negative thermoelement of greatly improved oxidation resistance. It is suited to use in both reducing and oxidizing atmospheres, where its stability of output is especially advantageous in fine wire applications at high temperatures. It is nominally 2.5 silicon-nickel.

Alumel 3-G-196 meets the accepted curve of emf versus platinum for Type K negative thermoelements at all temperatures from 0 to 1260°C (32 to 2300°F). It can be combined with either Chromel 3-G-345 or regular Chromel to form thermocouples meeting standard Type K thermocouple tolerances over the entire range from 0 to 1260°C (32 to 2300°F).

Type K thermocouples employing either or both special thermoelements can be used with conventional extension wires at no sacrifice in guaranteed accuracy of the thermocouple-extension wire combination.

3.3.4.1.5 Nickel-Chromium-Silicon versus Nickel-Silicon—

1. This combination resulted from many years of research in Australia and in the United States. Some of this work has been reported in NBS Monograph 161 [*3*].

2. The alloys in this combination have excellent resistance to preferential oxidation in the range 1000 to 1200°C.

3. The emf output differs from that of the Type K thermocouple but the curves have similar slopes over the elevated temperature range (see Fig. 3.9).

4. The short-term changes in the thermal emf outputs of 3.3 mm diameter nickel-chromium-silicon/nickel silicon and Type K thermocouples and of their individual thermoelements versus platinum on exposure in air at 1250°C are reported in NBS 161.

5. The long-term thermal emf drifts in 3.3 mm diameter nickel-chromium-silicon and Type K thermocouples and in their individual thermoelements versus platinum on exposure in air at 1000°C are also reported in NBS 161.

3.3.5 Nickel-Molybdenum Types

3.3.5.1 20 Alloy and 19 Alloy[10] *(nickel-nickel molybdenum alloys)—*

1. The 20 Alloy/19 Alloy thermocouple was developed for temperature sensing and control applications at elevated temperatures in hydrogen or other reducing atmospheres. The emf table of the 20 Alloy versus the 19 Alloy does not conform to the Type K or any existing base metal thermocouples designated by ASTM Specification E 230.
2. The 19 Alloy, which is the negative thermoelement, is essentially a nickel-1 percent molybdenum alloy. Its emf versus platinum values are somewhat more negative than those of the Type K negative thermoelement within the range of 0 to 1260°C (32 to 2300°F).
3. The 20 Alloy, which is the positive thermoelement, is essentially a nickel-18 percent molybdenum alloy. Its emf versus platinum values are less positive within the range of 0 to 260°C (32 to 500°F) than the Type K positive thermoelement, but more positive within the range of 260 to 1260°C (500 to 2300°F). Figure 3.10 and Table 3.17 show the characteristics of these alloys.
4. The 20 Alloy/19 Alloy thermocouple, when properly sealed in a protection tube, offers excellent emf stability at elevated temperatures in hydrogen or other reducing atmospheres.
5. The oxidation resistance of the 20 Alloy/19 Alloy thermocouple is not good. The 20 Alloy/19 Alloy thermocouples are not recommended for use in an oxidizing atmosphere above 649°C (1200°F).
6. 20 Alloy/19 Alloy extension wire should be used in connection with the 20 Alloy/19 Alloy thermocouple.

3.3.6 Tungsten-Rhenium Types

There are three tungsten-rhenium thermocouple systems available—tungsten versus tungsten-26 percent rhenium, doped tungsten-3 percent rhenium versus tungsten-25 percent rhenium and doped tungsten-5 percent rhenium versus tungsten-26 percent rhenium. The price of the first combination has the lowest cost. All have been employed to 2760°C (5000°F), but general use is below 2316°C (4200°F). Applications for these couples have been found in space vehicles, nuclear reactors, and many high-temperature electronic, thermoelectric, industrial heating, and structural projects. However, when employed in a nuclear environment, the effect of transmutation of the thermal emf of the couples should be considered.

The use of tungsten in certain applications as the positive element may pose a problem, since heating tungsten to or above its recrystallization

[10]The 20 Alloy versus 19 Alloy thermocouple was developed by the General Electric Company. Since 1962 the Wilbur B. Driver Company has been the sole manufacturer of the 20 Alloy and the 19 Alloy.

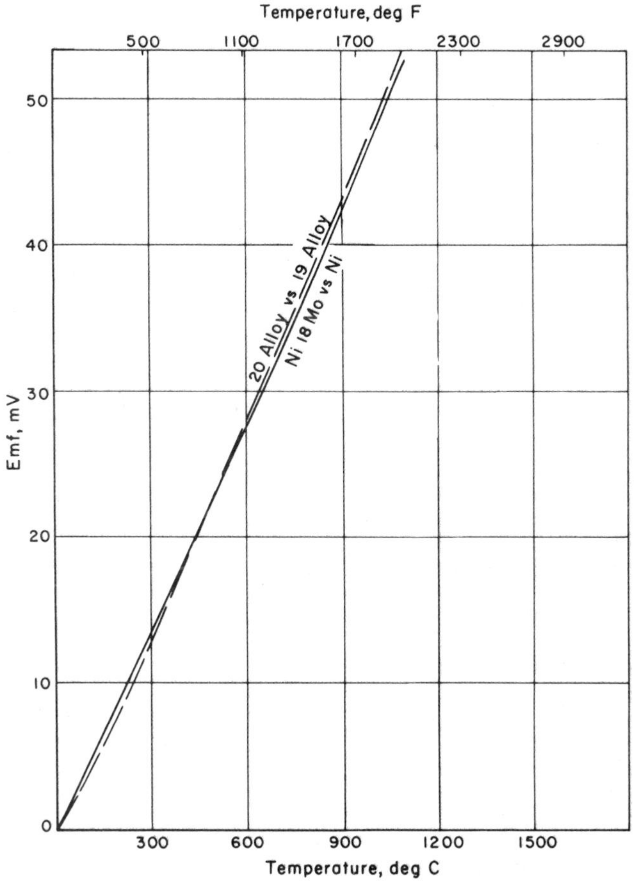

FIG. 3.10—*Thermal emf of nickel versus nickel-molybdenum thermocouples.*

temperature (approximately 1200°C) causes embrittlement resulting in loss of room-temperature ductility; an effect that is not experienced with the alloy leg containing high rhenium. With proper handling and usage this combination can be employed satisfactorily for long periods. One approach to the brittleness problem is to add rhenium to the tungsten thermoelement. Early research showed that the addition of 10 percent rhenium to the tungsten element did much to retain ductility after recrystallization. This much rhenium, however, greatly reduced the emf response for the thermocouple. Other techniques to retain room-temperature ductility are used by manufacturers; these include special processing and doping with the addition of 5 percent or less rhenium to the tungsten thermoelement.

Doping usually consists of using additives during the process of preparing

TABLE 3.17—*Physical data and recommended applications of the 20 Alloy/19 Alloy thermocouples.*

	20 Alloy/19 Alloy
Nominal operating temperature range, in:	
Reducing atmosphere (nonhydrogen)	1205°C (2200°F)
Wet hydrogen	1205°C (2200°F)
Dry hydrogen	1205°C (2200°F)
Inert atmosphere	1205°C (2200°F)
Oxidizing atmosphere	not recommended
Vacuum	1205°C (2200°F)
Maximum short time temperature	1260°C (2300°F)
Approximate microvolts per degree:	
Mean, over nominal operating range	55 μV/°C (31.0 μV/°F)
At top temperature of normal range	between the range of 1000 to 2300°F, 59 μV/°C (32.9 μV/°F)
Melting temperature, nominal:	
Positive thermoelement	1430°C (2600°F)
Negative thermoelement	1450°C (2640°F)
General stability with thermal cycling (good, fair, poor)	good
High temperature tensile properties (good, fair, poor)	good
Unaffected by mechanical working (good, fair, poor)	fair
Ductility (of most brittle thermal element) after use (good, fair, poor)	fair
Resistance to handling contamination (good, fair, poor)	good
Recommended extension wire:	
Positive conductor	20 Alloy
Negative conductor	19 Alloy

the tungsten powder and results in a unique microstructure in the finished wire. The additives essentially are eliminated during the subsequent sintering of the tungsten-rhenium powder compact. In fact, presently known analytical techniques do not disclose the presence of the additives above the background level of such substances normally present as impurities in non-doped tungsten or tungsten-rhenium alloys.

The emf response of tungsten-3 percent rhenium and tungsten-5 percent rhenium thermoelements used with thermoelements containing high percentages of rhenium is satisfactory. The thermoelectric power of the tungsten versus tungsten-26 percent rhenium, tungsten-3 percent rhenium versus tungsten-25 percent rhenium, and tungsten-5 percent rhenium versus tungsten-26 percent rhenium is comparable at lower temperatures, but drops off slightly for the latter two as the temperature is increased.

The tungsten thermoelement is not supplied to the user in a stabilized (recrystallized) condition; therefore, a small change in emf is encountered at the operating temperature. In the case of the doped tungsten-3 percent rhenium, doped tungsten-5 percent rhenium, tungsten-25 percent rhenium, and tungsten-26 percent rhenium thermoelements, these are supplied in a stabilized (recrystallized) condition.

All three thermocouple combinations are supplied as matched pairs guaranteed to meet the emf output of producer developed tables within ±1 percent. In addition, compensating extension wires are available for each of the three combinations with maximum service temperatures as high as 871°C (1600°F) for tungsten-5 percent rhenium versus tungsten-26 percent rhenium.

Important factors controlling the performance at high temperatures are: the diameter of the thermoelements (larger diameters are suitable for higher temperatures), the atmosphere (vacuum, high-purity hydrogen, or high-purity inert atmospheres required), the insulation, and sheath material. Some evidence is at hand, however, which indicates the possibility of selective vaporization of rhenium at temperatures of the order of 1900°C and higher when bare (unsheathed) tungsten-rhenium thermocouples are used in vacuum. For this reason, the vapor pressure of rhenium should be considered when a bare couple is used in a high vacuum at high temperatures. This, of course, is not a problem when these couples are protected with a suitable refractory metal sheath.

Figure 3.11 and Table 3.18 show the characteristics of these alloys.

3.4 Compatibility Problems at High Temperatures

In order for thermocouples to have a long life at high temperatures, it is necessary to limit reactions between the metals, the atmosphere, and the ceramic insulation. Such reactions may change the strength or corrosion-resistant properties of the alloys, the electrical output of the thermocouple, or the electrical insulation properties of the ceramic insulant.

At extremely high temperatures, reactions can be expected between almost any two materials. Table 3.19 has been included to show the temperatures at which such reactions occur between pairs of metallic elements. At lower temperatures, certain reactions do not occur and such as do occur, proceed at a slower rate. Because of potential reactions, it is important to identify the impurities and trace elements as well as the major constituents of the thermocouple components. The "free energy of formation" (Gibbs free energy) for the oxides of each element at the temperatures of interest, can be determined to predict possible oxide reactions. Other reactions may occur and attention should be given to the possible formation of the carbides, nitrides, etc. of the various elements.

Helpful data may be obtained from published reports, but, because of the importance of trace elements and impurities, the sources should be treated with caution. In some cases, the amount and types of impurities in the materials used were unknown.

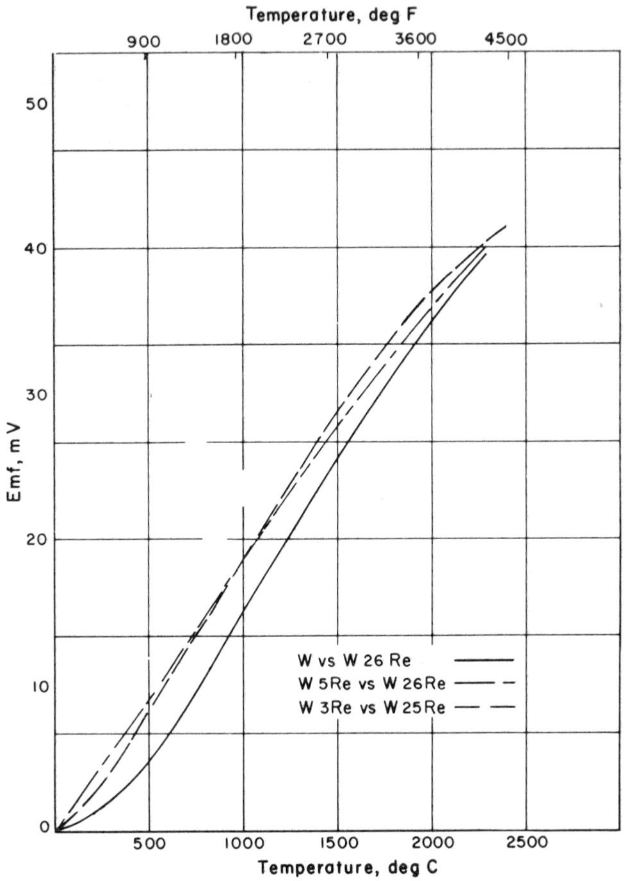

FIG. 3.11—*Thermal emf of tungsten-rhenium versus tungsten-rhenium thermocouples.*

Certain reactions may be somewhat self-limiting in that the reaction product provides a protective film against further reaction. However, spalling or chipping off of the reaction product may occur because of thermal or physical stress. Thus, the reaction rate and the use of the corrosion product as protection can be ascertained only if tested under the desired operating conditions and times.

The use of oxygen-gettering material should be considered in instances where oxygen is present in limited amounts [*4*]. This method has been employed with enclosed Type K thermocouples to limit the preferential oxidation of the positive thermoelement. A thin tube (or sliver) of titanium at the hottest location of the thermocouple has been used.

TABLE 3.18—*Tungsten-rhenium thermocouples.*

	W Versus W-26 Re	W-3Re Versus W-25 Re	W-5Re Versus W-26 Re
Nominal operating temperature range, in:			
Reducing atmosphere (nonhydrogen)	NR[a]	NR	NR
Wet hydrogen	NR	NR	NR
Dry hydrogen	2760°C (5000°F)	2760°C (5000°F)	2760°C (5000°F)
Inert atmosphere	2760°C (5000°F)	2760°C (5000°F)	2760°C (5000°F)
Oxidizing atmosphere	NR	NR	NR
Vacuum[b]	2760°C (5000°F)	2760°C (5000°F)	2760°C (5000°F)
Maximum short-time temperature	3000°C (5430°F)	3000°C (5430°F)	3000°C (5430°F)
Approximate microvolts per degree:			
Mean, over nominal operating range 0°C to 2316°C (32°F to 4200°F)	16.7/°C (9.3°F)	17.1/°C (9.5°F)	16.0/°C (8.9°F)
At top temperature of normal range 2316°C (4200°F)	12.1/°C (6.7/°F)	9.9/°C (5.5/°F)	8.8/°C (4.9/°F)
Melting temperature, nominal:			
Positive thermoelement	3410°C (6170°F)	3360°C (6080°F)	3350°C (6062°F)
Negative thermoelement	3120°C (5648°F)	3120°C (5648°F)	3120°C (5648°F)
Stability with thermal cycling	good	good	good
High-temperature tensile properties	good	good	good
Stability under mechanical working	fair	fair	fair
Ductility (of most brittle thermoelement after use)	poor	poor to good depending on atmosphere or degree of vacuum	poor to good depending on atmosphere or degree of vacuum
Resistance to handling contamination	good	good	good
Extension wire	available	available	available

[a] NR = not recommended.

[b] Preferential vaporization of rhenium may occur when bare (unsheathed) couple is used at high temperatures and high vacuum. Check vapor pressure of rhenium at operating temperature and vacuum before using bare couple.

TABLE 3.19—*Minimum melting temperatures of binary systems.*[a]

Element	W	Re	Os	Ta	Mo	Ir	Cb	Ru	Hf	Rh	V	Cr	Zr	Pt	Ti	Fe	Co	Ni	Si
Melting Point, °F	6170	5724	5432	5423	4730	4428	4425	4082	4030	3560	3452	3407	3353	3224	3034	2802	2723	2651	2588
Element:																			
Si	2550	~2057		2525	~2570	2588	2372				~2550	~2408	~2480	1526	2426	2192	2183	1770	
Ni	~2651	2651		2480	~2398		2145	2651	2102		2198	2444	~1764	~2651	1727	2588	~2646		
Co	~2696	~2723	~2723	2331	2444	~2460	2255	~2550	2214	~2550	2264	2552	1789	~2590	1860	~2696			
Fe	2775	~2802		~2462	2624		2480	2802	2372	~2750	2675	2743	1705	~2730	1985				
Ti	~3034	~3034		~3034	~3034	2687	~3034		2985		~2777	2525	~2876	2390					
Pt	~3224	~3224			3224	~3224	~3092			~3224		2552	2165						
Zr	~3020	2912		3308	2740		3164		3353		2246	~2372							
Cr	3038	3407		3092	3380		3020	2453			3182								
V	2974	3452		3308	3452		3290												
Rh					3524														
Hf	3488	~3416			3506														
Ru	4001			3578	3533		3056												
Cb	~4425	4415		~4425	~4250														
Ir																			
Mo	~4730	4424	~4406	~4730															
Ta	~5423	4874	~4388																
Os	4937																		
Re	~5070																		
W																			

[a] Adapted from: *Constitution of Binary Alloys* by Rodney P. Elliot, McGraw-Hill, New York, 1965.

3.5 References

[*1*] Moffat, R. J., "Understanding Thermocouple Behavior: The Key to Precision," Paper 68-628, Instrument Society of America, Vol. 5, Oct. 1968.

[*2*] Caldwell, F. R., "Thermocouple Materials," NBS Monograph 40, National Bureau of Standards, 1 March 1962.

[*3*] NBS Monograph 161, National Bureau of Standards.

[*4*] Neswald, R. G., "Titanium for Realists," *Space/Aeronautics*, Vol. 48, No. 5, Oct. 1967, pp. 90–99.

Chapter 4—Typical Thermocouple Designs and Applications

A complete thermocouple temperature sensing assembly, in accordance with the present state of the art, consists of one or more of the following:

A. Sensing Element Assembly—In its most basic form this assembly includes two dissimilar wires, supported or separated or both by electrical insulation and joined at one end to form a measuring junction. Such assemblies usually fall into one of three categories; (1) those formed from wires having nonceramic insulation, (2) those with hard-fired ceramic insulators, and (3) those made from sheathed, compacted ceramic-insulated wires. This chapter will deal only with the first two. See Chapter 5 for complete details on the latter.

B. Protecting Tube—Ceramic and metal protecting tubes serve the purpose of protecting the sensing element assembly from the deleterious effects of hostile atmospheres and environments. In some cases, two concentrically arranged protecting tubes may be used. The one closest to the sensing element assembly is designated the primary protecting tube, while the outer tube is termed the secondary protecting tube. Combinations such as an aluminum oxide primary tube and a silicon carbide secondary tube often are used to obtain the beneficial characteristics of the combination, such as resistance to cutting flame action and ability to resist thermal and mechanical shock.

C. Thermowell—More critical or demanding applications may require the use of specially machined and drilled solid bar stock called thermowells for not only protection of the thermocouple, but also to withstand high pressure or stresses or erosion or both caused by material flows within containment vessels. These drilled wells, as they are sometimes called, are machined to close tolerances, and highly polished to inhibit corrosion.

D. Terminations—Sensing element assembly wire terminations are made to:

(*a*) Terminals (usually screw type)
(*b*) Connection head
  1. General purpose type
  2. Screw cover type
  3. Open type

(*c*) Plug and jack quick-disconnect
(*d*) Military standard (MS) type of connector
(*e*) Miscellaneous connection devices such as crimp-on sleeves, transition fittings filled with epoxy or other potting materials, and so on.

E. Miscellaneous Hardware—These include:

(*a*) Pipe nipple or adapter—to join the protecting tube to the head
(*b*) Thermocouple gland—used primarily with sheathed, compacted ceramic-insulated thermocouple assemblies to serve the dual function of mounting and sealing-off pressure in the mounting hole (see Chapter 5). Such glands may also be used with small protecting tubes.
(*c*) Threaded bushing—welded or brazed to the protecting tube and screwed into the mounting hole.

4.1 Sensing Element Assemblies

Typical thermocouple element assemblies, shown in Fig. 4.1, A and B, illustrate common methods of forming the measuring junction, A—by twisting and welding (twisting provides added strength), and B—by butt-welding. C shows an assembly using nonceramic insulation such as asbestos or fiber glass. New ceramic fibers are available which extend the upper temperature limits of this form of insulation. D, E, and F show the use of various forms of hard fired ceramic insulators, double bore (D) fish-spine or ball and socket (E) and four-hole (F). Fish-spine provides flexibility, and four-hole provides for two independent sensing elements.

4.2 Nonceramic Insulation

The normal function of thermocouple insulation is to provide electrical separation for the thermoelements. If this function is not provided or is compromised in any way, the indicated temperature may be in error. Insulation is affected adversely by moisture, abrasion, flexing, temperature extremes, chemical attack, and nuclear radiation. Each type of insulation has its own limitations. A knowledge of these limitations is essential, if accurate and reliable measurements are to be made.

Some insulations have a natural moisture resistance. Teflon, polyvinyl chloride (PVC), and some forms of polyimides are examples of this group. With the fiber type insulations, moisture protection results from impregnating with substances such as wax, resins, or silicone compounds. Protection from abrasion and flexing is also generally provided by impregnating materials. However, one cycle over their rated temperatures will result in a deterioration of this protection. Once the impregnating materials have been exposed to temperatures at which they vaporize there is no longer moisture or mechanical protection.

The moisture penetration problem is not confined to the sensing end of the thermocouple assembly. For example, if the thermocouple passes through hot or cold zones or through an area which becomes alternately hot and cold, condensation may produce errors in the indicated temperature, unless adequate moisture resistance is provided.

After exposure to temperatures exceeding the limitations of the insulation, whole sections of the insulation can fall away resulting in bare wire and a possible "short." Thermocouples in this condition should not be used if any flexing or abrasion is expected. It is recommended that they be discarded or that the exposed portion of the thermocouple assembly be cut off and the measuring junction reformed.

A. Bare thermocouple element, twisted and welded.

B. Butt-welded thermocouple element.

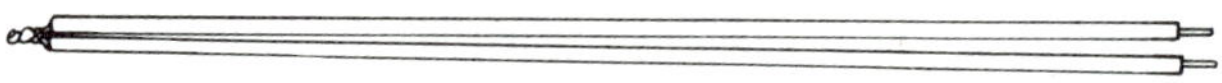

C. Thermocouple element, twisted and welded with asbestos insulation.

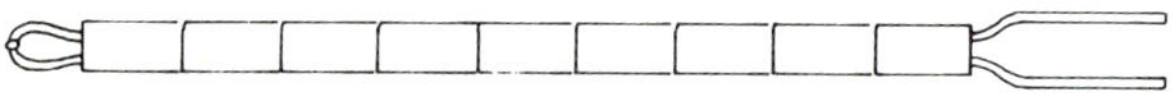

D. Butt-welded thermocouple element with double-bore insulators.

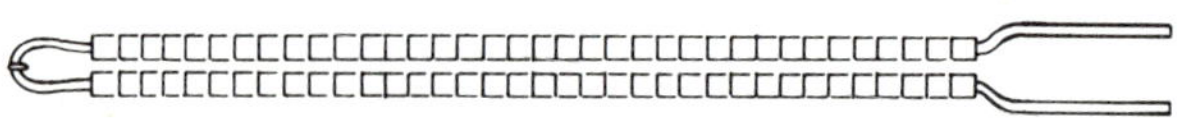

E. Butt-welded thermocouple element with fish-spine insulators.

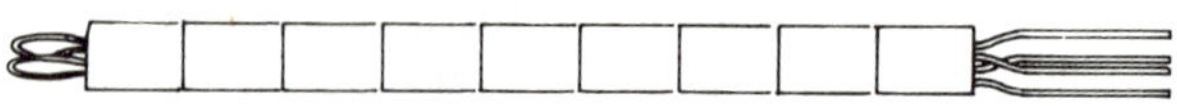

F. Two butt-welded thermocouple elements with 4-hole insulators.

FIG. 4.1—*Typical thermocouple element assemblies.*

Insulations are rated for a maximum temperature both for continuous usage and for a single exposure. It is imperative to observe the limits when selecting an insulating material. At elevated temperatures even those insulations which remain physically intact may become conductive. Under these conditions, the output of the thermocouple may be a function of the highest temperature to which the insulation is exposed, rather than the temperature of the measuring junction. The change in insulation resistance may be permanent if caused by deterioration of organic insulants or binders which leave a carbon residue. In considering the temperature to which the insulation is exposed, it should not be assumed that this is the temperature of the measuring junction. A thermocouple may be attached to a massive specimen which is exposed to a high-temperature source to achieve a rapid heating rate. Parts of the thermocouple wires not in thermal contact with the specimen can be overheated severely while the junction remains within safe temperature limits. With this in mind, high quality insulation should be used when rapid heating rates are expected. Very little factual information is available on actual deterioration rates and magnitudes, but the condition is real, so a conservative approach is a requisite of good engineering practice.

The basic types of flexible insulations for elevated temperature usage are fiber glass, fibrous silica, and asbestos.[1] Of the three materials, fibrous silica has the best high-temperature electrical properties, but, because this insulation normally is not impregnated, its handling and abrasion characteristics leave something to be desired. The next best high-temperature insulation is asbestos. Because this material has very poor mechanical properties a carrier fiber or an impregnating material is added. In some instances, this carrier is cotton or another organic compound which leaves a carbon residue after exposure to a temperature at which it burns. This results in a breakdown of electrical insulation. Asbestos loses its mechanical strength after exposure to elevated temperatures and may break away from the wire with little or even no handling. A more commonly used insulation is fiber glass. It can be impregnated to provide improved moisture and mechanical characteristics within the temperature limitations of the impregnating compound. The most frequently used type of fiber glass has an upper temperature limit of approximately 500°C (900°F) for continuous use. If one is willing to sacrifice the handling characteristics, nonimpregnated fiber glass insulations are available which withstand higher temperatures.

Modern technology has led to the development of ceramic fibers which markedly increase the upper use temperature of flexible insulations. These insulations, if properly applied and handled, will allow base metal thermocouples to be used to their upper use temperatures within their limits of exposure to the environment in which they are placed. For example, Type

[1] **Asbestos can be hazardous to health. Proper precaution should be exercised in its use.**

K thermocouples can be exposed to their stated upper temperature limit, and be adequately insulated provided the environmental factors do not otherwise affect the serviceability of the insulation. The same precautions must be observed with these thermocouples as with any other thermocouple installation. The user must be knowledgable about his application and the materials which he intends to use. These extremely high-temperature insulations should be employed without impregnants for maximum temperature service.

Chemical deterioration of insulation materials can produce a number of problems. If the environment reacts with the insulation, both the insulation and environment can be affected adversely. The insulation can be removed easily or become electrically conductive, and the process system can become contaminated. For example, some insulation materials are known to cause cracking in austenitic stainless steels.

In summary, an insulation should be selected only after considering possible exposure temperatures and heating rates, the number of temperature cycles, mechanical movement, moisture, routing of the thermocouple wire, and chemical deterioration (see Table 4.1).

Industry has established insulation color codes for standardized letter-designated thermocouple and extension wire types, as shown in Table 4.2.

4.3 Hard-Fired Ceramic Insulators

Hard-fired ceramic insulators most commonly used with bare thermocouple elements are mullite, aluminum oxide, and steatite, the latter being the most common material where fish-spine insulators are concerned. Single, double, and multibore insulators are available in a wide variety of sizes in both English and metric dimensions. Lengths commonly stocked by many suppliers are 1, 2, 3, 6, 12, 18, 24, and 36 in. Lengths to 72 in. are available on special order. The longer these insulators become, the more susceptible they are to breakage, so care must be exercised in handling.

It is advisable, especially in the case of precious metal thermocouple element assemblies, to keep the insulator in one piece to minimize contamination from the environment.

Hard-fired ceramic insulators are made in oval as well as circular cross-section examples of which are shown in Fig. 4.2. Properties of refractory oxides are tabulated in Table 4.3.

4.4 Protecting Tubes, Thermowells, and Ceramic Tubes

4.4.1 Factors Affecting Choice of Protection for Thermocouples

Thermocouples must be protected from atmospheres that are not compatible with the thermocouple alloys. Protecting tubes serve the double

TABLE 4.1—*Insulation characteristics.*

Insulation	Continuous Use Temperature Limit, °C (°F)	Single Exposure Temperature Limit, °C (°F)	Moisture Resistance	Abrasion Resistance
Cotton	95 (200)	95 (200)	poor	fair
Polyvinyl	105 (220)	105 (220)	excellent	good
Enamel and cotton	95 (200)	95 (200)	fair	fair
Nylon	125 (260)	125 (260)	poor	good
Teflon[a]	205 (400)	315 (600)	excellent	good
Polyimide	315 (600)	400 (750)	excellent	good
Teflon and fiber glass[b]	315 (600)	370 to 540 (700 to 1000)	excellent to 600°F	good
Fiber glass-varnish or silicone impregnation *E*[c]	480 (900)	540 (1000)	fair to 400°F, poor above 400°F	fair to 400°F, poor above 400°F
Fiber glass, nonimpregnated *S*[d]	540 (1000)	650 (1200)	poor	fair
Asbestos and fiber glass with silicone[e]	480 (900)	650 (1200)	fair to 400°F	fair to 400°F, poor above 400°F
Felted asbestos	540 (1000)	650 (1200)	poor	poor
Asbestos over asbestos	540 (1000)	650 (1200)	poor	poor
Refrasil[f]	870 (1600)	1100 (2000)	very poor	very poor
Ceramic fibers (for example, Nextel[g])	1000 (1830)	1370 (2500)	very poor	poor

[a] Trademark of the E. I. duPont de Nemours & Co.
[b] The Teflon vaporizes at 315°C (600°F) with possible toxic effects.
[c] E = electrical grade fiber glass.
[d] S = structural grade fiber glass.
[e] Individual wires are asbestos and overbraid is fiber glass.
[f] Trademark of the H. I. Thompson Co.
[g] Trademark of the 3M Corp.

TABLE 4.2—*Color code of thermocouple and extension wire insulations.*

Thermocouple Wire			Thermocouple Extension Wire		
	Insulation Color Code			Insulation Color Code	
Symbol	Single	Duplex	Symbol	Single[b]	Duplex
E		brown	EX		purple
EP	purple		EPX	purple	
EN	red		ENX	red with purple trace	
J		brown	JX		black
JP	white		JPX	white	
JN	red		JNX	red with white trace	
K		brown	KX		yellow
KP	yellow		KPX	yellow	
KN	red		KNX	red with yellow trace	
T		brown	TX		blue
TP	blue		TPX	blue	
TN	red		TNX	red with blue trace	
B	. . .[a]		BX		gray
BP	. . .[a]		BPX	gray	
BN	. . .[a]		BNX	red with gray trace	
R	. . .[a]		RX		green
RP	. . .[a]		RPX	black	
RN	. . .[a]		RNX	red with black trace	
S	. . .[a]		SX		green
SP	. . .[a]		SPX	black	
SN	. . .[a]		SNX	red with black trace	

NOTE—For a more complete explanation of color codes see ANSI MC96.1

[a] Normally noble metal thermocouples are not insulated with colored insulations. However, if they were, the color codes of the extension wire singles but with brown duplex would apply.

[b] The trace colors are recommended when duplex coverings are not present, but optional when duplex coverings are applied.

purpose of guarding the thermocouple against mechanical damage and interposing a shield between the thermocouple and its surroundings so as to maintain it as nearly as possible in its best working environment.

The conditions that must be excluded are: (*a*) metals (solid, liquid, or vapor) which, coming into contact with the thermocouple, would alter its chemical composition; (*b*) furnace gases and fumes which may attack the thermocouple materials (sulphur and its compounds are particularly deleterious); (*c*) materials such as silica and some of its metallic oxides, which, in contact with the thermocouple in a reducing atmosphere, are reduced, and combine with the thermocouple to attack it; and (*d*) electrolytes which would attack the thermocouple material.

The choice of the proper protecting tube is governed by the conditions of use and by the tolerable life of the thermocouple. On many occasions the strength of the protecting tube may be more important than the long-term thermoelectric stability of the thermocouple. On the other hand, gas tight-

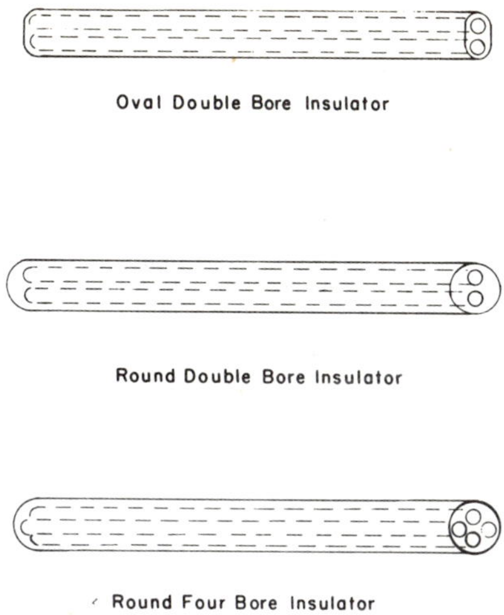

FIG. 4.2—*Cross-section examples of oval and circular hard-fired ceramic insulators.*

ness and resistance to thermal shock may be of paramount importance. In other cases, chemical compatibility of the protecting tube with the process may be the deciding factor. The problem of "green-rot" discussed in Section 3.1.1 also should be considered. In short, careful consideration must be given to each unique application.

As important as proper selection of the material is the cleanliness of the inside of the tube. Foreign matter, especially sulphur bearing compounds, can seriously affect the serviceability of the thermocouple as well as the tube itself.

4.4.2 Common Methods of Protecting Thermocouples

4.4.2.1 Protecting Tubes—By proper selection of material, metal protecting tubes can offer adequate mechanical protection for base metal thermocouples up to 1150°C (2100°F) in oxidizing atmospheres. It must be remembered that all metallic tubes are somewhat porous at temperatures exceeding 815°C (1500°F), so that, in some cases, it may be necessary to provide an inner tube of ceramic material otherwise damaging gaseous vapors may enter the tube and attack the thermocouple.

Typical use temperatures for protecting tubes are given next. These should be used only as a guide with careful selection of the proper material for each application.

(*a*) Carbon steels can be used to 540°C (1000°F) in oxidizing the atmospheres, and considerably higher if protected.

TABLE 4.3—*Properties*

	Composition	Porosity, volume %	Fusion Temperature, °C	Maximum Normal Use Temperature, °C	Density, Bulk (*b*), True (*t*), g/cm³
Sapphire crystal	99.9 Al_2O_3	0	2030	1950	3.97(*t*)
Sintered alumina	99.8 Al_2O_3	3 to 7	2030	1900	3.97(*t*)
Sintered beryllia	99.8 BeO	3 to 7	2570	1900	3.03(*t*)
Sintered calcia	99.8 CaO	5 to 10	2600	2000	3.32(*t*)
Chrome-alumina cermet (Haynes Stellite LT-1)	77 Cr, 23 Al_2O_3	2	1850	1300	5.9(*b*)
Sintered magnesia	99.8 MgO	3 to 7	2800	1900	3.58(*t*)
Sintered mullite	72 Al_2O_3, 28 SiO_2	3 to 10	1810	1750	3.03(*t*)
Sintered forsterite	99.5 Mg_2SiO_4	4 to 12	1885	1750	3.22(*t*)
Sintered spinel	99.8 $MgAl_2O_4$	3 to 10	2135	1850	3.58(*t*)
Sintered titania	99.5 TiO_2	3 to 7	1840	1600	4.24(*t*)
Sintered thoria	99.8 ThO_2	3 to 7	3050	2500	10.50(*t*)
Sintered yttria	99.8 Y_2O_3	2 to 5	2410	2000	4.50(*t*)
Sintered urania	99.8 UO_2	3 to 10	2800	2200	10.96(*t*)
Sintered stabilized zirconia	92 ZrO_2, 4 HfO_2, 4 CaO	3 to 10	2550	2200	5.6(*t*)
Sintered zircon	99.5 $ZrSIO_1$	5 to 15	2420	1800	4.7(*t*)
Silica glass	99.8 SiO_2	0	1710	1100	2.20(*t*)
Mullite porcelain	70 Al_2O_3, 27 SiO_2, 3 Mo + M_2O	2 to 10	1750	1400	2.8(*b*)
High alumina porcelain	90–95 Al_2O_3, 4–7 SiO_2, 1 to 4 Mo + M_2O	2 to 5	1800	1500	3.75(*b*)

[a] Kingery, W. D., "Oxides for High Temperature Applications," *Proceedings of the International Symposium on High Temperature Technology*. McGraw-Hill, New York, 1960.

(*b*) Austenitic stainless steels (AISI 300 series) can be used to 870°C (1600°F), in oxidizing atmospheres. Types 316, 317, and 318 can be used in some reducing atmospheres.

(*c*) Ferritic stainless steels (AISI 400 series) can, with proper selection, be used as high as 1093°C (2000°F) in both oxidizing and reducing atmospheres. Martensitic types (AISI 400 series), by contrast, are limited to lower temperatures for continuous use, with the top temperature being about 675°C (1250°F) for Type 410 stainless.

(*d*) High-nickel alloys, Nichrome,[2] Inconel,[3] etc., can be used to 1150°C (2100°F) in oxidizing atmospheres.

[2]Trademark of the Driver-Harris Company.
[3]Trademark of the International Nickel Company.

of refractory oxides. [a]

Specific Heat Capacity (cal/g/°C) 20 to 1000°C	Linear Expansion (10^{-6} in./in./°C) 20 to 1000°C	Thermal Conductivity (cal s^{-1} C^{-1} cm^{-2} cm) At 100°C	At 1000°C	Modulus of Rupture, psi At 20°C	At 1000°C	Modulus of Elasticity, 10^6 psi	Thermal Stress Resistance
0.26	8.6	0.072	0.019	40 000–150 000	30 000–100 000	55	very good
0.26	8.6	0.069	0.014	30 000	22 000	53	good
0.50	8.9	0.500	0.046	20 000	10 000	45	excellent
0.23	13.0	0.033	0.017	...	...	...	fair–poor
0.16	8.9	0.08	0.05	45 000	20 000	37.5	excellent
0.25	13.5	0.082	0.016	14 000	12 000	30.5	fair-poor
0.25	5.3	0.013	0.008	12 000	7 000	21	good
0.23	10.6	0.010	0.005	10 000	...	...	fair–poor
0.25	8.8	0.033	0.013	12 300	11 000	34.5	fair
0.20	8.7	0.015	0.008	8 000	6 000	...	fair–poor
0.06	9.0	0.022	0.007	12 000	7 000	21	fair–poor
0.13	9.3	(0.02)	...	...	...	...	fair–poor
0.06	10.0	0.020	0.007	12 000	18 000	25	fair–poor
0.14	10.0	0.005	0.005	20 000	15 000	22	fair–good
0.16	4.2	0.015	0.008	12 000	6 000	30	good
0.18	0.5	0.004	0.012	15 500	...	10.5	excellent
0.25	5.5	0.007	0.006	10 000	6 000	10	good
0.26	7.8	0.05	0.015	50 000	...	53	very good

As with carbon steels, the stainless steels and other alloy steels used for protecting tubes can be exposed to higher temperatures if properly protected.

The majority of metal protecting tubes are made from pipe sized tubing, cut to the proper length, and threaded on one end for attachment of the terminal head. The other end is closed by welding using a filler metal which is the same as the parent metal.

4.4.2.2 Thermowells—Where the thermocouple assembly is subject to high-pressure or flow-induced stresses or both, a drilled thermowell often is recommended. Although less expensive metal tubes, fabricated by plugging the end may satisfy application requirements, more stringent specifications usually dictate the choice of gun-drilled bar stock, polished and hydrostatically tested as a precaution against failure.

Examples of drilled thermowells are shown in Fig. 4.3.

4.4.2.3 Ceramic Tubes—Ceramic tubes are used usually at temperatures beyond the ranges of metal tubes although they are sometimes used at lower temperatures in atmospheres harmful to metal tubes.

The ceramic tube most widely used has a mullite base with certain additives to improve the mechanical and thermal shock properties. The upper temperature limit is 1650°C (3000°F).

Silicon carbide tubes are used as secondary protecting tubes. This material resists the cutting action of flames. It is not impermeable to gases and, where a dense tube is required, a nitride-bonded type material can be obtained with greatly reduced permeability.

Fused alumina tubes can be used as primary or secondary protecting tubes or both where temperatures to 1900°C (3450°F) are expected and a gas tight tube is essential. Fused alumina tubes and insulators should be used with platinum-rhodium/platinum thermocouples above 1200°C (2200°F) in order to ensure long life and attain maximum accuracy. The mullite types contain impurities which can contaminate platinum above

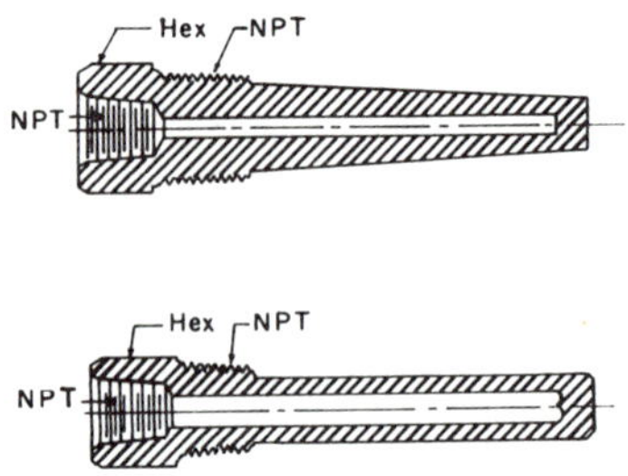

Straight and Tapered
Drilled Thermowells

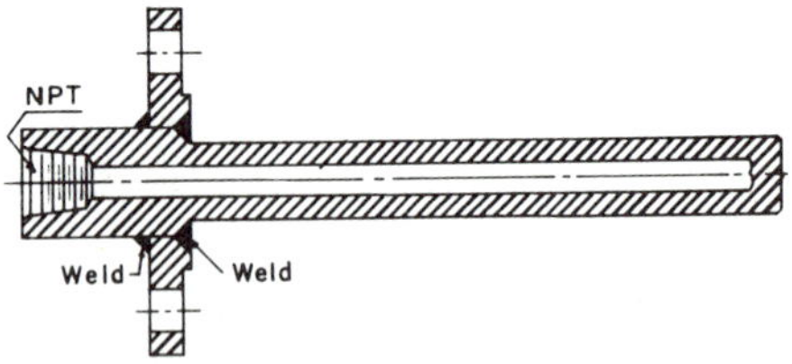

Flanged Thermowell

FIG. 4.3—*Examples of drilled thermowells.*

1200°C (2200°F). The alumina tubes are more expensive than the mullite base tubes but can be obtained impervious to most gases to 1815°C (3300°F).

4.4.2.4 Metal-Ceramic Tubes—"Cermets" are combinations of metals and metallic oxides which, after proper treatment, form dense, high-strength, corrosion-resistant tubes and are available for use to about 1425°C (2600°F) in most atmospheres.

4.5 Circuit Connections

To reduce costs and provide suitable insulation in the low-temperature parts of the thermocouple circuit, extension wire (see Section 3.2) is often used to extend the circuit to the measuring instrument. The intermediate junction between the thermocouple and extension wires must be held below the upper temperature limit of the extension wires (see Table 4.2) or considerable errors may be introduced.

Sheathed, mineral-insulated thermocouples can incorporate a transition fitting within which the extension wires are welded or brazed to the thermocouple wires. Where hard-fired ceramic insulators are used, the thermocouple wires usually terminate in a "head" which is attached to the insulator or to the protecting tube in which the insulator is housed. The interconnections are made by clamps or binding posts to permit easy replacement of the thermocouple. One assumes in such a component that, if a third metal is introduced into the circuit between the thermocouple and extension wires, no temperature gradient exists along its length. If this is so, in accordance with the law of intermediate metals, no error will result.

Industrial thermocouples often use heads of cast iron, aluminum, or die cast construction with screw covers and a temperature resistant gasket seal. Terminal blocks are commonly provided consisting of a ceramic or phenolic block and brass terminals. These assemblies are particularly likely to introduce errors due to temperature gradients, so they should be used with caution. Where phenolic terminal blocks are used, an upper temperature of about 170°C (340°F) applies.

To reduce the likelihood of thermal gradient errors, connectors of the quick disconnect type intended for use in thermocouple circuits have contacts made of thermoelectric materials matching the thermocouple conductors. Even with these connectors some errors can result when extreme gradients across the connectors are encountered. Quick disconnect connectors are more commonly used with sheathed ceramic insulated assemblies. See Chapter 5 for more details on their application.

Since any material inhomogenity in the circuit will introduce an error in the presence of a temperature gradient, best results will be obtained by avoiding extension wires, connectors, and terminal blocks. The thermocou-

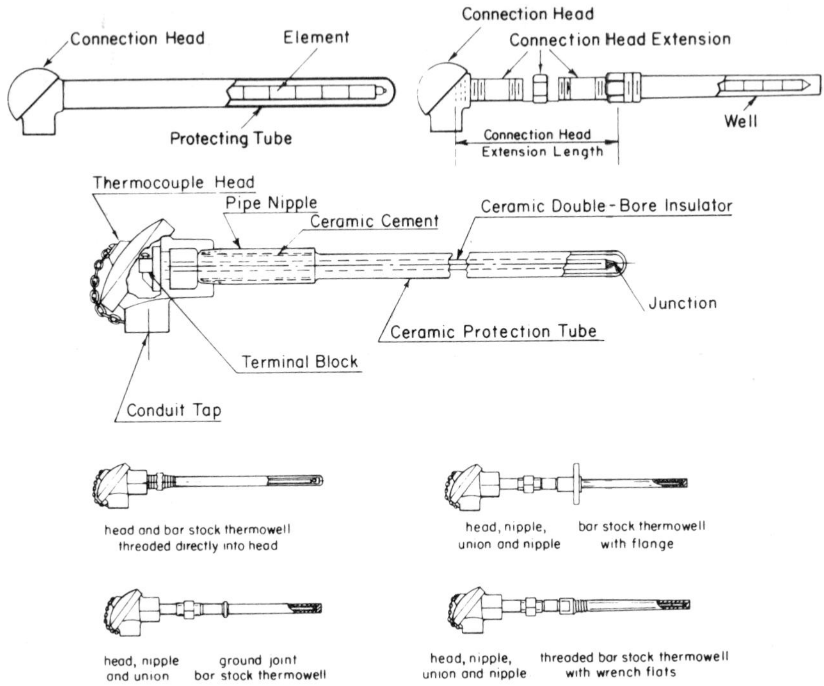

FIG. 4.4—*Typical examples of thermocouple assemblies with protecting tubes.*

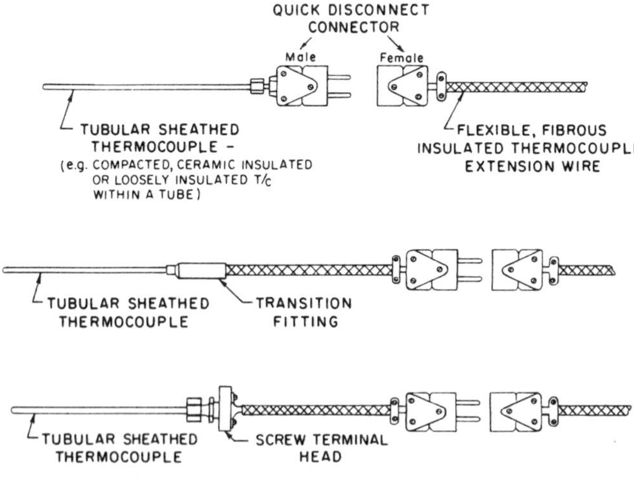

FIG. 4.5—*Typical examples of thermocouple assemblies using quick disconnect connectors.*

TABLE 4.4—*Selection guide for protecting tubes.*

Application	Protecting Tube Material
Heat treating:	
Annealing	
Up to 704°C (1300°F)	black steel
Over 704°C (1300°F)	Inconel 600,[a] Type 446 SS
Carburizing hardening	
Up to 816°C (1500°F)	black steel, Type 446 SS
1093°C (1500 to 2000°F)	Inconel 600, Type 446 SS
Over 1093°C (2000°F)	ceramic[b]
Nitriding salt baths	Type 446 SS
Cyanide	nickel (CP)
Neutral	Type 446 SS
High speed	ceramic[b]
Iron and steel:	
Basic oxygen furnace	quartz
Blast furnaces	
Downcomer	Inconel 600, Type 446 SS
Stove dome	silicon carbide
Hot blast main	Inconel 600
Stove trunk	Inconel 600
Stove outlet flue	black steel
Open hearth	
Flues and stack	Inconel 600, Type 446 SS
Checkers	Inconel 600, Cermets
Waste heat boiler	Inconel 600, Type 446 SS
Billet heating slab heating and butt welding	
Up to 1093°C (2000°F)	Inconel 600, Type 446 SS
Over 1093°C (2000°F)	silicon ceramic carbide[b]
Bright annealing batch	
Top work temperature	not required (use bare Type J thermocouple)
Bottom work temperature	Type 446 SS
Continuous furnace section	Inconel 600, ceramic[b]
Forging	silicon carbide, ceramic[b]
Soaking pits	
Up to 1093°C (2000°F)	Inconel 600
Over 1093°C (2000°F)	silicon ceramic carbide[b]
Nonferrous metals:	
Aluminum	
Melting	cast iron (white-washed)
Heat treating	black steel
Brass or bronze	not required (use dip-type thermocouple)
Lead	Type 446 SS, black steel
Magnesium	black steel, cast iron
Tin	extra heavy carbon steel
Zinc	extra heavy carbon steel
Pickling tanks	chemical lead
Cement:	
Exit flues	Inconel 600, Type 446 SS
Kilns, heating zone	Inconel 600
Ceramic:	
Kilns	ceramic[b] and silicon carbide[c]
Dryers	silicon carbide, black steel
Vitreous enamelling	Inconel 600, Type 446 SS

TABLE 4.4—*(Continued)*.

Application	Protecting Tube Material
Glass:	
Fore hearths and feeders	platinum thimble
Lehrs	black steel
Tanks	
Roof and wall	ceramic[a]
Flues and checkers	Inconel 600, Type 446 SS
Paper:	
Digesters	Type 316 SS, Type 446 SS
Petroleum:	
Dewaxing	Types 304, 310, 316, 321, 347 SS, carbon steel
Towers	Types 304, 310, 316, 321, 347 SS, carbon steel
Transfer lines	Types 304, 310, 316, 321, 347 SS, carbon steel
Factioning column	Types 304, 310, 316, 321, 347 SS, carbon steel
Bridgewall	Types 304, 310, 316, 321, 347 SS, carbon steel
Power:	
Coal-air mixtures	304 SS
Flue gases	black steel, Type 446 SS
Preheaters	black steel, Type 446 SS
Steel lines	Types 347 or 316 SS
Water lines	low carbon steels
Boiler tubes	Types 304, 309, or 310 SS
Gas producers:	
Producer gas	Type 446 SS
Water gas	
Carburetor	Inconel 600, Type 446 SS
Superheater	Inconel 600, Type 446 SS
Tar stills	low carbon steels
Incinerators:	
Up to 1093°C (2000°F)	Inconel 600, Type 446 SS
Over 1093°C (2000°F)	ceramic (primary) silicon carbide (secondary)[a]
Food:	
Baking ovens	black steel
Charretort, sugar	black steel
Vegetables and fruit	Type 304 SS
Chemical:	
Acetic acid	
10 to 50% 21°C (70°F)	Type 304, Hastelloy C,[d] Monel
50% 100°C (212°F)	Type 316, Hastelloy C,[d] Monel
99% 21 to 100°C (70 to 212°F)	Type 430, Hastelloy C,[d] Monel
Alcohol, ethyl, methyl	
21 to 100°C (70 to 212°F)	Type 304
Ammonia	
All concentration 21°C (70°F)	Types 304, 316 SS
Ammonium chloride	
All concentration 100°C (212°F)	Type 316 SS, Monel
Ammonium nitrate	
All concentration 21 to 100°C (70 to 212°F)	Type 316 SS
Ammonium sulphate, 10% to saturated, 100°C (212°F)	Type 316 SS

TABLE 4.4—*(Continued)*.

Application	Protecting Tube Material
Barium chloride, all concentration, 21°C (70°F)	Monel, Hastelloy C
Barium hydroxide, all concentration, 21°C (70°F)	low carbon steels
Barium sulphite	Nichrome,[e] Hastelloy C
Brines	Monel
Bromine	tantalum, Monel
Butadiene	Type 304 SS
Butane	Type 304 SS
Butylacetate	Monel
Butyl alcohol	copper, Type 304 SS
Calcium chlorate, dilute, 21 to 66°C (70 to 150°F)	
Calcium hydroxide	Type 304 SS
10 to 20% 100°C (212°F)	Type 304 SS, Hastelloy C
50% 100°C (212°F)	Type 316 SS, Hastelloy C
Carbolic acid, all, 100°C (212°F)	Type 316 SS
Carbon dioxide, wet or dry	2017-T4 aluminum, Monel, nickel
Chlorine gas	
Dry, 21°C (70°F)	Type 316 SS, Monel
Moist, −7 to 100°C (20 to 212°F)	Hastelloy C
Chromic acid, 10 to 50% 100°C (212°F)	Type 316 SS, Hastelloy C (all concentrations)
Citric acid	
15% 21°C (70°F)	Type 304 SS, Hastelloy C (all concentrations)
15% 100°C (212°F)	Type 316 SS, Hastelloy C (all concentrations)
Concentrated, 100°C (212°F)	Type 316 SS, Hastelloy C (all concentrations)
Copper nitrate	Types 304 SS, 316 SS
Copper sulphate	Types 304 SS, 316 SS
Cresols	Type 304 SS
Cyanogen gas	Type 304 SS
Dow therm[f]	low carbon steels
Ether	Type 304 SS
Ethyl acetate	Monel, Type 304 SS
Ethyl chloride, 21°C (70°F)	Type 304 SS, low carbon steel
Ethyl sulphate, 21°C (70°F)	Monel
Ferric chloride, 5% 21°C (70°F) to boiling	tontalum, Hastelloy C
Ferric sulphate, 5% 21°C (70°F)	Type 304 SS
Ferrous sulphate, dilute, 21°C (70°F)	Type 304 SS
Formaldehyde	Types 304 SS, 316 SS
Formic acid, 5% 21 to 66°C (70 to 150°F)	Type 316 SS
Freon	Monel
Gallic acid, 5% 21 to 66°C (70 to 150°F)	Monel
Gasoline, 21°C (70°F)	Type 304 SS, low carbon steel
Glucose, 21°C (70°F)	Type 304 SS
Glycerine, 21°C (70°F)	Type 304 SS
Glycerol	Type 304 SS
Hydrobromic acid, 98% 100°C (212°F)	Hastelloy B
Hydrochloric acid	
1%, 5% 21°C (70°F)	Hastelloy C
1%, 5% 100°C (212°F)	Hastelloy B
25% 21 to 100°C (70 to 212°F)	Hastelloy B
Hydrofluoric acid, 60% 100°C (212°F)	Hastelloy C, Monel
Hydrogen peroxide, 21 to 100°C (70 to 212°F)	Types 316 SS, 304 SS
Hydrogen sulphide, wet and dry	Type 316 SS

TABLE 4.4—*(Continued).*

Application	Protecting Tube Material
Iodine, 21°C (70°F)	tantalum
Lactic acid	
5% 21°C (70°F)	Type 304 SS, 316 SS
5% 66°C (150°F)	Type 316 SS
10% 100°C (212°F)	tantalum
Magnesium chloride	
5% 21°C (70°F)	Monel, nickel
5% 100°C (212°F)	nickel
Magnesium sulphate, hot and cold	Monel
Muriatic acid, 21°C (70°F)	tantalum
Naptha, 21°C (70°F)	Type 304 SS
Natural gas, 21°C (70°F)	Types 304 SS, 316 SS, 317 SS
Nickel chloride, 21°C (70°F)	Type 304 SS
Nickel sulphate, hot and cold	Type 304 SS
Nitric acid	
5% 21°C (70°F)	Types 304 SS, 316 SS
20% 21°C (70°F)	Types 304 SS, 316 SS
50% 100°C (70°F)	Types 304 SS, 316 SS
50% 100°C (212°F)	Types 304 SS, 316 SS
65% 100°C (212°F)	Type 316 SS
Concentrated, 21°C (70°F)	Types 304 SS, 316 SS
Concentrated, 100°C (212°F)	tantalum
Nitrobenzene, 21°C (70°F)	Type 304 SS
Oleic acid, 21°C (70°F)	Type 316 SS
Oleum, 21°C (70°F)	Type 316 SS
Oxalic acid	
5% hot and cold	Type 304 SS
10% 100°C (212°F)	Monel
Oxygen	
21°C (70°F)	steel
Liquid	SS
Elevated temperatures	SS
Palmitic acid	Type 316 SS
Pentane	Type 340 SS
Phenol	Types 304 SS, 316 SS
Phosphoric acid	
1%, 5% 21°C (70°F)	Type 304 SS
10% 21°C (70°F)	Type 316 SS
10% 100°C (212°F)	Hastelloy C
30% 21 to 100°C (70 to 212°F)	Hastelloy B
85% 21 to 100°C (70 to 212°F)	Hastelloy B
Picric acid, 21°C (70°F)	Type 304 SS
Potassium bromide, 21°C (70°F)	Type 316 SS
Potassium carbonate, 1% 21°C (70°F)	Types 304 SS, 316 SS
Potassium chlorate, 21°C (70°F)	Type 304 SS
Potassium hydroxide	
5% 21°C (70°F)	Type 304 SS
25% 100°C (212°F)	Type 304 SS
60% 100°C (212°F)	Type 316 SS
Potassium nitrate	
5% 21°C (70°F)	Type 304 SS
5% 100°C (212°F)	Type 304 SS
Potassium permanganate, 5% 21°C (70°F)	Type 304 SS
Potassium sulphate, 5% 21°C (70°F)	Types 304 SS, 316 SS
Potassium sulphide, 21°C (70°F)	Types 304 SS, 316 SS

TABLE 4.4—*(Continued)*.

Application	Protecting Tube Material
Propane	Type 304 SS, low carbon steel
Pyrogallic acid	Type 304 SS
Quinine bisulphate, dry	Type 316 SS
Quinine sulphate, dry	Type 304 SS
Seawater	Monel or Hastelloy C
Salicylic acid	nickel
Sodium bicarbonate	
All concentration, 21°C (70°F)	Type 304 SS
5% 66°C (150°F)	Types 304 SS, 316 SS
Sodium carbonate, 5% 21 to 66°C (70 to 150°F)	Types 304 SS, 316 SS
Sodium chloride	
5% 21 to 66°C (70 to 150°F)	Type 316 SS
Saturated 21 to 100°C (70 to 212°F)	Type 316 SS, Monel
Sodium fluoride, 5% 21°C (70°F)	Monel
Sodium hydroxide	Types 304 SS, 316 SS, Hastelloy C
Sodium hypochlorite, 5% still	Type 316 SS, Hastelloy C
Sodium nitrate, fused	Type 316 SS
Sodium peroxide	Type 304 SS
Sodium sulphate, 21°C (70°F)	Types 304 SS, 316 SS
Sodium sulphide, 21°C (70°F)	Type 316 SS
Sodium sulphite, 30% 66°C (150°F)	Type 304 SS
Sulphur dioxide	
Moist gas, 21°C (70°F)	Type 316 SS
Gas, 302°C (575°F)	Types 304 SS, 316 SS
Sulphur	
Dry molten	Type 304 SS
Wet	Type 316 SS
Sulphuric acid	
5% 21 to 100°C (70 to 212°F)	Hastelloy B, 316 SS
10% 21 to 100°C (70 to 212°F)	Hastelloy B
50% 21 to 100°C (70 to 212°F)	Hastelloy B
90% 21°C (70°F)	Hastelloy B
90% 100°C (212°F)	Hastelloy D
Tannic acid, 21°C (70°F)	Type 304 SS, Hastelloy B
Tartaric acid	
21°C (70°F)	Type 304 SS
66°C (150°F)	Type 316 SS
Toluene	2017-T4 aluminum, low carbon steel
Turpentine	Types 304 SS, 316 SS
Whiskey and wine	Type 304 SS, nickel
Xylene	copper
Zinc chloride	Monel
Zinc sulphate	
5% 21°C (70°F)	Types 304 SS, 316 SS
Saturated, 21°C (70°F)	Types 304 SS, 316 SS
25% 100°C (212°F)	Types 304 SS, 316 SS

[a] Trademark of the International Nickel Co.
[b] Due to susceptability to cracking, sudden thermal shocks should be avoided.
[c] Due to susceptability to cracking, sudden thermal shocks should be avoided.
[d] Trademark of the Cabot Corp.
[e] Trademark of the Driver-Harris Co.
[f] Trademark of the Dow Chemical Corp.

ple wires should be run directly to the reference junction after which copper wires are interconnected for extension to the measuring instrument.

4.6 Complete Assemblies

Figures 4.4 and 4.5 show complete assemblies of the components which have been described in the foregoing sections. Many other combinations are possible. Manufacturers' catalogs may be consulted for other varieties and details.

4.7 Selection Guide for Protecting Tubes

The following information in Table 4.1 has been extracted from various manufacturers' literature. It is offered as a guide to the selection of protecting tubes. Caution should be exercised in applying this information to specific situations.

4.8 Bibliography

White, F. J., "Accuracy of Thermocouples in Radiant Heat Testing," *Experimental Mechanics.* Vol. 2, July 1962, p. 204.

Baker, H. D., Ryder, E. A., and Baker, N. H., *Temperature Measurement in Engineering.* Vol. 1, Wiley, New York, 1953.

McLaren, E. H. and Murdock, E. G., "New Considerations on the Preparation, Properties, and Limitations of the Standard Thermocouple for Thermometry," *Temperature, Its Measurement and Control in Science and Industry,* Reinhold, New York, Vol. 4, Part 3, 1972, p. 1543.

Benedict, R. P., "*Fundamentals of Temperature, Pressure, and Flow Measurements,* Second Edition, Wiley, New York, 1977.

Chapter 5—Sheathed, Ceramic-Insulated Thermocouples

5.1 General Considerations

Compacted ceramic insulated themocouple material consists of three parts as shown in Fig. 5.1. This type of thermocouple is in common use because it isolates the thermocouple wires from environments that may cause rapid deterioration and provides excellent high-temperature insulation for thermocouple wires. The sheath can be made of a metal compatible with the process in which it is being used and provides mechanical protection. The material is easy to use because it forms easily, retains the bent configuration, and is readily fabricated into finished thermocouple assemblies.

5.2 Construction

All compacted types of thermocouples are made by similar processes. They begin with matched thermocouple wires surrounded by noncompacted ceramic insulating material held within a metal tube. By drawing, swaging, or other mechanical reduction processes the tube is reduced in diameter, and the insulation is compacted around the wires. Several options are available depending upon the material combinations selected for the temperature measurement application.

A ductile sheath and brittle refractory metal wire combination requires a design wherein the starting tubing diameter is only slightly larger than the finished size and only large enough on the inside diameter to accommodate a crushable preformed ceramic insulator with the wire strung through the insulators. This combination then would be reduced to the final diameter by one of the compaction methods usually in a single reduction pass. The process is such that the wire is neither elongated nor reduced in diameter in recognition of its brittle nature at room temperature.

A brittle sheath/brittle wire combination does not lend itself to a compacted insulation design and, therefore, is assembled as a tube-insulator-wire combination without the subsequent sheath reduction. Ductile wire/ductile sheath combinations cover the widest range of commonly used materials and offer the producer the widest choice of design approaches.

Nominal physical dimensions of sheathed ceramic insulated ther-

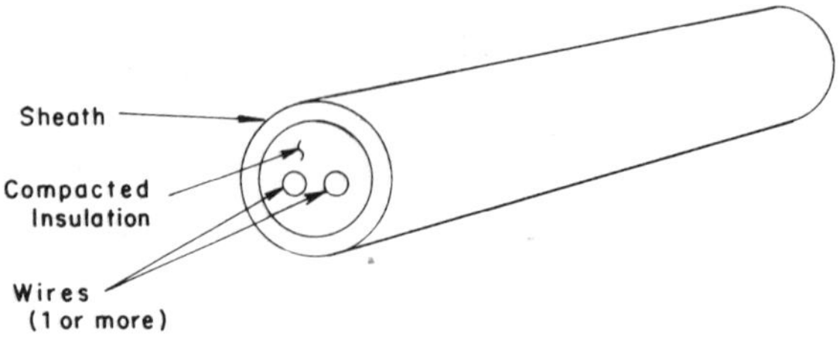

FIG. 5.1—*Compacted ceramic insulated thermocouple showing its three parts.*

mocouples are shown in Fig. 5.2. The ratios of sheath outside diameter to wire size and to sheath wall thickness offer a balance between maximum wall thickness (for protection of the sheath compacted insulant) and suitable insulation spacing for effective insulation resistance at elevated temperature. It is estimated that 90 percent of all sheathed thermocouples produced to date have used magnesium oxide (MgO) as the insulation material. Magnesium oxide is popular as a thermocouple insulator because of its overall compatibility with standard thermoelements and sheathing materials, its relative low cost, and its availability.

Aluminum, beryllium and thorium oxide insulations are also available from suppliers for use with certain wire and sheath combinations. The latter two materials are usually combined with refractory metal sheaths and thermoelements.

Because many applications of ceramic insulated thermocouples are at temperatures above 200°C (400°F), much attention must be given to cleanliness and chemical and metallurgical purity of the components. Great pains and expense having been taken to control the purity of the insulation during fabrication, the user would be foolish to abuse the sheathed material during assembly or application of the finished thermocouple. A warning heeded by the sailors and frontiersmen of early America can be applied today to the manufacture and use of sheathed thermocouples: Keep your powder clean and dry!

5.3 Insulation

For most practical purposes the sheathed thermocouple material should have a minimum insulation resistance of 100 megohms. This is readily obtained using dry, uncontaminated compacted ceramic. The capture of oil, oil vapors, moisture, perspiration, and lint during manufacture can cause low insulation resistance. The hygroscopic nature of the insulants, especially MgO, and capillary attraction cause rapid absorption of moisture through exposed ends of the sheath. Also, the insulation resistance of all ceramics,

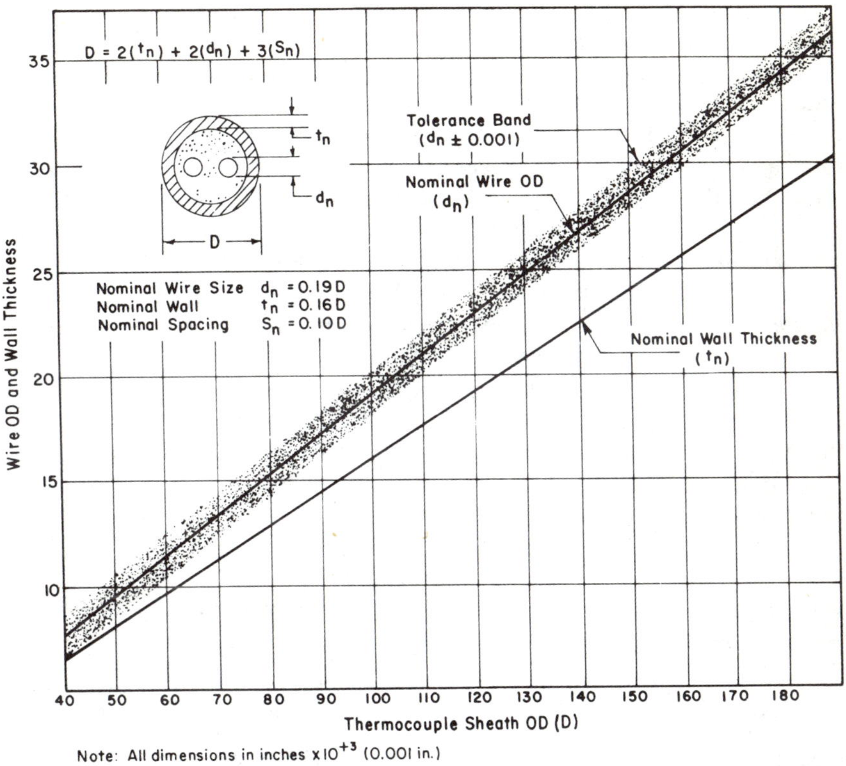

FIG. 5.2—*Nominal thermocouple sheath outside diameter versus internal dimensions.*

compacted and uncompacted, reduces with an increase in temperature, especially if contaminated.

When purity has been maintained so that the insulation resistance is greater than 1000 megohms, special techniques are required to maintain these values. Such material exposed to 21°C (70°F) air and a relative humidity above 50 percent will experience a degradation of insulation resistance to less than 0.1 megohm in 15 min. Higher humidity will cause a more rapid degradation. The following precautions should be exercised when handling compacted ceramic insulated thermocouples:

1. Do not leave an end exposed for periods longer than one minute. Immediately seal the end, preferably immediately after heating to expel any moisture.
2. Expose ends only in a region of low relative humidity.
3. Store sealed assembly in a warm (above 38°C (100°F)) and dry (relative humidity less than 25 percent) area.

Once a surface of the compacted oxide insulator is exposed to normal ambient humidity, moisture enters and diffuses inward through the insulator. The rate of diffusion depends on the humidity, the insulation temperature, and the material and degree of compaction of the oxide. As indicated, it is usually rapid. In fact, the ultimate purity can be only maintained if all operations are performed in an inert-gas dry box [*1*].[1] After having been absorbed and diffused, the moisture is very difficult to remove. Heat added to drive out the moisture will only increase its rate of reaction (oxidation) with the sheath wall and the thermoelements. Even though the insulation resistance may be restored, the damage has been done. Failure to recognize the effect and nature of this phenomenon is responsible for the poor performance reported by many users of sheathed thermocouples.

In the special case of a brief exposure of high-density material, limited success has been achieved by moving a moderate heat source 200 to 300°C (400 to 600°F) toward the exposed end and immediate sealing with epoxy.

Some of the characteristics of the more common compacted ceramics are shown in Tables 5.1 and 5.2.

TABLE 5.1—*Characteristic of insulating materials used in ceramic-packed thermocouple stock.*

Insulator	Minimum Purity, %	Melting Point, °C (°F)	Approximate Usable Temperature, °C (°F)
Magnesia (MgO)	99.4	2790 (5050)	1650 (3000)
Alumina (Al_2O_2)	99.5	2010 (3650)	1540 (2800)
Zirconia (ZrO_2)	99.4	2480 (4500)	650 (1200)
Beryllia (BeO)	99.8	2510 (4550)	2315 (4200)
Thoria (ThO_2)	99.5	3290 (5950)	2760 (5000)

TABLE 5.2—*Thermal expansion coefficient of refractory insulating materials and three common metals.*

Material	Average Coefficient of Expansion $\times 10^{-4}$ (25 to 700°C), °C^{-1}
Copper	16.5
Stainless steels	13.9 to 16.4
Aluminum	9.6
Magnesium oxide	12.9
Beryllium oxide	8.1
Aluminum oxide	7.1 to 8.0
Zirconium oxide	4.2 to 5.2

[1]The italic numbers in brackets refer to the list of references appended to this chapter.

5.4 Wire

The characteristics of the wires are no different from those of other thermocouples described in this manual. Since they are contained in a protective sheath and firmly held, small diameter wires can be exposed to high temperatures for long periods of time with less deterioration than that experienced by bare wire of the same size, provided they are properly fabricated and sealed.

5.5 Sheath

The sheath material performs several functions in the ceramic-insulated thermocouple system:

1. It holds the ceramic in compaction.
2. It shields the ceramic and thermoelements from the environment.
3. It furnishes mechanical strength to the assembly.

Since there is no sheath material suitable for all environments, a wide variety of different materials are offered. Table 5.3 shows a group of materials that can be used as sheaths and some of their characteristics. For maximum life a sheath diameter should be selected that offers the heaviest wall.

5.6 Combinations of Sheath, Insulation, and Wire

Precautions to be observed in selecting combinations are:

1. Be sure the sheath material will survive the environment. Remember that a thin sheath will deteriorate more rapidly than a thick one.
2. Be sure the assembly of sheath and wire is suitably annealed for optimum sheath life and stability of wire calibration.
3. Select a sheath and wire material of similar coefficients of linear expansion. For example: Platinum-rhodium and platinum have about one half the expansion of stainless or Inconel; a grounded hot junction may pull apart. The usual solution is to use hard-fired insulators in place of crushable and perhaps isolate the junction. Another solution is to use a platinum-rhodium sheath.
4. Select a sheath and wire material of similar annealing characteristics.
5. Do not use Types K, J, or E with an aluminum or copper sheath.

Table 5.4 indicates the compatibility of wire and sheath materials.
Table 5.5 gives dimensions of typical ceramic-packed stock.

5.7 Characteristics of the Basic Material

1. The sheath can usually be bent around a mandrel twice the sheath diameter without damage.

TABLE 5.3—*Sheath materials of ceramic-packed thermocouple stock and some of their properties.*

Material	Melting Point, °F	Recommended Maximum in Air, °F	Recommended: Operating Atmosphere	Recommended: Continuous Maximum Temperature, °F	Tensile Strength,[a] psi: At 200°F	Tensile Strength,[a] psi: At 1600°F
Stainless steel:						
304	2 560	1 920	ORNV	1 650	68 000	
309			ORNV	2 000		
310	2 560	2 000	ORNV	2 100	87 000	23 000
316	2 500	1 650	ORNV	1 700	75 000	23 000
321	2 550	1 650	ORNV	1 600	70 000	17 000
347	2 600	1 680	ORNV	1 600	75 000	
430	2 700	1 550	ORNV	1 200		
446	2 700	2 000	ORNV	2 000		
Inconel	2 550	2 000	ONV[b]	2 100	93 000	5 000
Inconel X	2 620	2 500	ONV[b]	2 200	150 000	11 000
Incoloy	2 500	1 640			77 000	3 000
Hastelloy X	2 350	2 300			106 000	7 000
Hastelloy C	2 310	1 820			136 000	64 000
Haynes 25	2 425	1 820			147 000	13 000
Hastelloy B	2 375	1 400			125 000	51 000
Monel	2 460	1 640				
Chromel	2 600	2 100	ONV		90 000	21 000
Copper	1 980	600	ORNV	600		
Brass	1 850	700				
Aluminum	1 220	800	ORNV	700		
Nichrome	2 550	2 200		2 000		
Alumel	2 550	2 100	ONV		82 000	19 000
Nickel	2 647	1 100				
Iron	2 798	600				
Zircaloy	3 350	1 400				
Platinum	3 217	3 000	ON[b]	3 000		
Pt-Rh 10%	3 362	3 100	ON	3 100		
Niobium	4 474	1 600	VN	3 800	110 000	
Molybdenum	4 730	400	VNR		137 000	30 000
Molybdenum disilicized		3 100	ON	3 000		
Molybdenum chromalized		3 100	ON	3 000		
Tantalum	5 425	750	V	5 000	96 000	22 000
Titanium	3 035	600	VN	2 000		

NOTE—Symbols describing atmospheres are: O = oxidizing; R = reducing; N = neutral; V = vacuum. 32°F = 0°C

[a] Scales readily in oxidizing atmosphere.

[b] Very sensitive to sulfur corrosion.

[c] After exposure to temperature of 100 h except for stainless steels; Haynes 25, W, Mo, Ta, and Nb.

TABLE 5.4—*Compatibility of wire and sheath material* [6].

Wire	Sheath													
	304	310	316	321	347	440	Platinum Alloys	Hastelloy X	Copper	Aluminum	Inconel 600	Inconel 702	Tantalum	Niobium Alloys
Type K	1	1	1	1	1	1	1	3	4	4	1	1	3	3
Type J	1	1	1	1	1	1	2	4	4	4	3	3	4	4
Type T	1	1	1	1	1	1	2	4	1	4	3	3	4	4
Type E	1	1	1	1	1	1	1	1	4	4	1	1	3	3
Types B, R, and S	1	1	1	1	1	1	1	1	4	4	1	1	4	4
Tungsten rhenium alloys	3	3	3	3	3	3	3	1	4	4	3	3	3	3
Iridium rhodium alloys	3	3	3	3	3	3	3	1	4	4	3	3	3	3
Copper	3	4	3	3	3	4	4	4	1	2	3	4	2	2
Nickel	1	1	1	1	1	1	1	1	4	4	1	1	3	3
Aluminum	4	4	4	4	4	4	4	4	4	1	4	4	4	4
Nichrome[a]	1	1	1	1	1	1	1	1	4	4	1	1	3	3

NOTE—1. Easy manufacturing and good operational compatibility.
2. Easy manufacturing but poor operational compatibility.
3. Difficult manufacturing but good operational compatibility.
4. Difficult manufacturing and poor operational compatibility.

[a] Trademark of the Driver-Harris Co.

TABLE 5.5—*Dimensions and wire sizes of typical ceramic-packed material.*

(*a*)

Sheath Outside Diameter, in.	Outside Diameter Tolerance, ±in.	Nominal Wall Thickness, in.	Approximate Wire, B&S gage	Nominal Production Length, ft
0.010	0.001	0.0015	48	250
0.020	0.001	0.003	38	250
0.032	0.001	0.005	36	250
0.040	0.001	0.007	34	250
0.062	0.002	0.010	29	150
0.090	0.002	0.014	26	125
0.125	0.002	0.018	24	100
0.188	0.003	0.025	18	60
0.250	0.003	0.032	17	40
0.313	0.003	0.040	16	40
0.375	0.003	0.049	14	30
0.430	0.003	0.065	13	30
0.500	0.003	0.070	12	30

(*b*)

Sheath Diameter, in.	Nominal Conductor Diameters, in.			
	1-Wire	2-Wire	3-Wire	4-Wire
0.313	0.064	0.051	0.040	0.040
0.250	0.051	0.040	0.032	0.032
0.188	0.040	0.032	0.022	0.022
0.125	0.032	0.022	0.011	0.011
0.042	0.022	0.011	0.006	0.006
0.040	0.011	0.006	...	...
0.025	0.006	0.004	...	...

NOTE—1 in. = 25.4 mm; 1 ft = 0.3048 m.

2. The life of material having a diameter of 0.81 mm (0.032 in.) or less may be limited by grain growth in the sheath wall.
3. Four wires in 1.57 mm (0.062 in.) sheath diameter and smaller are not practical to handle.
4. Two wires in 0.81 mm (0.032 in.) sheath diameter and smaller are difficult to handle but are used in laboratory environments.

Stock material and completed thermocouples are usually supplied to the end user in the fully annealed state with proper metallurgical grain size and no surface corrosion. Improper handling can easily destroy this condition. When wire is delivered coiled, it should not be uncoiled until needed for fabrication. Repeated or excessive bending will affect the annealed state. Sometimes the wire has been further heat treated after solution annealling, to control or stabilize calibration. If this is the case, further heat treatment to remove cold work will destroy these characteristics.

5.8 Testing

Many tests have been devised to evaluate sheathed ceramic insulated thermocouples. These tests cover the physical and metallurgical properties of the sheath and thermoelements, the electrical properties of the insulation, and the thermoelectric properties of the thermoelements.

ASTM has issued a specification E 235 [*2*] on sheathed thermocouples for nuclear and other high reliability applications which covers Type K thermoelements in various sheath materials, with alumina or magnesia insulations and alternative thermoelements. In Aerospace Information Report 46 [*3*] the Society of Automotive Engineers has covered the preparation and use of Type K thermocouples for turbojet engines, while the military has issued Mil T 22300 (Ships) and Mil T 23234 (Ships) for nuclear application of the sheathed thermocouples. In addition many government and commercial laboratories, thermocouple fabricators, and thermocouple users have issued specifications covering specific tests and test procedures for evaluating sheathed thermocouples.

Scadron [*4*] outlined some of the tests and tolerances that are applied for the verification of physical, electrical, and thermoelectric integrity of the sheathed thermocouple.

Most manufacturers' catalogs cover some aspects of the tests, guarantees, and procedures used to verify product integrity. Fenton et al [5] were concerned with inhomogeneity in bare and sheathed thermoelements and developed procedures and apparatus for testing for drift and inhomogeneity. National Bureau of Standards Circular 590 [*6*] covers thermocouple testing. While this publication applies to the bare thermoelements it is applicable to the electrical characteristics of the metal sheathed ceramic insulated material.

Table 5.6 shows various characteristics, tests, and the source of testing procedure which are applicable to sheathed ceramic-insulated thermocouples.

5.9 Measuring Junction

Numerous variations in measuring junction construction are possible with this type of material. The application dictates the most desirable method.

1. *Exposed or Bare Wire Junction* (Fig 5.3)—In this type of a junction the sheath and insulating material are removed to expose the thermocouple wires. These wires are joined to form a measuring junction. The junction may be of the twist-and-weld or butt-weld type. While the thermocouple will have a fast response, the exposed ceramic is not pressure tight and will pick up moisture, and the wires are subject to mechanical damage and are exposed to the environment.

2. *Grounded Junction* (Fig 5.4)—A closure is made by welding in an inert

TABLE 5.6—*Various characteristics tests and the source of testing procedure applicable to sheathed ceramic-insulated thermocouples.*

Characteristics	Tests	Procedure	Comments
		PHYSICAL AND METALLURGICAL	
1. OD of sheath	ring gage, micrometer, or met mounts for high precision	good mechanical practice or transverse section per ASTM Method E 3	tolerances per Table 29 sample preparation: requires care to prevent smear
2. Roundness of sheath	rotate for total indicated reading	Vee block and dial indicator	important for tube fitting use
3. Sheath surface finish	profilometer or roughness gage	ANSI Standard B46.1	also check clean finish for welding and brazing
4. Sheath wall thickness	thread gage micrometer or metallurgical mount	transverse section per ASTM Method E 3	concerns machining, bending, welding, life: can be checked only at ends
5. Insulation spacing	comparator or metallurgical mount	transverse section per ASTM Method E 3	see electrical tests
6. Insulation compactness	(A) tap test, (B) helium pressure, (C) dye absorption, (D) compaction density	(A) supplier catalog, (B) MIL-T-22300, (C) and (D) ASTM Method D 2771	relative values must be related to application
7. Wire diameter	micrometer or metallurgical mount	visual or transverse section per ASTM Method E 3	±1 wire gage size (Table 29)
8. Wire surface and roundness	micrometer and visual	3-point diameter check	10% out of round max showing slight embedment of insulating material into wire
9. Ductility and formability	bend and visual or metallurgical mount	ASTM E 235	bend 90 deg only for ease of making metallurgical mount
10. Sheath integrity	(A) water immersion, (B) hydrostatic, (C) helium leak, (D) dye penetrant	(A) water soak with ends protected [18], (B) ASTM Method E 165, (C) ASTM Method E 235, (D) ASTM Method E 235	(A) may use hydrostatic pressure or steam for some applications: seal weld to protect insulation
11. Junction integrity	radiographs	ASTM Method E 235	application for junction and 4 in. of sheath
12. Material compatability	thermal cycling	ASTM Method E 235	thermal expansion for grounded junction integrity check

13. Metallurgical integrity	metallurgical mount	ASTM Method E 235	checks grain size
14. Response time	thermal response test	gas: SAE 46, liquid [*24*]	assesses thermal mass and welding technique
15. Uniformity of properties	Tests 1, 2, 3, 10, 11, and 13 full length	as noted	use sampling procedure for large runs
16. Stability under service conditions	application experience or life tests	as required	check with other users in same field
	CONSTANT TEMPERATURE ELECTRICAL		
17. Insulation resistance	megaohm meter between wires and sheath and wires	ASTM Method E 235	50 V for OD less than 0.061, 500 V for greater than 0.062
18. Electrical resistance (loop)	Wheatstone bridge	Chapter 3	tolerances ±1 gage size based on wire type, length and dia: also checks continuity
19. High temperature insulation resistance	megohm meter or Wheatstone bridge	immerse full length except for end in temperature chamber: test with ungrounded junction or before junction is made	relative to operational requirement, effective means of testing insulation quality at expected operating temperature of thermocouple
20. Dielectric strength	high potential generator	ac voltage (current limited) applied at room temperature for 1 min 80 V per mil of insulation [*32*]	seldom used check for contamination and voids in insulation: voltage usually 500 V for 0.062 OD and larger
	ISOTHERMAL ELECTRICAL		
21. Capacitance	capacitance bridge		test is usually not applied to thermocouples
22. Uniformity along length	Test 17, 18, 19, 20, 21 for full length	check samples or full lengths coiled if necessary	all tests are seldom necessary for single application
23. Stability under service conditions	application experience or life tests	as required	check for experience by other users

TABLE 5.6—*(Continued).*

Characteristics	Tests	Procedure	Comments
	THERMOELECTRIC PROPERTIES		
24. Emf versus temperature	potentiometer and heat or cold source	Chapter 6	test at one or more points up to maximum use temperature
25. Homogeneity	heat source and galvanometer	NBC Circular 590 and Ref *6*	max of ±100 μV for J, K, T; ±25 μV for B, R, S thermoelements
26. Uniformity full length	Test 24 and 25 for full length	applying sampling procedure for large quantities	random sampling is most economical approach

FIG. 5.3—*Exposed or bare wire junction.*

FIG. 5.4—*Grounded junction.*

atmosphere so that the two thermocouple wires become an integral part of the sheath weld closure. Thus, the wires are "grounded" to the sheath. This will give a slower response than exposed wire but the insulation is pressure tight, and the wires are protected from mechanical damage and from the environment.

3. *Ungrounded or Isolated Junction* (Fig. 5.5)—This type is similar to the grounded junction except that the thermocouple wires are first made into a junction which is then insulated from the sheath and the sheath closure. The closure is formed by welding without touching the thermocouple wires. Thus, the thermocouple is "ungrounded" to the sheath material. This junction has slower response than the grounded junction but is still pressure tight and protected from mechanical damage and from the environment. The strain due to differential expansion between wires and sheath may be reduced.

4. *Reduced Diameter Junction* (Fig. 5.6)—This may be either grounded or insulated and is used where fast response is required, and a heavier sheath or wires are desired for strength, life, or lower circuit resistance over the balance of the unit.

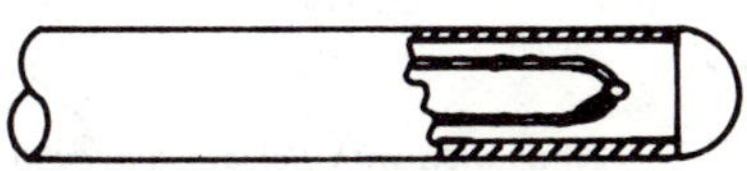

FIG. 5.5—*Ungrounded or isolated junction.*

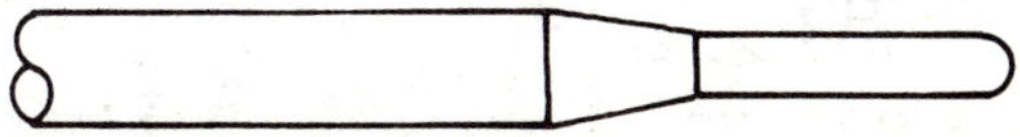

FIG. 5.6—*Reduced diameter junction.*

5.10 Terminations

There are numerous ways in which thermocouples of this type can be treated at the reference end. The most common treatments are as follows:

1. *Wires Exposed and Sealed*—The sheath and the insulation are removed leaving the bare thermocouple wires exposed for a specific length. The insulating material is then sealed with epoxy to inhibit moisture absorption.
2. *Transition Fitting* (Fig 5.7)—The terminal end is provided with a fitting wherein the thermocouple wires are joined to more suitable wires. In this fitting the necessary sealant for the mineral insulant is also provided. One commercial machine produces a molded, thermoplastic transition.
3. *Terminals or Connectors*—Various types of fittings are available to facilitate external electrical connections. These include screw terminal heads, open or enclosed, and plug or jack connections (Fig. 5.8).

5.11 Installation of the Finished Thermocouple

Many types of installation are possible with sheathed thermocouples. Typical installations are shown in Figs. 5.9, 5.10, and 5.11. Other special flanges and adaptors are available from many thermocouple manufacturers.

5.12 Sheathed Thermocouple Applications

Application information for sheathed thermocouples has been well covered in the literature. Many suggested applications are made in the various suppliers catalogs. The Sannes article [7] on the application of the smaller

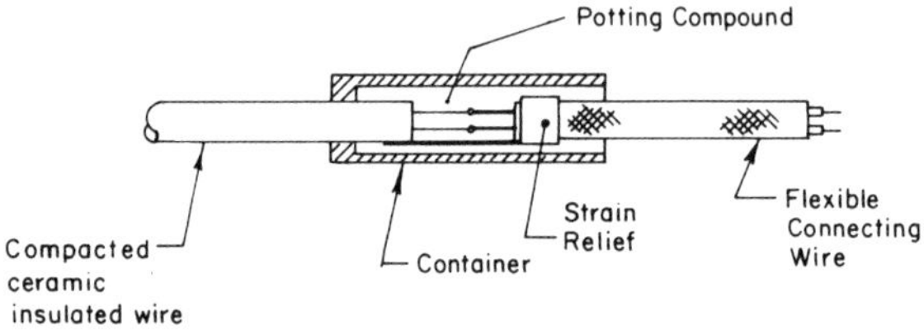

FIG. 5.7—*Termination with flexible connecting wires.*

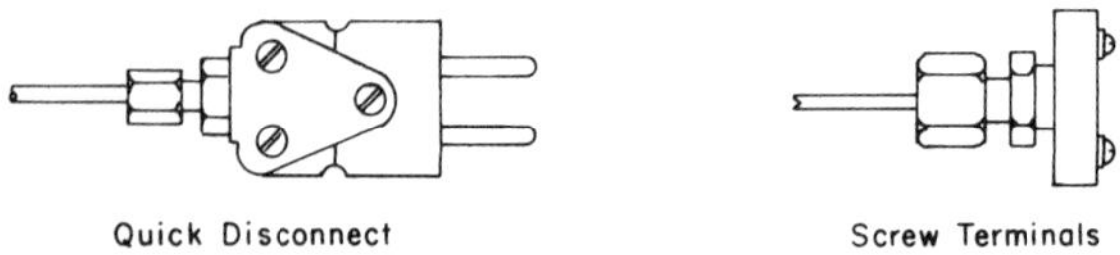

FIG. 5.8—Quick disconnect and screw terminals.

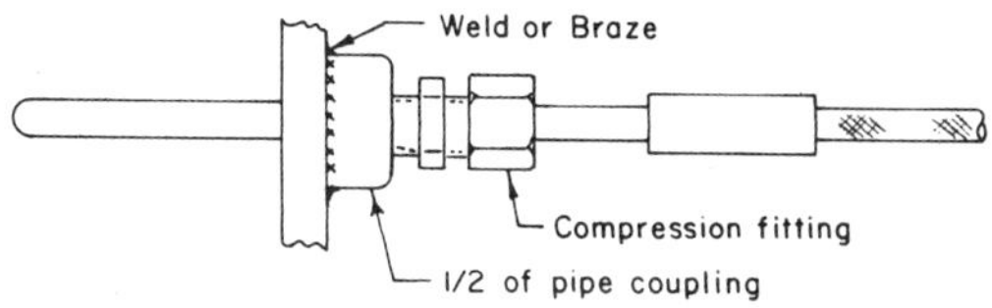

FIG. 5.9—*Fittings to adapt into process line [up to 3.48 × 10⁴ kPa (5000 psi)].*

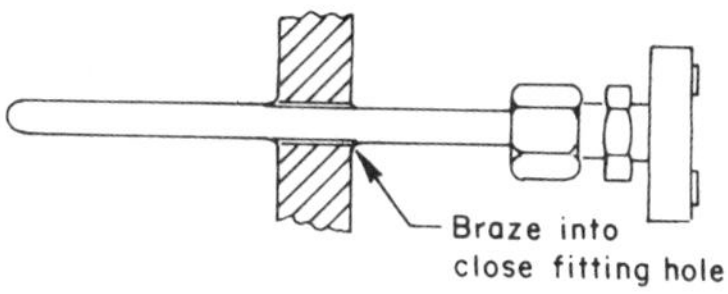

FIG. 5.10—*Braze for high pressure operation [up to 6.89 × 10⁵ kPa (100 000 psi)].*

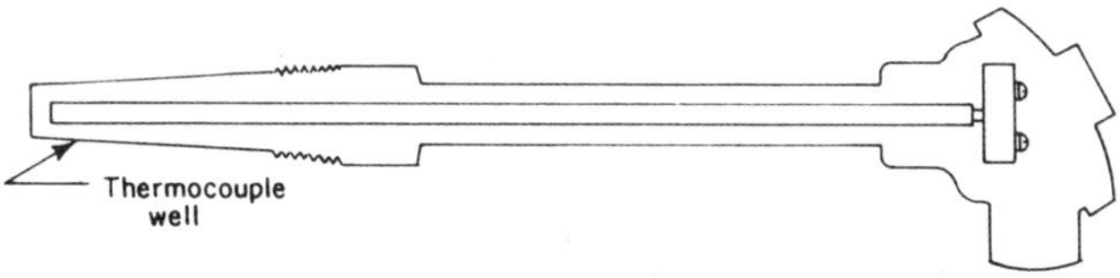

FIG. 5.11—*Thermocouple in thermowell.*

sheath diameters is useful. Chapter 9.2 of this manual is an excellent reference on surface temperature measurement. Gas stream performance is well documented [*8–10*] because of the applications to jet engines and rockets.

Nuclear reactor applications have used sheathed thermocouples extensively for monitoring and control. Johannessen [*12*] discussed the reliability assurance of such applications.

5.13 References

[*1*] Bliss, P., "Fabrication of High-Reliability Sheathed Thermocouples," *Temperature, Its Measurement and Control in Science and Industry*, Vol. 4, Part 3, Section VIII-C, June 1971.

[*2*] Bliss, P., "Fabrication of High-Reliability Sheathed Thermocouples," *Temperature, Its Measurement and Control in Science and Industry*, Vol 4, Part 3, Section VIII-C, Reinhold, New York, June 1971.

[*3*] Committee AE-2, Society of Automotive Engineers, "The Preparation and Use of Chromel-Alumel Thermocouples for Turbojet Engines," Aerospace Information Report No. 46, 15 March 1956.

[*4*] Scadron, L., "Ceramic Insulated Thermocouples," *Instruments and Control Systems*, May 1961.

[*5*] Fenton, A. W., Dacey, R., and Evans, E. J., "Thermocouples: Instabilities of Seebeck Coefficient," TRG Report 1447 (R), United Kingdom Atomic Energy Authority, 1967.

[*6*] Roeser, W. F. and Lonberg, S. T., "Methods of Testing Thermocouples and Thermocouples Materials," NBS Circular 590, National Bureau of Standards, 6 Feb. 1958.

[*7*] Sannes, T., "Tiny Thermocouples," *Instrumentation Technology*, March 1967.

[*8*] Faul, J. C., "Thermocouple Performance in Gas Streams," *Instruments and Control Systems*, Vol. 35, Dec. 1962.

[*9*] Moffat, R. J., "Designing Thermocouples for Response Rate," *Transactions*, American Society of Mechanical Engineers, Feb. 1958.

[*10*] Ihnat, M. E. and Hagel, W. D., "A Thermocouple System for Measuring Turbine Inlet Temperatures," *Transactions*, American Society of Mechanical Engineers, *Journal of Basic Engineering*, March 1960.

[*11*] Johannessen, H. G., "Reliability Assurance in Sheathed Thermocouples for Nuclear Reactors," High Temperature Thermometry Seminar, TID7586, Oak Ridge National Laboratory, Oak Ridge, Tenn., Aug. 1960.

Chapter 6—Emf Measurements

6.1 General Considerations

The basic principle of thermoelectric thermometry is that a thermocouple develops an emf which is a function of the difference in temperature of its measuring junction and reference junction. If the temperature of the reference junction is known, the temperature of the measuring junction can be determined by measuring the emf generated in the circuit. The use of a thermocouple in temperature measurements, therefore, requires the use of an instrument capable of measuring emf.

There are three types of emf measuring instruments in use in industry: deflection meters (millivoltmeters), digital voltmeters, and potentiometers. Because of its limitations, the deflection meter is not used for precise measurements. Where precision and accuracy are required in the measurement of thermal emfs, either the digital voltmeter or the potentiometer is used.

Digital voltmeters can be used in most industrial applications and in making standard measurements. Digital voltmeters are very high impedance devices so that the readings are essentially independent of external circuit resistance.

When the greatest accuracy is required in measuring emfs, the potentiometer is often used. Readings are completely free from uncertainties arising from changing circuit resistance. Choice of range can be quite wide due to the very high sensitivity capability.

6.2 Deflection Millivoltmeters

The deflection meter consists of a galvanometer with a rigid pointer which moves over a scale graduated in millivolts or degrees. The galvanometer indicates by its deflection the magnitude of the current passing through it, and if the circuit in which it is placed includes a thermocouple, it measures the current I generated by the thermocouple in the circuit. If the circuit has a resistance R and the emf is E, by Ohm's law, $E = RI$. If R is kept constant, I is proportional to E, then the scale can be calibrated in terms of millivolts rather than in milli- or microamperes. This calibration holds as long as R remains constant. Any change in R introduces an error in the indicated value of E. Changes in resistance may result from changes in temperature of the thermocouple or its extension wires or of the copper galvanometer coil, from corrosion of the thermocouple wires, from changes in the depth of immersion of the thermocouple, or from changes in contact resistance at switches or binding posts.

In general, the reference junctions of the thermocouple measuring circuit are located near the meter movement. The effect of temperature changes on these junctions is compensated for by a bimetallic spiral attached to one of the control springs of the pointer. This system maintains accurate meter calibration during changes in ambient temperature.

In spite of its limitations, the deflection meter serves a very useful purpose in a great variety of industrial measurements of temperature where the precision and accuracy required are not of a high order. Its insensitivity to small a-c signals may prove an advantage.

6.3 Digital Voltmeters

Two kinds of digital meters are generally available: (1) the integrating type and (2) the successive approximation type. The integrating type compares the value of the emf being measured to an internal reference voltage standard by timed charging and discharging of a reference capacitor. Measurements are relatively slow, on the order of a second, but are relatively unaffected by the presence of some types of electrical noise.

The successive approximation type digital meter operates by first estimating a voltage and comparing it with that being measured. The error is noted by the circuit, and a second approximation is made and compared, etc., until the estimated voltage equals the unknown. Solid-state circuitry enables this type of meter to display a reading within milliseconds. It is, however, susceptible to errors from electrical noise interference.

Both types of meters may display voltage, or a linearizing circuit can be included to provide readings directly in temperature.

Typical uncertainty of instruments reading in volts is 0.01 percent of range ± 1 digit. If readout is directly in temperature, uncertainty might be plus or minus several degrees for a broad range or less than $\pm \frac{1}{2}$ deg for a short range.

6.4 Potentiometers

6.4.1 Potentiometer Theory

Accurate measurement is usually a matter of comparing an unknown quantity against a known quantity or standard—the more direct the comparison, the better. Accurate weighing, for example, often is accomplished by direct comparison against standard weights using a mechanical balance. If the measured weights are too heavy for direct comparison, lever arms may be used to multiply the forces.

The potentiometer, as the term is used here, serves a similar function in the measurement of voltage, and, in fact, may be called a "voltage balance," the standard voltage being furnished by a standard cell, the "lever" being resistance ratios, and the galvanometer serving as the balance indicator. Since no current is drawn from the standard cell or the measured source at balance, the measurement is independent of external circuit resistance, except to the extent that this affects galvanometer or balancing mechanism sensitivity.

6.4.2 Potentiometer Circuits

Figure 6.1 shows a simple potentiometer circuit which includes a resistor R, a standard cell S, a battery B, a galvanometer G, and a rheostat r. R may be a calibrated slidewire, with a known resistance, and R' a fixed resistor such that R plus R' is some simple multiple of the emf of S. If the emf of S is taken as 1.019 V, the sum of R plus R' may be chosen as 101.9. If the switch K is turned to the standard cell position, the galvanometer, in general, will deflect. The setting of rheostat r, is adjusted until the galvanometer remains at rest when K is closed. Then the drop of potential of the battery current through R and R' is exactly equal to the emf of S, so that if the current in R and R', $T - 1.019/101.9$ A, or 10 mA, through each ohm of R there is then a drop of potential 0.010 V. If R is 20 ohms the total drop through the slidewire is 0.2 V.

Now if K is turned to the "thermocouple" position, a setting may be found for the sliding contact on R at which there will be no deflection of the galvanometer. At this position, the drop of potential through R of the contact is equal to the emf of the thermocouple. If balance occurs at the 5 ohm point, the emf of the thermocouple is $0.01 \times 5 = 0.05$ V $= 50$ mV.

In this measurement the galvanometer has been used only as a means of detecting the presence of a current, and readings are made only when no perceptible current is passing through it. Therefore, it is not calibrated, and it is only called upon to indicate on balance with sufficient sensitivity to give the desired precision of setting of the sliding contact. An increase in the resistance of the thermocouple circuit can increase only the limits of positions of the contact between which there is no perceptible deflection, but does not affect the position of balance, nor the measured value of the emf. Since the galvanometer is used only to indicate the existence and direction of current, it is not necessary to design it to give a linear relationship between current and deflection. Zero stability is not extremely important since it is only necessary to look for departure from any equilibrium position when the key is closed to detect a need for adjustment and the direction in which it should be made. The galvanometer may be of the suspension type or an electronic null balance detector.

The calibration of an instrument of this type is stable since resistors and slidewires can be made with a high degree of stability. The emf of an un-

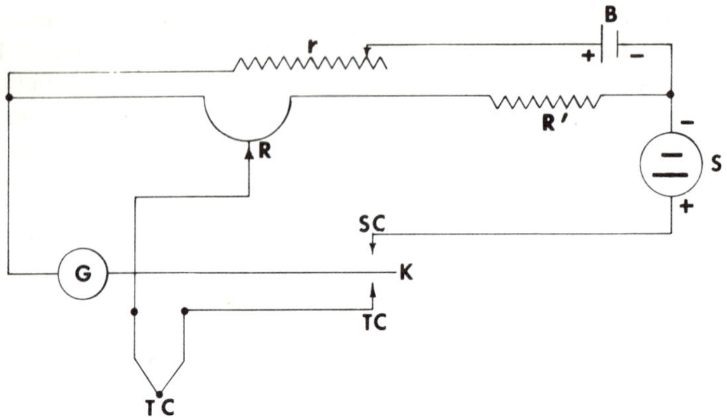

FIG. 6.1—*A simple potentiometer circuit.*

saturated standard cell, as used for potentiometric work, does not change more than about 0.01 percent per year and has a negligible temperature coefficient.

The usefulness of an elementary potentiometer of this sort, having its entire range across the slidewire, is limited by the resolution possible in setting and reading the position of the contact on *R*.

6.4.3 *Types of Potentiometer Instruments*

6.4.3.1 Laboratory High Precision Type—The six-dial potentiometer meets the most exacting requirements for standardizing, research and testing laboratories. It is particularly useful for precise temperature measurements in studies of specific heats, melting and boiling points, and for calibrating less precise potentiometers. It also is used for precise measurements such as required in calorimetry cryogenics. It measures emfs in the range of 0 to 1.6 V. It is essentially free of thermal emfs, and transients. Overall limits of error are typically 0.5 mV or less up to 1000 mV, and 5 mV or less up to 50 000 mV.

6.4.3.2 Laboratory Precision Type—The three-dial, manually operated potentiometer, also is used for precise measurements of voltage. As applied to the measurement of thermal emfs, its characteristics are such as to make it the most generally used of all the potentiometers in this field when accuracy and convenience of measurement are demanded. It is not portable in the sense in which the term is applied to self-contained potentiometers, since to attain the results of which it is capable, some parts of the circuit such as the null detector and the reference voltage source must be mounted separately. Overall limits of error are typically 1 μV or less up to 1000 μV, and 12 μV or less up to 50 000 μV.

6.4.3.3 Portable Precision Type—This potentiometer is a self-contained, three-dial, manually balancing instrument. It is used for precision checking of thermocouple pyrometers in laboratory and plant and for general temperature measurements. The measuring range typically extends from −10.1 to 100.1 mV, readable to 1 μV, and typical limits of error are: (*a*) ±(0.03 percent of reading plus 3 μV) without internal reference junction compensation and (*b*) ±(0.03 percent of reading plus 6 μV) with reference junction compensation.

An example of (*a*) when using an external reference junction (see Chapter 7) is as follows:

> A Type K thermocouple develops 41.269 mV at 1000°C (1832°F). The thermocouple error (±3/4 percent) can be 7.5°C; the instrument error will be 0.0003 × 41.269 + 0.003 = 0.015 mV or 0.4°C.

The reference junction compensation (*b*) is provided by an additional slidewire. This slidewire is used to set the reference junction voltage to the value which corresponds to the temperature of the reference junction (obtain millivolt value from reference tables. Chapter 10).

The portable precision potentiometer is designed to measure thermocouple emfs with a precision adequate for all but the more refined laboratory applications. It includes a built-in lamp and scale galvanometer.

6.4.3.4 Semiprecision Type—Potentiometers are available which are similar to those used for precision potentiometry except for limits of error. They are essentially of the two-dial type with self-contained galvanometer, standard cell, and battery. They are used primarily for general temperature measurements by means of thermocouples, checking thermocouple pyrometers in laboratory and plant, and as a calibrated source of voltage.

One model has two ranges: −10.1 to 0 mV and 0 to 100.1 mV, it is readable to 1 μV. The limit of error is ±(0.05 percent of reading +20 μV) without reference junction compensator and ±(0.05 percent of reading +40 μV) with reference junction compensator.

6.4.3.5 Recording Type—Recording potentiometers are used widely for industrial process temperature measurement and control. These instruments balance automatically. When reliability and convenience are the prime concern, such instruments are satisfactory. However, because of potential inaccuracies in charts caused by printing limitations and humidity effects and practical limits on chart widths and scale lengths resulting in inadequate readability, they are not generally used where precision is the criterion.

6.5 Voltage References

Both digital meters and potentiometers require a standard voltage reference for comparison.

Digital voltmeters and moderate precision potentiometers generally use the Zener diode, although many potentiometers still use Weston-type standard cells. High-precision potentiometers use standard cells. Two types of cells are available: saturated and unsaturated.

The saturated cell containing undissolved cadmium sulfate crystals is used in standardizing laboratories such as the National Bureau of Standards to represent the value of the volt. A typical cell has a potential of 1.01864 V at 20°C reproducible to a few microvolts. This voltage is constant for a long period of time. (As long as 60 years or more is possible with proper care of the cell.) These cells have a substantial but predictable negative temperature coefficient ranging from about $-40\ \mu V/°C$ at 20°C to $-70\ \mu V/°C$ at 37°C. It is, therefore, necessary to control their temperature environment very precisely, confining their usefulness to the laboratory.

Unsaturated cells have a very low temperature coefficient but a less stable voltage output. Normal voltage drift rates can be expected to be -20 to -40 μV/year. The temperature coefficient is only a few microvolts/°C between 15 and 45°C and may be either positive or negative, depending upon the chemical composition of the particular cell. Useful life of these cells may be ten years or more. As the cells age, the emf decreases. When it drops to about 1.0130 V, the potential becomes erratic; the temperature coefficient increases and excessive temperature emf hysteresis occurs.

A standard cell must be never short circuited, nor should its emf be measured with a voltmeter. In precise measurements, the balances should be made with a resistance of at least 10 000 ohms in series with the cell until the balance is well within the range of the detector scale.

For more complete information concerning standard cell characteristics and care, refer to *NBS Monograph No. 84.*

6.6 Reference Junction Compensation

Temperature measurement with thermocouples always requires the knowledge of the reference junction temperature. It can be either held at a constant known temperature or allowed to follow ambient temperature with the necessary compensation applied (see Chapter 7).

Chapter 7—Reference Junctions

7.1 General Considerations

A thermocouple circuit is by its nature a differential measuring device, producing an emf which is a function of the temperatures of its two junctions. One of these junctions is at the temperature which is to be measured and is referred to as the measuring junction. The other junction is maintained at a known temperature and is referred to as the reference junction. (In a practical thermocouple circuit (see Section 2.3) copper wires are often connected to the thermocouple alloy conductors at the reference junction. The term reference junctions will be used to refer to this situation.) If these junctions are both at the same temperature, the presence of the copper "intermediate metal" introduces no change in the thermocouple's emf. If they are not at the same temperature the analysis of the circuit is complicated. Moffat [*1*][1] gives a helpful analytical method.

The Seebeck coefficient (thermoelectric power) of many common thermocouples is approximately constant from the ice point to the upper temperature limit of the materials (see Section 3.1.1). For such thermocouples, an uncertainty in the temperature of the reference junction will reflect a similar uncertainty in the deduced temperature of the measuring junction. However this situation does not exist for all thermocouple pairs. Notable exceptions occur in the case of the high rhodium-in-platinum alloy thermocouples [*2,3*]. In particular, if the reference junction of a platinum-30 percent rhodium versus platinum-6 percent rhodium (Type B) thermocouple lies within the range 0 to 50°C (32 to 122°F), a 0°C (32°F) reference junction may be assumed, and the error will not exceed 3 μV. This represents about 0.3°C (0.5°F) error in high-temperature measurements [*4*].

7.2 Reference Junction Techniques

Three basic methods are used to take account of the reference junctions of the thermocouple circuit: (1) the junction is maintained at a fixed temperature, (2) the temperature of the reference junction is allowed to vary, and a compensating emf is introduced into the circuit or accounted for by calculation, (3) the temperature of the reference junction is allowed to vary, and

[1] The italic numbers in brackets refer to the list of references appended to this chapter.

compensation is made by a mechanical adjustment of the readout instrument.

Some variations of these techniques will be described in the following sections. Sources of error which are common to several techniques will be discussed in Section 7.3.

7.2.1 Fixed Reference Temperature

7.2.1.1 Triple Point of Water—A cell can be constructed in which there is an equilibrium between ice, water, and water vapor [*5*]. The temperature of this triple point is 0.01°C on the International Practical Temperature Scale of 1968, and it is reproducible to about 0.0001°C. Williams [*6*] describes a commercially available cell which is not affected by factors such as air saturation and pressure which can cause several millikelvins fluctuation in the temperature of an ice bath. To utilize such precision, extreme attention to immersion error and galvanic error is required, and the measurement system must be of the highest quality.

7.2.1.2 Ice Point—An ice bath, consisting of a mixture of melting, shaved ice and water, forms an easy way of bringing the reference junctions of a thermocouple to 0°C (32°F). If a proper technique is used, the uncertainty in the reference junction temperature can be made negligibly small. With extreme care the ice point can be reproduced to 0.0001°C [*7*].

A recommended form of ice bath is described in Ref *8* and is shown in Fig. 7.1. Using the illustrated construction, with 14 gage B&S iron or nickel-base alloy wires and 20 gage B&S copper and noble metal wires, immersed at least 110 mm (4.5 in.) in the water-ice mixture, Caldwell [*9*] found the immersion error (see Section 7.3.1) to be less than 0.05°C. Smaller diameter wires would reduce the error. If large conductors must be used, Finch [*10*] describes a technique which minimizes the error.

To avoid the use of mercury[2] in the glass tubes, the tubes may be filled with moisture-free oil. Kerosine, transformer oil and silicone oil are all suitable [*11-13*].[3] Moisture-proof insulation should extend beneath the surface of the oil to the ends of the wires where they may be connected by any means which ensures a low resistance contact (for example, welding, soldering, crimping, etc.). If the glass tubes are immersed 200 mm (8 in.) in the ice-water mixture and the oil extends to within a few millimeters of the ice-water surface, the immersion error will be negligible. Sutton [*11*] plots the error as a function of the wire material, diameter, and immersion depth.

If the ice bath is improperly used, serious errors can result. The largest error which is likely to occur arises due to melting of the ice at the bottom of a

[2] Mercury can be hazardous to health.

[3] Care should be taken to prevent oil contamination of that part of the thermocouples which will be exposed to high temperatures.

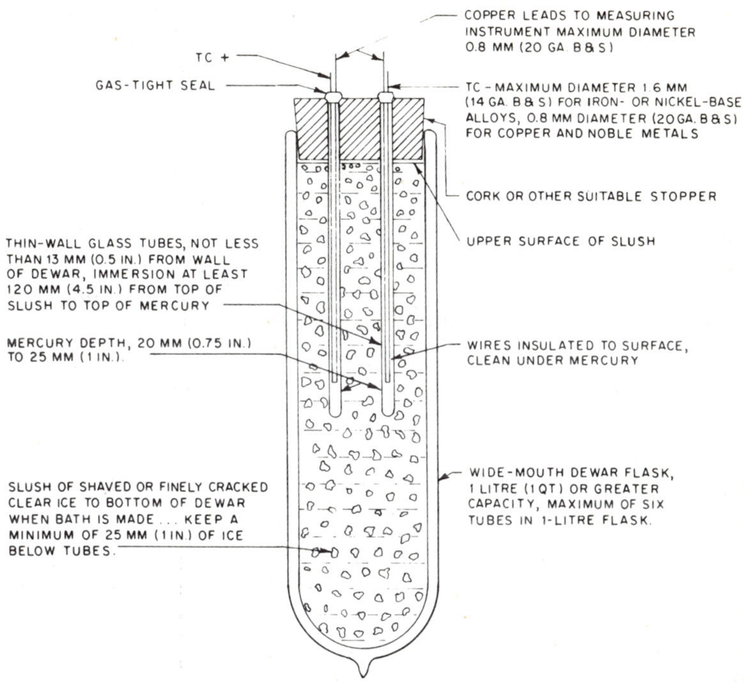

FIG. 7.1—*Recommended ice bath for reference junction.*

bath until the reference junctions are below the ice level and surrounded by water alone. This water may be as much as 4°C above the ice point. While an ice bath is being used excess water should be removed periodically and more ice added, so that the ice level is maintained safely below the reference junctions [*8,12*].

If the ice used to prepare the bath has been stored in a freezer at a temperature below 0°C, it may freeze the surrounding water and remain at a temperature below 0°C for some time. To avoid the condition, the ice should be shaved rather than crushed and thoroughly wetted with water before placing it in the vacuum bottle [*11-13*].

If appreciable concentrations of salts are present in the water used to make the ice bath, the melting point can be affected. McElroy [*14*] investigated various combinations of tap and distilled water to determine the error introduced. He observed bath temperatures of +0.013°C using distilled water with tap ice and −0.006°C using tap water and tap ice. It is probable that the use of tap water will not introduce significant errors unless accuracies better than 0.01°C are required [*9*].

Another possible source of error is galvanic action which is discussed in Section 7.3.2.

Although the ice bath is an easy way of achieving a convenient fixed point for the reference junctions in the laboratory, the need for constant attention makes it unsuitable for industrial application where some form of automatic reference junction is desirable.

7.2.1.3 Automatic Ice Point—The development of the thermoelectric refrigerator (Peltier cooler) has enabled the production of practical devices in which an equilibrium between ice and water is constantly maintained [*15, 16*]. Since water expands 9 percent on freezing, the change of volume may be used to control the heat transfer. Substances which undergo a phase change do so without change of temperature. This is an important advantage since the system behaves as if it had an infinite thermal capacity.

This device can provide a reference medium which is maintained at a precise temperature, but careful design is necessary if this precision is to be utilized fully. The system is subject to immersion error (Section 7.3.1) and galvanic error (7.3.2) due to condensation.

Some commercially available devices provide wells into which the user may insert reference junctions formed from his own calibrated wire. Others are provided with many reference junction pairs brought out to terminals which the user may connect into his system. The latter type are subject to the wire matching error (Section 7.3.3).

If the potential errors have been successfully overcome, the error introduced into a system by these devices may be less than 0.1°C.

Automatic ice points can be used to further advantage in conjunction with a "zone box" (see 7.2.2.1).

7.2.1.4 Constant Temperature Ovens—A thermostatically controlled oven provides a way of holding a reference junction at an approximately constant temperature. The major advantage of this method is that one oven can be used with a large number of circuits while maintaining isolation between the circuits without the need for providing separate power supplies for each circuit.

The oven must be maintained at a temperature above the highest anticipated ambient temperature. The need for temperature uniformity within the oven and a precise temperature control system are inherent complications, in addition to the immersion error (Section 7.3.1) and the wire matching error (Section 7.3.3).

Ovens are available with rated accuracies ranging from 0.1 to 0.05°C. A common commercially used junction temperature is 66°C (150°F), although other values may be specified.

Elevated temperature reference junctions can be used with instruments or tables based on 0°C reference junctions, provided an appropriate correction is applied.

A device using two ovens is available which makes this correction automatically. To generate the necessary emf (for example, 2.662 mV for a Type K system with a 66°C (150°F) reference junction temperature) an additional internal thermocouple is connected in series with the external thermocouple circuit. This internal thermocouple has its reference junction in the first oven and its measuring junction in the second oven. The second oven is maintained at a suitable temperature above that of the first so that the required emf is generated. The external thermocouple circuit then behaves as if its reference junction were at 0°C. The manufacturers' stated accuracies of these devices are similar to those of single oven units.

7.2.2 Electrical Compensation

If the temperature of the reference junction of a thermocouple is allowed to change, the output emf will vary in accordance with the Seebeck coefficient of the couple at the reference temperature. A compensating circuit containing a source of current, a combination of fixed resistors, and a temperature sensitive resistor (TSR) can be designed which will have a similar variation of emf as the temperature of the TSR is varied. If the reference junction of the thermocouple is coupled thermally to the TSR and the compensating circuit is connected in series with the thermocouple so that its temperature-variable emf opposes that caused by the reference junction, the thermocouple behaves as if the reference junction temperature were held constant. In addition, by suitable choice of the fixed resistors, any fixed reference junction temperature may be simulated. Since the TSR is at the temperature of the reference junction, no warmup or stabilization time is involved. Muth [*17*] has given a more extended description of these circuits.

The disadvantages of this arrangement include the need for a stable power source for each thermocouple circuit, the difficulty of exactly matching the Seebeck coefficient over an extended temperature range, and the addition of series resistance in the thermocouple circuit.

This principle is used in almost all self-balancing recording thermocouple potentiometers. Here the power source already is present as part of the potentiometer circuit. The Seebeck coefficient is matched adequately to allow the accuracy of the entire instrument to be typically 0.25 percent of full scale over a reasonable range of ambient temperatures.

Electric reference junction compensators are also available as small circuit modules with self-contained battery power sources or for connection to a-c power. A typical specification for a battery powered unit is compensation to ±0.3°C over an ambient of 13 to 35°C (55 to 90°F). Improved specifications are quoted for more elaborate devices.

Some portable manually balanced potentiometers are provided with a thermometer to read the reference junction temperature and an adjustable circuit calibrated in millivolts or temperature. The control must be set manually to add the required emf to simulate an ice-point reference junction temperature.

7.2.2.1 Zone Box—In systems employing large numbers of thermocouples, an alternative method of dealing with the reference junctions is particularly useful. All of the thermocouples are routed to a device called a zone box where each thermocouple conductor is joined to a copper wire which is routed to the emf measuring instrument [*18, 19*].

Within the zone box all of the reference junctions between the thermocouples and the copper wires are insulated electrically but kept in good thermal contact with each other and with a transducer which measures the temperature within the zone box. This transducer can be a thermocouple of the same type as the measuring thermocouples with its reference junction in an automatic ice point or any of the other devices described in Section 7.2.1. The output of this thermocouple can be added to that of the measuring thermocouples, either electrically or in a data-collection computer. Several thermocouples are sometimes used in the zone box to monitor its temperature uniformity.

Alternatively, the zone box temperature can be measured by other transducers such as thermistors or resistance thermometers which need no reference temperature. In this case the measuring instrument determines the necessary correction to the thermal emfs, based on the temperature of the measuring-thermocouple reference junctions in the zone box. Several variations of this technique are described in Ref *19*.

The advantage of the zone box is its simplicity. It generally requires no heaters, controls or power supplies. Since it is approximately at the ambient temperature, the immersion error (Section 7.3.1) can be made negligible with a moderate amount of thermal insulation if care is taken to avoid locations having extreme thermal gradients. The wire matching error (Section 7.3.3) should receive attention because the reference junctions are at ambient temperature and the calibration error at this temperature must be accounted for.

7.2.2.2 Extended Uniform Temperature Zone—To reduce the length of thermocouple wires required to reach a zone box, Sutton [*11*] extended the zone by the use of a piped water system. The system operates at ambient temperature. By circulating water with a low power pump, uniformity within 0.1°C between the first and last reference junction is achievable. The zone length may extend to many tens of meters.

7.2.3 Mechanical Reference Compensation

To complete this account of methods used to compensate for the temperature of thermocouple reference junctions, a device used on millivolt pyrometers must be included. The millivolt pyrometer measures the thermal emf of a thermocouple circuit by measuring the current produced in a circuit of fixed and known resistance. The current operates a galvanometer with a rigid pointer which moves over a scale graduated in degrees [*10*]. The reference junction is at the temperature of the instrument, and hence the available thermal emf is a function of the temperature of the instrument. Compensation is often accomplished by attaching one of the hairsprings of the D'Arsonval galvanometer movement to a bimetallic thermometer element so that the electrical zero of the instrument is adjusted to correspond to the temperature of the instrument. This system is subject to the wire matching error (Section 7.3.3.), but the precision of the pyrometer seldom justifies making corrections.

7.3 Sources of Error

Several sources of error which may disturb the control or measurement of the reference junction temperature are discussed in this section.

7.3.1 Immersion Error

Whenever reference junctions are being maintained at a temperature which differs from the ambient, heat is transferred between the reference temperature medium (oven, ice bath, etc.) and the ambient via the electrical insulation which separates the junctions from the medium and via the wires which emerge from the reference junctions. Thus the temperature of the junctions always differs from that of the reference medium to a greater or lesser degree. Caldwell [*9*] gives data which allow the error to be estimated for the standard type of ice bath. For other situations the error may be calculated by methods outlined in Refs *20* and *21*, if the coefficients governing heat flow between the medium, the wires, and the ambient can be evaluated.

With careful design the immersion error usually can be made negligible.

7.3.2 Galvanic Error

If water is allowed to contact the thermocouple alloy and copper wires of the reference junction, a galvanic cell may be set up, causing voltage drops which disturb the thermal emfs. If the reference junction is at a temperature below the dew point, the water may appear due to condensation. Insulation on both wires and precautions to avoid the accumulation of water in contact with the wires normally will prevent this error [*9,14*].

7.3.3. Wire Matching Error

Thermocouple wire normally is calibrated with its reference junction at 0°C, and corrections are determined to enable accurate measurements at elevated temperatures. The calibration deviation at ambient temperature is seldom of interest and usually is not determined.

Many reference junction devices are equipped with thermocouple alloy wires, and provision is made for interconnection with the user's thermocouples at ambient temperatures. If the calibration of the wire supplied with the device differs from that of the thermocouples at the ambient temperature, a significant error can result due to the interconnection of the wires. This source of error often is overlooked, since it is assumed that if the interconnection of both wires of a thermocouple pair occurs at the same temperature no error is introduced. The existence of this error is visualized easily if the circuit is analyzed using Moffat's method [*1*].

A simple correction for the wire matching error can be made if the ambient temperature deviation of the wire supplied with the reference junction device is known and the user's thermocouples are calibrated at ambient temperature.

The wire matching error is avoided in those reference junction devices in which the user's wire is used to form the reference junctions.

7.4 References

[*1*] Moffat, R. J., "The Gradient Approach to Thermocouple Circuitry," *Temperature, Its Measurement and Control in Science and Industry*, Vol. 3, Part 2, Reinhold, New York, 1962, pp. 33-38.

[*2*] Caldwell, F. R., "Thermocouple Materials," *Temperature, Its Measurement and Control in Science and Industry*, Vol. 3, Part 2, Reinhold, New York, 1962, pp. 81-134.

[*3*] Zysk, E. D., "Noble Metals in Thermometry, Recent Developments," *Technical Bulletin*, Englehard Industries, Inc., Vol. 5, No. 3, 1964.

[*4*] Burns, G. W. and Gallagher, J. S., "Reference Tables for the Pt-30 Percent Rh Versus Pt-6 Percent Rh Thermocouple," *Journal of Research, NBS-C Engineering and Instrumentation*, Vol. 70C, No. 2, 1966.

[*5*] Stimson, H. F., "Precision Resistance Thermometry and Fixed Points," *Temperature, Its Measurement and Control in Science and Industry*, Vol. 2, Reinhold, New York, 1955, pp. 141-168.

[*6*] Williams, S. B., "Triple Point of Water Temperature Reference," *Instruments and Control Systems*, Vol. 33, 1960.

[*7*] Thomas, J. L., "Reproducibility of the Ice Point," *Temperature, Its Measurement and Control in Science and Industry*, Vol. 1, Reinhold, New York, 1941, pp. 159-161.

[*8*] ASTM E 563, "Standard Recommended Practice for Preparation and Use of Freezing Point Reference Baths," American Society for Testing and Materials.

[*9*] Caldwell, F. R., "Temperatures of Thermocouple Reference Junctions in an Ice Bath," *Journal of Research, NBS Engineering and Instrumentation*, Vol. 69C, No. 2, 1965.

[*10*] Finch, D. I., "General Principles of Thermoelectric Thermometry," *Temperature, Its Measurement and Control in Science and Industry*, Vol. 3, Part 2, Reinhold, New York, 1962, pp. 3-32.

[*11*] Sutton, G. R., "Thermocouple Referencing," *Temperature Measurement, 1975;* Conference Series No. 26, Chapter 4, Institute of Physics, London, pp. 188-194.

[*12*] Hollander, B., "Referencing Thermocouple Junctions," *Instruments and Control Systems*, March 1973, pp. 38-39.

[*13*] McElroy, D. L., and Fulkerson, W., in "Temperature Measurement and Control," *Techniques of Metals Research*, Vol. 1, Part 1, Chapter 2, *Temperature Measurement and Control*, R. F. Bunshah, Ed., Interscience, New York, 1969, pp. 175-177.

[*14*] McElroy, D. L., "Progress Report I. Thermocouple Research Report for the Period, November 1, 1956 to October 31, 1957," ORNL-2467, Oak Ridge National Laboratory, Oak Ridge, Tenn..

[*15*] Morgan, W. A., "Close Temperature Control of Small Volumes, A New Approach," Preprint Number 11.7-2-64, Instrument Society of America, 1964.

[*16*] Feldman, C. L., "Automatic Ice-Point Thermocouple Reference Junction," *Instruments and Control Systems*, Jan. 1965, pp. 101-103.

[*17*] Muth, S., Jr., "Reference Junctions," *Instruments and Control Systems*, May 1967, pp. 133–134.

[*18*] Roeser, W. F., "Thermoelectric Thermometry," *Temperature, Its Measurement and Control in Science and Industry*, Vol. 1, Reinhold, New York, 1941, p. 202.

[*19*] Claggett, T. J., "External Thermocouple Compensation," *Instrumentation*, Vol. 18, No. 2, Second Quarter, 1965.

[*20*] Baker, H. D., Ryder, M. E., and Baker, M. A., *Temperature Measurement in Engineering*, Vol. 1, Section 7, Wiley, New York, 1953.

[*21*] Bauerle, J. E., "Analysis of Immersed Thermocouple Error," *Review of Scientific Instruments*, Vol. 32, No. 3, March 1961, pp. 313–316.

Chapter 8—Calibration of Thermocouples

The calibration of a thermocouple consists of the determination of its electromotive force (emf) at a sufficient number of known temperatures so that, with some accepted means of interpolation, its emf will be known over the entire temperature range in which it is to be used. The process requires a standard thermometer to indicate temperatures on a standard scale, a means for measuring the emf of the thermocouple, and a controlled environment in which the thermocouple and the standard can be brought to the same temperature. Some of the more commonly used techniques for accomplishing such calibrations will be discussed in this chapter.

Much of this material is based upon National Bureau of Standards Circular 590, *Methods of Testing Thermocouples and Thermocouple Materials*, and the calibration methods appearing in *Temperature Measurement* of the American Society of Mechanical Engineers Performance Test Codes, PTC 19.3.

8.1 General Considerations

8.1.1 Temperature Scale

The International Practical Temperature Scale of 1968 (IPTS-68) is realized and maintained by the National Bureau of Standards to provide a standard scale of temperature for use by science and industry in the United States. This scale [*1,2*][1] was adopted by the International Committee of Weights and Measures at its meeting in October 1968 and was amended in 1975 [*3*]. The scale has general international acceptance, and it replaces the International Practical Temperature Scale of 1948 [*4*]. The IPTS-68 distinguishes between the International Practical Kelvin Temperature with the symbol T_{68} and the International Practical Kelvin Temperature with the symbol t_{68}; the relation between T_{68} and t_{68} is

$$t_{68} = T_{68} - 273.15 \text{ K}$$

The units of T_{68} and t_{68} are the kelvin, symbol K, and degree Celsius, symbol °C (formerly called centigrade). It is common practice to express values of temperature on the IPTS-68 in Kelvin temperatures below 0°C and in Celsius

[1]The italic numbers in brackets refer to the list of references appended to this chapter.

temperatures above 0°C. Values of Fahrenheit temperature, symbol °F, are obtained from the conversion formula

$$°F = 9/5 \times °C + 32$$

The IPTS-68 is a practical, standard, empirical scale chosen in such a way that temperatures measured on it closely approximate thermodynamic temperatures. It is a practical scale because the value for a given temperature is more easily and reproducibly determined on it than on the thermodynamic scale. The scale is based on eleven reproducible equilibrium states (defining fixed points) to which numerical values are assigned, and on formulas establishing the relationship between values of International Practical Temperature and the indications of standard instruments calibrated at the defining fixed points.

The defining fixed points are established by realizing specified equilibrium states between phases of pure substances. These fixed points and the exact numerical values assigned to them are given in Table 8.1.

The standard instrument used from 13.81 K (−259.34°C) to 630.74°C is the platinum resistance thermometer. The thermometer resistor must be annealed pure platinum that is mounted in a strain-free configuration. The resistance ratio $W(T_{68})$ of the platinum resistance thermometer, defined by

$$W(T_{68}) = R(T_{68})/R(273.15\ \mathrm{K})$$

where $R(T_{68})$ is the resistance at T_{68} and $R(273.15\ \mathrm{K})$ is the resistance at 273.15 K, must not be less than 1.39250 at $T_{68} = 373.15$ K. From 630.74 to 1064.43°C the standard instrument is the platinum-10 percent rhodium versus platinum thermocouple. Above 1064.43°C, the IPTS-68 is defined by the Planck law of radiation with 1064.43°C as the reference temperature, but the type of standard instrument to be used for the measurement is not specified. However, visual and photoelectric optical pyrometers are the instruments usually used to realize the temperature scale in this region.

The procedures for interpolation lead to a division of the scale into four parts.

(*a*) From 13.81 to 273.15 K (−259.34 to 0°C) the temperature T_{68} is defined by the formula

$$W(T_{68}) = W_{\mathrm{CCT\text{-}68}}(T_{68}) + \Delta W(T_{68})$$

where $W(T_{68})$ is the resistance ratio of the platinum resistance thermometer and $W_{\mathrm{CCT\text{-}68}}(T_{68})$ is the resistance ratio given by a reference function [*3*]. The deviations $\Delta W(T_{68})$ are calculated from measured values of $W(T_{68})$ at the defining fixed points and the corresponding values of $W_{\mathrm{CCT\text{-}68}}(T_{68})$. To obtain $\Delta W(T_{68})$ at intermediate temperatures interpolation formulas are used. For this purpose the range between 13.81 and 273.15 K is divided into four parts; 13.81 to 20.28 K, 20.28 to 54.361 K, 54.361 to 90.188 K, and 90.188

TABLE 8.1—*Defining fixed points of* The International Practical Temperature Scale of 1968.[a]

Equilibrium State	Assigned Value of International Practical Temperature	
	T_{68}(K)	t_{68}(°C)
Equilibrium between the solid, liquid and vapor phases of equilibrium hydrogen (triple point of equilibrium hydrogen)	13.81	−259.34
Equilibrium between the liquid and vapor phases of equilibrium hydrogen at a pressure of 33 330.6 N/m^2 (25/76 standard atmosphere)	17.042	−256.108
Equilibrium between the liquid and vapor phases of equilibrum hydrogen (boiling point of equilibrium hydrogen)	20.28	−252.87
Equilibrium between the liquid and vapor phases of neon (boiling point of neon)	27.102	−246.048
Equilibrium between the solid, liquid and vapor phases of oxygen (triple point of oxygen)	54.361	−218.789
Equilibrium between the solid, liquid and vapor phases of argon (triple point of argon)[b]	83.798	−189.352
Equilibrium between the liquid and vapor phases of oxygen (boiling point of oxygen)	90.188	−182.962
Equilibrium between the solid, liquid and vapor phases of water (triple point of water)	273.16	0.01
Equilibrium between the liquid and vapor phases of water (boiling point of water)	373.15	100
Equilibrium between the solid and liquid phases of tin (freezing point of tin)[c]	505.1181	231.9681
Equilibrium between the solid and liquid phases of zinc (freezing point of zinc)	692.73	419.58
Equilibrium between the solid and liquid phases of silver (freezing point of silver)	1 235.08	961.93
Equilibrium between the solid and liquid phases of gold (freezing point of gold)	1 337.58	1 064.43

[a]Except for the triple points and one equilibrium hydrogen point (17.042 K) the assigned values of temperature are for equilibrium states at a pressure p_0 = 1 standard atmosphere (101 325 N/m^2). In the realization of the fixed points small departures from the assigned temperatures will occur as a result of the differing immersion depths of thermometers or the failure to realize the required pressure exactly. If due allowance is made for these small temperature differences, they will not affect the accuracy of realization of the Scale. The magnitudes of these differences are given in section 111 of Ref *3*.

[b]The triple point of argon may be used as an alternative to the condensation point of oxygen.

[c]The freezing point of tin (t' = 231.9292°C, see Equation in subsection (*b*) on next page) may be used as an alternative to the boiling point of water.

to 273.15 K. In each part $\Delta W(T_{68})$ is given by a polynomial [*3*] in T_{68}. The constants in the polynomials are determined from values of $\Delta W(T_{68})$ at the fixed points and the condition that values of $d\,\Delta W(T_{68})/d\,T_{68}$ given by joining polynomials must be equal at the junctions of the temperature ranges.

(*b*) From 0 to 630.74°C t_{68} is defined by

$$t_{68} = t' + 0.045 \left(\frac{t'}{100°C}\right)\left(\frac{t'}{100°C} - 1\right)\left(\frac{t'}{419.58°C} - 1\right)\left(\frac{t'}{630.74°C} - 1\right) °C$$

where t' is defined by the formula

$$t' = \frac{1}{\alpha}\left[W(t') - 1\right] + \delta\left(\frac{t'}{100°C}\right)\left(\frac{t'}{100°C} - 1\right) °C$$

where $W(t') = R(t')/R(0°C)$. The constants $R(0°C)$, α, and δ are determined by measurement of the resistance of the platinum resistance thermometer at the triple point of water, the boiling point of water (or the freezing point of tin[2] and the freezing point of zinc. The preceding formula for t' may be also expressed as

$$W(t') = 1 + At' + Bt'^2$$

where

$$A = \alpha\left(1 + \frac{\delta}{100°C}\right) \quad \text{and} \quad B = -10^{-4}\,\alpha\delta\,°C^{-2}$$

(*c*) From 630.74°C to the gold point (1064.43°C) the temperature t_{68} is defined by the formula

$$E(t_{68}) = a + bt_{68} + ct_{68}^2$$

where $E(t_{68})$ is the electromotive force of a standard platinum-10 percent rhodium versus platinum thermocouple when one of its junctions is at 0°C and the other is at the temperature t_{68}. The constants *a*, *b*, and *c* are calculated from values of *E* at 630.74°C ± 0.2°C, as determined by a platinum resistance thermometer, and at the freezing points of silver and gold.

For the foregoing interpolation formula to be valid, the platinum thermoelement of the standard thermocouple must meet certain purity requirements and the values of *E* at 630.74°C, the silver point and the gold point must also satisfy certain criteria [*3*]. Not all Type S thermocouples will meet these requirements, but suitable thermocouples may be obtained commercially.

(*d*) Above the gold point the temperature T_{68} is defined by the formula

$$\frac{L_\lambda(T_{68})}{L_\lambda[T_{68}(\text{Au})]} = \frac{\exp\left[\dfrac{c_2}{\lambda T_{68}(\text{Au})}\right] - 1}{\exp\left[\dfrac{c_2}{\lambda T_{68}}\right] - 1}$$

[2]See note *c*, Table 8.1.

where $L_\lambda(T_{68})$ and $L_\lambda(T_{68}(Au))$ are the spectral concentrations at temperature T_{68} and at the freezing point of gold, $T_{68}(Au)$, of the radiance of a blackbody at the wavelength λ; c_2 is defined to be 0.014388 metre kelvin and λ is in metres (exp $[x]$ is defined to be e^x).

In addition to the defining points of the Scale, given in Table 8.1 certain other points may be useful for calibration purposes. Some of these and their reported temperatures are given in Table 8.2. Except for the triple point of benzoic acid, each temperature is for a system in equilibrium under a pressure of 1 standard atm.

8.1.2 Reference Thermometers

Any one of several types of thermometers, calibrated in terms of the IPTS, may be used as a reference thermometer for the calibration of thermocouples. The choice will depend upon the temperature range covered, whether a laboratory furnace or a stirred liquid bath is used, the accuracy expected of the calibration, or in cases where more than one type will suffice, the convenience or preference of the calibrating laboratory.

8.1.2.1 Resistance Thermometers—The standard platinum resistance thermometer is the most accurate reference for use from approximately −260°C (−436°F) to 630°C (1166°F). In cases where an uncertainty approaching 0.1°C is necessary at temperatures below −56°C (−69°F) or

TABLE 8.2—*Secondary reference points.*[a]
The pressure[b] is 1 standard atm, except for the triple point of benzoic acid.

	°C
Boiling point of normal hydrogen	−252.753
Boiling point of nitrogen	−195.806
Sublimation point of carbon dioxide	−78.476
Freezing point of mercury	−38.836
Freezing point of water	0.00
Triple point of benzoic acid	122.37
Freezing point of indium	156.634
Freezing point of bismuth	271.442
Freezing point of cadmium	321.108
Freezing point of lead	327.502
Freezing point of antimony	630.755
Freezing point of aluminum	660.46
Freezing point of copper	1 084.88
Freezing point of nickel	1 455
Freezing point of palladium	1 554
Freezing point of platinum	1 769

[a] Values are extracted from Table 6 of Ref *3*.
[b] See Ref *3* for information on the effect of pressure variation.

above about 200°C (392°F) there are few alternatives to the use of resistance thermometers as a reference.

8.1.2.2 Liquid-in-Glass Thermometers—This type of thermometer may be used from approximately −183°C (−297°F) to 400°C (752°F), or even higher with special types. Generally, the accuracy of these thermometers is less below −56°C (−69°F) where organic thermometric fluids are used, and above 300°C (572°F) where instability of the bulb glass may require frequent calibration. Specifications for ASTM liquid-in-glass thermometers are given in Ref *5*.

8.1.2.3 Types E and T Thermocouples—Either of these types of thermocouples may be used down to a temperature of −183°C(−297°F) or lower, but the attainable accuracy may be limited by the accuracy of the emf measurements and the inhomogeneity of the wire at low temperatures. The stability of the larger sizes of wire is greater than that of smaller wires under the same conditions. Twenty-four gage wire is a useful compromise between the lesser stability of smaller wire and the greater thermal conduction (greater required depth of immersion) of larger wire. Recommended upper limits are 250°C (482°F) for the Type E and 200°C (392°F) for Type T.

8.1.2.4 Types R and S Thermocouples—A Type S or R thermocouple is the most satisfactory reference thermometer for use in the range from 630°C (1166°F) up to about 1200°C (2192°F). Its use may be extended down to room temperature if it is desired to use the same reference over a wide range, but its sensitivity falls off appreciably as temperatures below 200°C (392°F) are reached. Twenty-four gage wire most commonly is used for these types of thermocouples.

8.1.2.5 High Temperature Standards—The IPTS-68 above 1064°C (1947°F) is defined by the Planck law of radiation, and the scale is usually realized by means of an optical pyrometer. The optical pyrometer [*6*], sighted on a blackbody cavity built into the calibration furnace, therefore, can serve as a reference thermometer for all temperatures above 1064°C (1947°F). On the other hand, thermocouples, calibrated on the optical pyrometer scale, can be used themselves as references. The Type B thermocouple [*7*] is useful up to about 1700°C (3092°F). Tungsten rhenium alloys can be used to higher temperatures, but the optical pyrometer more commonly is used.

8.1.3 Annealing

Practically all base-metal thermocouple wire is annealed or given a "stabilizing heat treatment" by the manufacturer. Such treatment generally is considered sufficient, and seldom is it found advisable to further anneal the wire before testing.

Although new platinum and platinum-rhodium thermocouple wire as sold by some manufacturers already is annealed, it has become regular practice in many laboratories to anneal all Types R, S, and B thermocouples, whether new or previously used, before attempting an accurate calibration. This is accomplished usually by heating the thermocouple electrically in air. The entire thermocouple is supported between two binding posts, which should be close together, so that the tension in the wires and stretching while hot are kept at a minimum. The temperature of the wire is determined most conveniently with an optical pyrometer.[3]

There are some questions as to the optimum temperature and length of time at which such thermocouples should be annealed to produce the most constant characteristics in later use [*8*] and as to whether annealing for more than a few minutes is harmful or beneficial. Most of the mechanical strains are relieved during the first few minutes of heatings at 1400 to 1500°C (2552 to 2732°F), but it has been claimed that the changes in the thermal emf of a couple in later use will be smaller if the wires are heated for several hours before calibration and use. The principal objection to annealing thermocouples for a long time at high temperatures, aside from the changes in emf taking place, is that the wires are weakened mechanically as a result of grain growth. It has been found that annealing at temperatures much above 1500°C (2732°F) produces changes in the emf and leaves the wire very weak mechanically. In addition, rapid cooling [*8*] of the wires following annealing at elevated temperatures should be avoided since it can substantially alter the emf. Except for work of the highest precision, cooling at a rate not exceeding about 400°C/min is satisfactory. The National Bureau of Standards has adopted the procedure of annealing Types R, S, and B thermocouples for 45 min at 1450°C (2642°F), followed by 30 min at 750°C (1382°F) and then slow cooling to room temperature.

It has not been demonstrated conclusively that Types R, S, and B thermocouples after contamination can be improved materially in homogeneity by prolonged heating in air, although it is logical to suppose that certain impurities can be driven off or, through oxidation, rendered less detrimental.

8.1.4 Measurement of Emf

One of the factors in the accuracy of the calibration of a thermocouple is the accuracy of the instrument used to measure the emf. Fortunately in most instances, an instrument is available whose performance is such that the accuracy of the calibration need not be limited by the accuracy of the emf measurements. For work of the highest accuracy it is advisable to use a potentiometer of the type designed by Diesselhorst [*9*], White [*10*], or Wenner [*11*], in which there are no slidewires and in which all the settings are

[3]The ordinary portable type of optical pyrometer is very satisfactory for this purpose. As commonly used, the magnification is too low for sighting on an object as small as the wires of noble-metal thermocouples, but this is remedied easily by lengthening the telescope tube or using an objective lens of shorter focal length.

made by means of dial switches. However, for most work on which an accuracy of 5 μV will suffice, slidewire potentiometers of the laboratory type are sufficiently accurate.

Electronic digital voltmeters and analog-to-digital converters that are of potentiometric or other high-impedence design and that possess sufficient accuracy may be also used. Such instruments permit fast readings of a large number of thermocouples. For a more detailed consideration of emf measurements see Chapter 6.

8.1.5 Homogeneity

The emf developed by a thermocouple made from homogeneous wires will be a function of the temperature difference between the measuring and the reference junction. If, however, the wires are not homogeneous, and the inhomogeneity is present in a region where a temperature gradient exists, extraneous emfs will be developed, and the output of the thermocouple will depend upon factors in addition to the temperature difference between the two junctions. The inhomogeneity of the thermocouple wire, therefore, is an important factor in accurate measurements.

Thermocouple wire now being produced is usually sufficiently homogeneous in chemical composition for most purposes. Occasionally inhomogeneity in a thermocouple may be traced to the manufacturer, but such cases are rare. More often it is introduced in the wires during tests or use. It usually is not necessary, therefore, to examine new thermocouples for inhomogeneity, but thermocouples that have been used for some time should be so examined before an accurate calibration is attempted.

While rather simple methods are available for detecting thermoelectric inhomogeneity, no satisfactory method has been devised for quantitatively determining it or the resulting errors in the measurement of temperatures. Abrupt changes in the thermoelectric power may be detected by connecting the two ends of the wire to a sensitive galvanometer and slowly moving a source of heat, such as a bunsen burner or small electric furnace, along the wire. This method is not satisfactory for detecting gradual changes in the thermoelectric power along the length of the wire. Inhomogeneity of this nature may be detected by doubling the wire and inserting it to various depths in a uniformly heated furnace, the two ends of the wire being connected to a galvanometer as before. If, for example, the doubled end of the wire is immersed 250 mm (10 in.) in a furnace with a sharp temperature gradient so that two points on the wire 500 mm (20 in.) apart are in the temperature gradient, the emf determined with the galvanometer is a measure of the difference in the thermoelectric properties of the wire at these two points.

After reasonable homogeneity of one sample of wire has been established, it may be used in testing the homogeneity of similar wires by welding the two together and inserting the junction into a heated furnace. The resulting emf at various depths of immersion may be measured by any convenient method. Other similar methods have been described for detecting inhomogeneity [*12*].

Tests such as those described previously will indicate the uncertainty in temperature measurements due to inhomogeneity in the wires. For example, if a difference in emf of 10 μV is detected along either element of a platinum-rhodium couple by heating various parts of the wire to 600°C (1112°F), measurements made with it are subject to an uncertainty of the order of 1 deg below 500°C, or 2 deg to 1200°C. Similarly, if an emf of 10 μV is detected along an element of a base-metal couple with a source of heat at 100°C, measurements made with the thermocouple are subject to an uncertainty of the order of 0.2°C at this temperature. The effects of inhomogeneity in both wires may be either additive or subtractive, and, as the emf developed along an inhomogeneous wire depends upon the temperature distribution, it is evident that corrections for inhomogeneity are impracticable if not impossible.

8.1.6 General Calibration Methods

The temperature-emf relation of a homogeneous [*13*] thermocouple is a definite physical property and therefore does not depend upon the details of the apparatus or method employed in determining this relation. Consequently, there are numerous methods of calibrating thermocouples, the choice of which depends upon the type of thermocouple, temperature range, accuracy required, size of wires, apparatus available, and personal preference. However, the emf of a thermocouple with its measuring junction at a specified temperature depends upon the temperature difference between its measuring and reference junctions. Therefore, whatever method of calibration is used, the reference junction must be maintained constant at some known temperature (see Chapter 7), and this temperature must be stated as a necessary part of the calibration results.

Thermocouple calibrations are required with various degrees of accuracy ranging from 0.1 to 5 or 10°C. For an accuracy of 0.1°C, agreement with the IPTS-68 and methods of interpolating between the calibration points become problems of prime importance, but for an accuracy of about 10°C comparatively simple methods of calibration usually will suffice. The most accurate calibrations in the range −260°C (−436°F) to 300°C (572°F) are made by comparing the couples directly with a standard platinum-resistance themometer in a stirred liquid bath. In the range 300 to 630°C (572 to 1166°F) (and below if a platinum-resistance thermometer and stirred liquid bath is not available) thermocouples are calibrated most accurately at the freezing or boiling points of pure substances. Between 630 and 1064°C (1166 and 1947°F), the Type S thermocouple calibrated at 630.74°C and the freezing points of gold and silver, serves to define the IPTS-68, and other types of thermocouples are calibrated most accurately in this range by direct comparison with a Type S thermocouple calibrated as specified. Other thermocouples may be calibrated just as accurately at the fixed points as the Type S thermocouple, but interpolated values at intermediate points may depart slightly from the IPTS-68. Above 1064°C (1947°F), the most basic calibrations are made by observing the emf when one junction of the thermocouple is in a blackbody furnace, the temperature of which is measured with an optical pyrometer. However, the difficulties encountered in bringing a blackbody furnace to a uniform temperature make the direct comparison of these two types of instruments by no means a simple matter.

Although the Type S thermocouple serves to define the IPTS-68 only in the range 630.74 to 1064.43°C, this type of thermocouple calibrated at fixed points is used extensively both above and below this range as a reference thermometer in the calibration of other thermocouples. For most industrial purposes a calibration accurate to 2 or 3°C in the range room temperature to 1200°C (2192°F) is sufficient. Other thermocouples can be calibrated by comparison with such a Type S reference thermocouple almost as accurately as the calibration of the reference thermocouple is known. However, it might be pointed up that outside the range 630.74 to 1064.43°C any type of thermocouple suitable for the purpose, and calibrated to agree with the resistance thermometer or optical pyrometer in their respective ranges, has as much claim to yielding temperatures on the IPTS-68 as the Type S thermocouple. In fact, at the lower temperatures certain types of base-metal couples are definitely better adapted for precise measurements.

The calibration of couples then may be divided into two general classes, depending upon the method of determining the temperature of the measuring junction: (1) calibration at fixed points and (2) calibration by comparison with standard instruments such as thermocouples, resistance thermometers, etc.

In order to obtain the high accuracies referred to previously and usually associated with calibrations at fixed points, it is necessary to follow certain prescribed methods and to take the special precautions described in detail in the following paragraphs, but for an accuracy of about 5°C the more elaborate apparatus to be described need not be employed.

8.1.7 Calibration Uncertainties

The several factors which contribute to the uncertainties in the emf versus temperature relationship for a particular thermocouple as determined by calibration may be grouped into two kinds; those influencing the observations at calibration points, and those arising from any added uncertainty as a result of interpolation between the calibration points. Errors from either of these sources of uncertainty can be reduced materially, within limits, through use of well designed equipment and careful techniques; hence, the required accuracy should be clearly understood when choosing calibration facilities.

Estimates of the uncertainties in calibrating homogeneous thermocouples by different techniques are given in Tables 8.3, 8.4, 8.5, 8.6, and 8.7. The estimates assume that reasonable care is exercised in the work. More or less accurate results are possible using the same methods, depending upon soundness of the techniques used. While excessive care is a waste when relatively crude measurements are sufficient, it should be emphasized that inadequate attention to possible sources of error is more often found to be the practice than the converse. In the following some of the important considerations associated with the various calibration methods are emphasized briefly.

8.1.7.1 Uncertainties Using Fixed Points—The equilibrium temperatures listed in Table 8.2 are sufficiently exact, and the materials are readily available in high enough purity, that accurate work can be done using these fixed points with no significant error being introduced by accepting the temperatures listed. Using freezing points, however, good designs of freezing point cells and furnaces are important for controlling the freezes and for providing sufficient immersion for the thermocouple, if the full potential of the method is to be realized.

Although uncertainties of the order of ±1°C in the temperatures are assigned to the freezing points (and hence by implication to the melting points) of palladium and platinum, these contribute in only a minor way overall uncertainties of calibrations using freezing point techniques.

8.1.7.2 Uncertainties Using Comparison Methods—The accuracy attained at each calibration point using the comparison method will depend upon the degree to which the reference thermometer and the test thermocouple are maintained at the same temperature and the accuracy of the reference thermometer used. Comparison measurements made in stirred liquid baths usually present no special problems provided that sufficient immersion is used. Because of the high-thermal conductivity of copper, special attention

TABLE 8.3—*Calibration uncertainties using fixed point techniques.*

Type	Temperature Range, °C	Calibration Points[a]	Calibration Uncertainty: At Observed Points, °C	Calibration Uncertainty: Of Interpolated Values, °C
S	0 to 1100	Zn, Sb[b], Ag, Au	0.2	0.3
R	0 to 1100	Sn, Zn, Al, Ag, Au	0.2	0.5
B	600 to 1100	Al, Ag, Au	0.2	0.5
E	0 to 870	Sn, Zn, Al	0.2	0.5
J	0 to 760	Sn, Zn, Al	0.2	1.0
K	0 to 1100	Sn, Zn, Al, Ag, Au	0.2	1.0

[a]Metal freezing points.
[b]Temperature measured by standard platinum resistance thermometer.

TABLE 8.4—*Calibration uncertainties using comparison techniques in laboratory furnaces (Types R or S standards).*

Type	Temperature Range, °C	Calibration Points[a]	Calibration Uncertainty: At Observed Points, °C	Calibration Uncertainty: Of Interpolated Values, °C
R or S	0 to 1100	about every 100°C	0.3	0.5
B	600 to 1100	about every 100°C	0.3	0.5
E	0 to 870	about every 100°C	0.5	0.5
J	0 to 760	about every 100°C	0.5	1.0
K	0 to 1100	about every 100°C	0.5	1.0

TABLE 8.5—*Calibration uncertainties using comparison techniques in stirred liquid baths.*

Type	Temperature Range, °C	Calibration Points	Type of Standard[a]	Calibration Uncertainty: At Observed Points, °C	Calibration Uncertainty: Of Interpolated Values, °C
E	−196 to 425	about every 100°C	PRT	0.1	0.2
	−196 to 435	about every 50°C	PRT	0.1	0.1
	−196 to 435	about every 50°C	E or T	0.2	0.2
	−56 to 200	about every 50°C	LIG	0.1	0.1
T	−196 to 250	about every 100°C	PRT	0.1	0.2
	−196 to 250	about every 50°C	PRT	0.1	0.1
	−196 to 250	about every 50°C	E or T	0.2	0.2
	−56 to 200	about every 50°C	LIG	0.1	0.1

[a]PRT = standard platinum resistance thermometer; E or T = Type E or T thermocouple; and LIG = liquid-in-glass thermometer.

TABLE 8.6—*Calibration uncertainties: tungsten-rhenium type thermocouples.*

Calibration Uncertainty: At Observed Points		Of Interpolated Values[a]
Gold (1064.43°C)	±0.5°C	1000 to 1455°C, ±2.7°C
Nickel (1455°C)	±3.5°C	1455 to 1554°C, ±4.0°C
Palladium (1554°C)	±3.0°C	1554 to 1768°C, ±4.0°C
Platinum (1769°C)	±3.0°C	1768 to 2000°C, ±7.0°C
Rhodium (1963°C)	±5.0°C	

[a]These values apply only when all five observed points are taken.

TABLE 8.7—*Calibration uncertainties using comparison techniques in special furnaces (optical pyrometer standard).*

Type	Temperature Range, °C	Calibration Uncertainty: At Observed Points, °C	Calibration Uncertainty: Of Interpolated Values,[a] °C
IrRh versus Ir[b]	1000 to 1300	2	3
IrRh versus Ir[b]	1300 to 1600	3	4
IrRh versus Ir[b]	1600 to 2000	5	8
W versus WRe[c]	1000 to 1300	2	3
W versus WRe[c]	1300 to 1600	3	4
W versus WRe[c]	1600 to 2000	5	8
R, S, or B	1100 to 1450	2	3
B	1450 to 1750	3	5

[a]Using difference curve from reference table with calibration points spaced every 200°C.

[b]40Ir60Rh versus Ir, 50Ir50Rh versus Ir, or 60Ir10Rh versus Ir.

[c]W versus 74W26Re, 97W3Re versus 75W25Re, or 95W5Re versus 74W26Re.

should be given to the problem of immersion when calibrating Type T thermocouples.

As higher and higher temperatures are used, the difficulties of maintaining the test and reference thermocouples at the same measured temperature are magnified whether a tube furnace, an oven with moderating block, or whatever means is used for maintaining the desired temperature. In addition, at temperatures of about 1500°C (2732°F), and higher, the choice of insulating materials becomes very important (see Chapter 4). Special attention must be paid to possible errors arising from contamination from the insulators or protection tube and from electrical leakage.

When an optical pyrometer is used as the reference thermometer, a good blackbody must be used, and the design must be such that the test thermocouple is at the same temperature as the blackbody.

8.2 Calibration Using Fixed Points

The indications of the Type S thermocouple calibrated at 630.74°C and the silver and gold points, as mentioned in Section 8.1.1, define the IPTS-68 between 630.74°C (1167°F) and 1064.43°C (1948°F). If such a thermocouple is calibrated also at the zinc point, a reference thermometer will result which is accurate to about 0.3°C in the range 0 to 1100°C (32 to 2012°F). While the fixed-point calibration is prescribed for defining the IPTS-68, similar methods are also useful in the calibration of other types of thermocouples. Fixed points can be used with various degrees of accuracy, ranging from 0.1 to 5°C, for the calibration of various types of thermocouples in the range −260°C (−436°F) to the melting point of platinum at 1769°C (3216°F). Some of the fixed points for which values have been determined accurately are listed in Section 8.1.1, Table 8.2. Because of experimental difficulties, fixed points at temperatures higher than the freezing point of copper usually are realized as melting points rather than freezing points, as described later.

8.2.1 Freezing Points

The emf developed by a homogeneous thermocouple at the freezing point of a metal is constant and reproducible if all of the following conditions are fulfilled: (1) the thermocouple is protected from contamination: (2) the thermocouple is immersed in the freezing-point sample sufficiently far to eliminate heating or cooling of the measuring junction by heat flow along the wires and protection tube; (3) the reference junctions are maintained at a constant and reproducible temperature; (4) the freezing-point sample is pure, and (5) the metal is maintained at essentially a uniform temperature during freezing.

Techniques for achieving these conditions are well developed [*12,14,15*]. Many of the metals listed in Table 8.2 of Section 8.1.1 are available commercially in high purity (99.999 percent or better) and can be used assuming the freezing point temperatures given in the table. It is essential, however, that protecting tubes and crucibles be chosen of such material (see Section 4) that the pure metals will not be contaminated. Copper and silver must be protected from oxygen contamination [*12*] and it is also advisable to protect aluminum and antimony; this is done usually by using covered crucibles and covering the freezing point metals with powdered graphite or by sealing the crucible in a glass tube that contains a nonoxidizing gas such as argon or helium. The choice of a suitable furnace is also important. The furnace must provide uniform heating in the region of the freezing point sample, and have adequate controls to bring the sample slowly into its freeze. Complete units consisting of freezing point sample, crucible, and furnace are available commercially. Freezing point standard samples of tin, lead, zinc, aluminum, and copper may be purchased from the National Bureau of Standards.

8.2.2 Melting Points

The emf of a thermocouple at the melting point of a metal may be determined with the same apparatus as that described above for freezing points, but the use of the freeze is usually more satisfactory. Melting points are used to advantage, however, when only a limited amount of material is available or at high temperatures where experimental techniques with freezing points are difficult. To apply this method [*16–19*], a short length of metal whose melting point is known is joined between the end of the two wires of the thermocouple and placed in an electrically heated furnace the temperature of which is raised slowly. When the melting point of the metal is reached, the emf of the thermocouple remains steady for a few minutes and then drops to zero as the fused metal drops away from the junction. With good technique[4] the method can result in accuracies comparable to those with which the IPTS-68 is realized above 1064°C by optical pyrometry.

8.3 Calibration Using Comparison Methods

The calibration of a thermocouple by comparison with a reference thermometer [*20*] is sufficiently accurate for most purposes and can be done conveniently in most industrial and technical laboratories. The success of this method usually depends upon the ability of the observer to bring the measuring junction of the thermocouple to the same temperature as the actuating element of the reference thermometer, such as the measuring junction of a standard thermocouple or the bulb of a resistance or liquid-in-glass thermometer. The accuracy obtained is further limited by the accuracy of the reference thermometer. Of course, the reference junction temperature must be known, but this can be controlled, as described in Chapter 7. The method of bringing the measuring junction of the thermocouple to the same temperature as that of the actuating element of the reference thermometer depends upon the type of thermocouple, type of reference thermometer, and the method of heating.

8.3.1 Laboratory Furnaces

The calibration procedure consists of measuring the emf of the thermocouple being calibrated at selected calibration points, the temperature of each point being measured with a reference thermometer. The number and choice of calibration points will depend on the type of thermocouple, the temperature range covered, and the accuracy required (see Sections 8.1.6 and 8.4).

8.3.1.1 Noble-Metal Thermocouples—Such thermocouples usually may be calibrated at temperatures from ambient up to 1200°C by comparison with either a Type S or Type R reference thermocouple in electrically heated furnaces. Above 1200°C (2192°F) the Type B thermocouple is a preferred reference thermometer because of its greater stability at high temperatures. This thermocouple may be used to 1700°C (3092°F) or higher.

One method for the comparison of two such thermocouples is based upon the simultaneous reading of the emf of the reference and the test thermocouple without waiting for the furnace to stabilize at any given temperature. The measuring junctions are maintained always at close to the same temperature

[4]This method is not well adapted to metals that oxidize rapidly, and, if used with materials whose melting temperature is altered by the oxide, the metal should be melted in a neutral atmosphere.

by welding them into a common bead or by wrapping them together with platinum wire or ribbon. A separate potentiometer is used to measure each emf, one connected to each thermocouple, and each potentiometer is provided with a reflecting galvanometer. The two spots of light are reflected into a single scale, the galvanometers being set in such a position that the spots coincide at the zero point on the scale when the circuits are open, and therefore also when the potentiometers are set to balance the emf of each thermocouple. Simultaneous readings are obtained by setting one potentiometer to a desired value and adjusting the other so that both spots of light pass across the zero of the scale together as the temperature of the furnace is raised or lowered.

By making observations first with a rising and then with a falling temperature, the rates of rise and fall being approximately equal, and taking the mean of the results found, several minor errors such as those due to differences in the periods of the galvanometers, etc., are eliminated or greatly reduced.

This method is particularly adapted to the calibration of thermocouples at any number of selected points. For example, if it is desired to determine the temperature of a thermocouple corresponding to 10.0 mV, this emf is set up on the potentiometer connected to the thermocouple, the emf of the reference thermocouple observed as desired above, and the temperature obtained from the emf of the reference. If it is desired to determine the emf of a thermocouple corresponding to 1000°C (1832°F), the emf of the reference thermocouple corresponding to this temperature is set up on the potentiometer connected to the reference thermocouple, and the emf of the thermocouple being calibrated is observed directly with the second potentiometer. To reduce the time required to calibrate by this method the furnace should be so constructed that it will heat or cool rapidly. Fast response is obtained in one furnace design which employs a nickel-chromium tube as the heating element [*14*].

A similar furnace using a silicon carbide tube as the heating element can be used to extend the calibration range upward [*7*]. At temperatures above 1064°C (1947°F) the IPTS-68 is defined in terms of ratios of radiation (Section 8.1.1) usually measured with a visual or a photoelectric optical pyrometer. If the test thermocouple is inserted into the back of a blackbody cavity built into the furnace, a pyrometer may be used directly as the reference thermometer. Alternatively, the Type B thermocouple can be used as the reference thermometer after it has been calibrated against a pyrometer.

The thermocouples are insulated and protected by suitable ceramic tubes (Chapter 4). It is essential that good insulation be maintained between the two potentiometers and thermocouple circuits except at the point where the junctions are welded together. The reference junctions are maintained at a known temperature (Chapter 7).

Variations of the two potentiometer method may be used to automate the calibration process when the thermocouple being calibrated and the reference thermocouple are of the same type [*20*]. If the emf of the reference is read with one potentiometer and the emf difference between the reference and the unknown are read with the second potentiometer, the calibration data may be recorded automatically [*21,22*].

If two potentiometers are not available for taking simultaneous readings, the furnace may be brought to essentially a constant temperature, and the emf of each thermocouple read alternately on one instrument [*20*].

When the thermocouples are calibrated by welding or wrapping the junctions together, both would be expected to be close to the same temperature even when the temperature of the furnace is changing. If it is necessary or advisable to calibrate the thermocouples without removing them from the protecting tubes, then the junctions of the thermocouple being tested and that of the reference thermocouple should be brought as close together as possible in a uniformly heated portion of the furnace. In this case it is necessary that the furnace be brought to approximately a constant temperature before taking observations.

There are a number of other methods of heating and of bringing the junctions to approximately the same temperature, for example, inserting the thermocouples properly protected into a bath of molten metal or into holes drilled in a large metal block. The block of metal may be heated in a muffle furnace or, if made of a good thermal conductor such as copper, may be heated electrically. Tin, which has a low melting point, 232°C (450°F), and low volatility, makes a satisfactory bath material. The thermocouples should be immersed to the same depth with the junctions close together. Ceramic tubes are sufficient protection, but to avoid breakage by thermal shock when immersed in molten metal it is preferable to place them inside of secondary tubes of iron, nickel-chromium, graphite, or similar material. In all of these methods, particularly in those cases in which the junctions of the thermocouples are not brought into direct contact, it is important that the depth of immersion be sufficient to eliminate cooling or heating of the junctions by heat flow along the thermocouple and the insulating and protecting tubes. This can be determined by observing the change in the emf of the thermocouple as the depth of immersion is changed slightly. If proper precautions are taken, the accuracy yielded by any method of heating or bringing the junctions to the same temperature may be as great as that obtained by any other method.

8.3.1.2 Base-Metal Thermocouples—The methods of testing base-metal thermocouples above room temperature are generally the same as those just described for testing noble-metal thermocouples with the exception, in some cases, of the methods of bringing the junctions of the reference and the thermocouple being tested to the same temperature and the methods of protecting platinum-rhodium reference thermocouples from contamination. One arrangement of bringing the junction of a platinum-rhodium reference thermocouple to the same temperature as that of a large base-metal thermocouple for accurate calibration is to insert the junction of the reference thermocouple into a small hole (about 1.5 mm (0.06 in.) in diameter) drilled in the hot junction of the base-metal thermocouple. The platinum-rhodium reference thermocouple is protected by ceramic tubes to within a few tenths of an inch of the hot junction, and the end of the ceramic tube is sealed to the thermocouple by pyrex glass or by a small amount of kaolin and water-glass cement. This prevents contamination of the reference thermocouple, with the exception of the small length of about 2.5 mm (0.1 in.) which is necessarily in contact with the base-metal thermocouple. If the furnace is uniformly heated in this region (and it is of little value to make such a test unless it is) contamination at this point will not cause any error. If the wire of the reference thermocouple becomes brittle at the junction, this part of the wire may be cut off and enough wire drawn through the softened seal to form a new junction. The seal should be examined after each test and remade if it does not appear to be good. More than one base-metal thermocouple may be welded together and the hole drilled in the composite junction. The thermocouples should be clamped in place so that the junctions remain in contact. If two potentiometers are used for taking simultaneous readings, the temperature of the furnace may be changing as much as a few degrees per minute during an observation, but if a single instrument is used for measuring the emf, the furnace temperature should be maintained practically constant during observations. When wires, insulators, and protection tubes are large, tests should be made to ensure that the depth of immersion is sufficient.

8.3.2 Stirred Liquid Baths

At temperatures below 620°C (1148°F) stirred liquid baths provide an efficient medium for bringing a thermocouple and a reference thermometer to the same temperature.

Water, petroleum oils, or other organic liquids, depending upon temperature range, are commonly used bath media. Molten salts or liquid tin are used at temperatures higher than are suitable for oil. Baths suitable for this work are described in Ref *23*.

Base-metal thermocouples, either bare wire or insulated, may be calibrated accurately in such baths. Usually no special preparation of the thermocouple will be required other than to insert it to the bottom of a protection tube for immersion in the liquid bath. Borosilicate glass tubing, such as Pyrex glass, is convenient for use up to 500°C (932°F). Vitreous silica or ceramic tubing may be used to 620°C (1148°F). The tube should be closed at the immersed end and of an internal diameter such as to permit easy insertion of the thermocouple or thermocouples to be calibrated but no larger than necessary. Unfavorable heat transfer conditions in an unnecessarily large diameter tube will require a greater depth of immersion in the bath than would a close fitting tube. If a bare wire thermocouple is being calibrated, the wires must be provided with electrical insulation over the length inserted in the protection tube. Sheathed thermocouples may be immersed directly in the bath liquid in cases where the sheath material will not be attacked by the liquid. Salt baths for use at high temperature must be provided with suitable wells into which the thermocouple protection tubes and reference thermometers may be inserted for protection from the molten salt.

The reference thermometer may be a calibrated thermocouple inserted in the protection tube with the thermocouple being calibrated, or it may be a liquid-in-glass thermometer or a resistance thermometer immersed in the bath close to the thermocouple protection tube. The choice of a reference thermometer will be governed principally by the degree of uncertainty which can be tolerated.

8.3.3 Fixed Installations

After thermocouples have been used for some time at high temperatures, it is difficult, if not impossible, to determine how much the calibrations are in error by removing them from an installation and testing in a laboratory furnace. The thermocouples are usually heterogeneous after such use and in such a condition that the emf developed by the thermocouples depends upon the temperature distribution along the wires [*24*]. If possible, such a thermocouple should be tested under the same conditions and in the same installation in which it is used. Although it is not usually possible to obtain as high a precision by testing the thermocouple in place as is obtained in laboratory tests, the result is far more useful in the sense of being representative of the behavior of the thermocouple [*19*]. The calibration is accomplished by comparing the thermocouple with a reference thermocouple.

In this case, as in the calibration of any thermocouple by comparison methods, the main objective is to bring the measuring junction to the same temperature as that of the thermocouple being tested. One method is to drill a hole in the furnace, flue, etc., at the side of each thermocouple permanently installed, large enough to permit insertion of the reference thermocouples. The hole is kept plugged, except when tests are being made. The reference thermocouple is inserted through this hole to the same depth as the thermocouple being tested with the measuring junction ends of the protecting tubes as close together as possible. Preferably a potentiometer should be used to measure the emf of the reference thermocouple.

In many installations the base-metal thermocouple and its protecting tube are mounted inside another protecting tube of iron, fire clay, carborundum, or some other refractory which is permanently cemented or fastened into the furnace wall. Frequently there is room to insert a reference thermocouple in this outer tube alongside of the fixed thermocouple. A third method, much less satisfactory, is to wait until the furnace, flue, etc., have reached a constant temperature and make observations with the thermocouple being tested, then remove this thermocouple and insert the reference thermocouple to the same depth.

If desired, comparisons can be made, preferably by either of the first or second methods at several temperatures, and a curve obtained for each permanently installed thermocouple showing the necessary corrections to be applied to its readings. Although testing a thermocouple at one temperature yields some information, it is not safe to assume that the changes in the emf of the thermocouple are proportional to the temperature or to the emf. For example, it has been observed that a thermocouple which had changed in use by the equivalent of 9°C at 315°C (16°F at 599°F) had changed only the equivalent of 6°C at 1100°C (11°F at 2012°F).

It may be thought that the method of calibrating thermocouples under working conditions is unsatisfactory because, in most furnaces used in industrial processes, large temperature gradients exist, and there is no certainty that the reference thermocouple is at the same temperature as the thermocouple being tested. This objection, however, is not serious, because if temperature gradients do exist of such a magnitude as to cause much difference in temperature between two similarly mounted thermocouples located close together, the reading of the reference thermocouple represents the temperature of the fixed thermocouple as closely as the temperature of the latter represents that of the furnace.

Another advantage of calibrating thermocouples in the same installation in which they are used is that the thermocouple, extension wires, and indicator are tested as a unit and under the conditions of use.

8.4 Interpolation Methods

An experimental thermocouple calibration consists of a series of voltage measurements determined at a finite number of known temperatures. If a test thermocouple were compared with a standard temperature instrument at 100 temperatures within a 5°C (10°F) range, there would be little need for interpolation between the calibration points. However, if from 4 to 10 calibration points are all that can be afforded in a given range of interest, then what is needed to characterize an individual thermocouple is a continuous relation, by means of which temperatures can be approximated with a minimum uncertainty from voltage measurements at intermediate levels. Efforts to obtain such a continuous relation appear thwarted from the start because of the small number of discrete calibration points available. However, interpolation between the calibration points is possible since the emf changes only slowly and smoothly with temperature.

One can present raw calibration data directly in terms of temperature (T) and voltage (E_{couple}), on a scale so chosen that the information appears well represented by a single curve (see Fig. 8.1) or by a simple mathematical

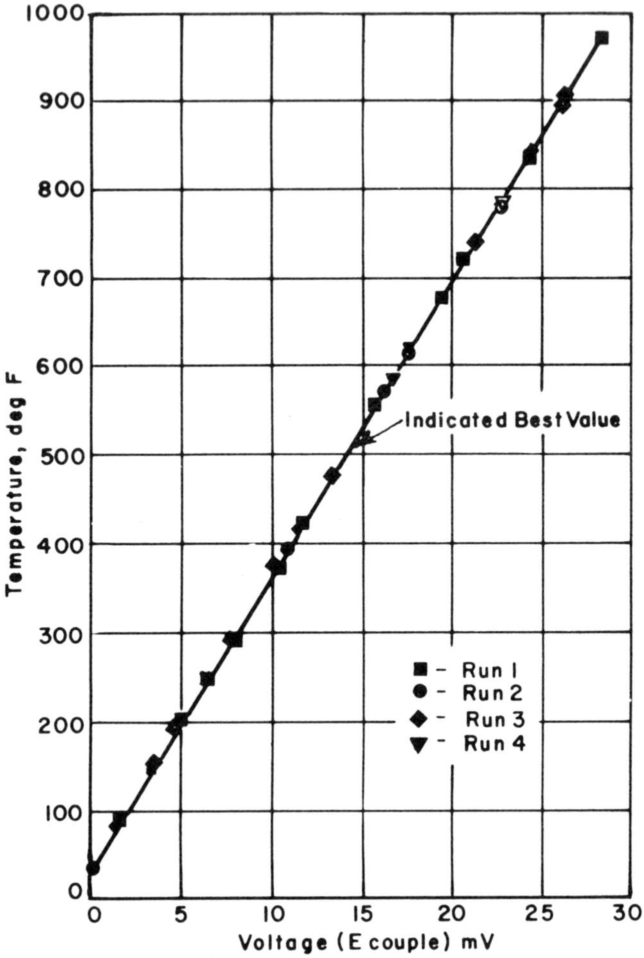

FIG. 8.1—*Temperature emf plot of raw calibration data for an iron/constantan thermocouple.*

equation. For example, for the highest accuracy in the range 630 to 1064°C with the Type S thermocouple, the method is that prescribed in Ref *12*. An equation of the form $e = a + bt + ct^2$, is used where a, b, and c are constants determined by calibration at the freezing points of gold, silver, and antimony. By calibrating the thermocouple also at the freezing point of zinc and using an equation of the form $e = a + bt + ct^2 + dt^3$, the temperature range can be extended down to 400°C without introducing an uncertainty of more than 0.1°C in the range 630 to 1064°C. By calibrating the thermocouple at freezing points of gold, antimony, and zinc and using an equation of the form $e = a + bt + ct^2$, a calibration is obtained for the range 400 to 1100°C, which agrees with IPTS-68 to 0.5°C. However, in general, this practice of directly representing thermocouple characteristics does not yield results within the required limits of uncertainty.

A better method[5] is based on the use of differences between observed values and values obtained from standard reference tables. Such reference tables and the mathematical means for generating them are presented in Chapter 10 of this Manual. The data of Fig. 8.1 are replotted in Fig. 8.2 in terms of differences from the proper reference table. The maximum spread between points taken at the same level (replication), but obtained in random order with respect to time and level (randomization) is taken as the uncertainty envelope. This information, taken from Fig. 8.2 is plotted in Fig. 8.3, and constitutes a vital bit of information about the particular thermocouple and the calibration system. In lieu of an experimental determination of the uncertainty, one must rely on judgment or on the current literature for this information.

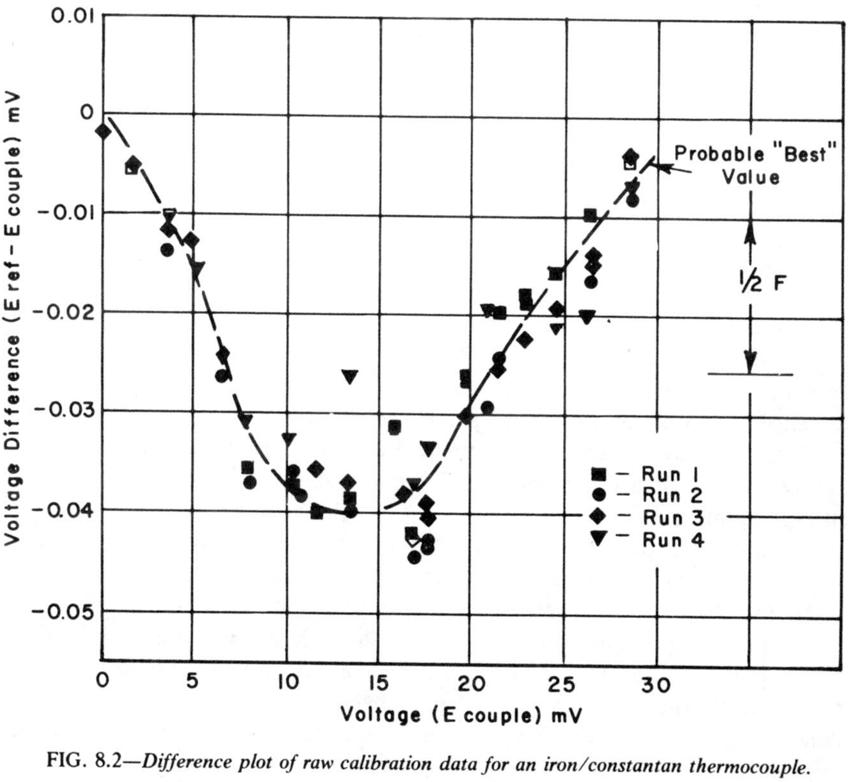

FIG. 8.2—*Difference plot of raw calibration data for an iron/constantan thermocouple.*

[5]Much of the material in this section is based on Refs *25* to *27*.

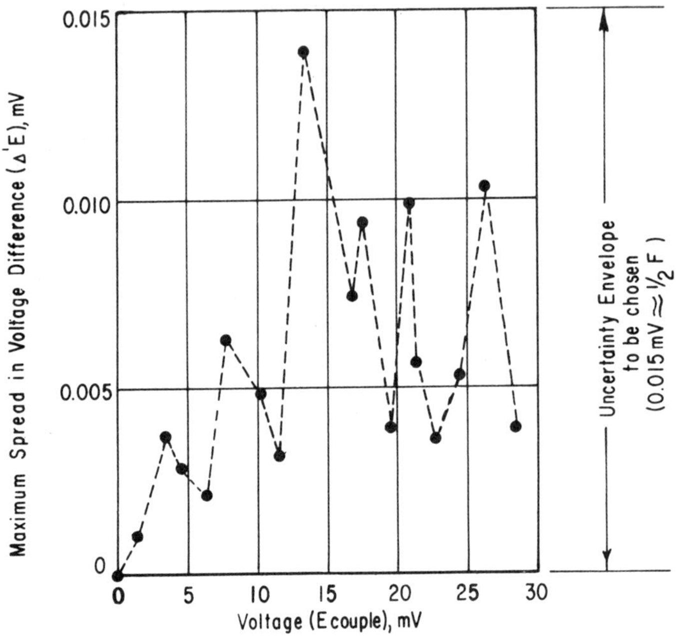

FIG. 8.3—*Typical determination of uncertainty envelope (from data of Fig. 8.2).*

Usually, only a single set of calibration points is available. Typical points would be those taken from one run shown in Figs. 8.1 or 8.2, and these are shown in Fig. 8.4 together with four of the many possible methods for representing the thermocouple difference characteristic. Although at first it appears that the most probable relation characterizing a given thermocouple is sensibly indeterminate from a single set of calibration points, it is an important fact that all experimental points must be continued within the uncertainty interval when the uncertainty interval is centered on the most probable interpolation equation.

Making use of this principle, together with the fact that overall experimental uncertainties are minimized by use of the least squares technique, one starts the search for the most probable interpolation equation by passing a least squares equation of the first degree through the experimental data. A check is then made to ascertain whether all experimental points are contained within the uncertainty envelope which is centered on the linear interpolation equation (see Fig. 8.5). One proceeds, according to the results of the foregoing check, to the next highest degree equation, stopping at the lowest degree least squares equation which satisfies the uncertainty requirements. For the example given here, a third degree interpolation equation is required

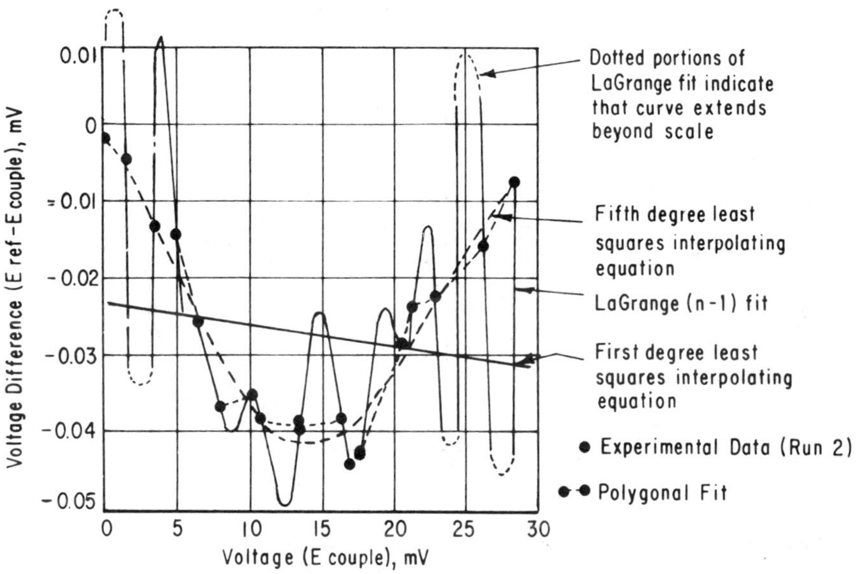

FIG. 8.4—*Various possible empirical representations of the thermocouple characteristic (based on a single calibration run).*

(see Fig. 8.6). By obtaining voltage differences from the least squares fit of any set of calibration points, the uncertainty in the thermocouple difference characteristic will be within one half the uncertainty interval. Generally, the form of the uncertainty envelope and the degree of the most probable least squares interpolation equation are strongly dependent on the amount of calibration data available and on the temperature range under consideration. It is recommended that the number of distinct calibration points available should be at least 2 (degree +1). The factor two is arrived at from numerical analysis reasoning. A distinct calibration point is defined arbitrarily as one which is separated, temperaturewise, from all other points in the set by as much as one tenth the difference in temperature between the maximum and minimum temperatures of the particular run. The choice of one tenth presupposes a maximum practical degree of four for the least squares interpolation equation, in keeping with the low degree requirement of numerical analysis. Indeed, if the data cannot be represented by a fourth degree interpolation equation, one should increase the uncertainty interval and start the fitting procedure again.

Thus, in general, by using the proper reference table in conjunction with a difference curve, greater precision in temperature determination by means of thermocouples can be obtained from a given number of calibration points than from the use of the calibration data alone.

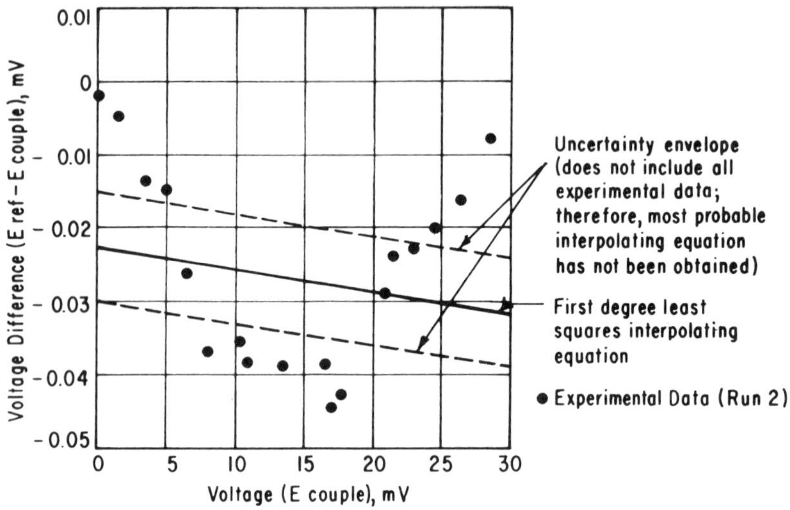

FIG. 8.5—*Uncertainty envelope method for determining degree of least squares interpolating equation for a single calibration run (linear).*

8.5 Single Thermoelement Materials

The standard method provided by ASTM for evaluating the emf characteristics of single thermoelement materials used in thermocouples for temperature measurement is designated as E 207 and is entitled "Standard Method of Thermal Emf Test of Single Thermoelement Materials by Comparison with a Secondary Standard of Similar Emf-Temperature Properties." The method covers the determination of the thermoelectromotive force of single thermoelement materials (thermoelements), against standard platinum, the cold junction being at the ice point, by comparison to the thermoelectromotive force of a working standard thermoelement of similar emf-temperature properties independently standardized with respect to the same standard platinum.

Summary of Method—The test thermoelements are welded to the working standard to form the test thermocouple. The method involves measuring the small electromotive force developed between the test thermoelement and the secondary working standard having emf-temperature properties similar to the thermoelement being evaluated. The thermoelectromotive force of the test thermoelement then is determined by algebraically adding this measured small emf to the known emf of the secondary working standard referenced to the standard platinum. The testing circuit is shown schematically in Fig. 8.7. Since the thermal emfs (against any reference material) of the test and standard thermoelements are similar, it is unnecessary to accurately control or

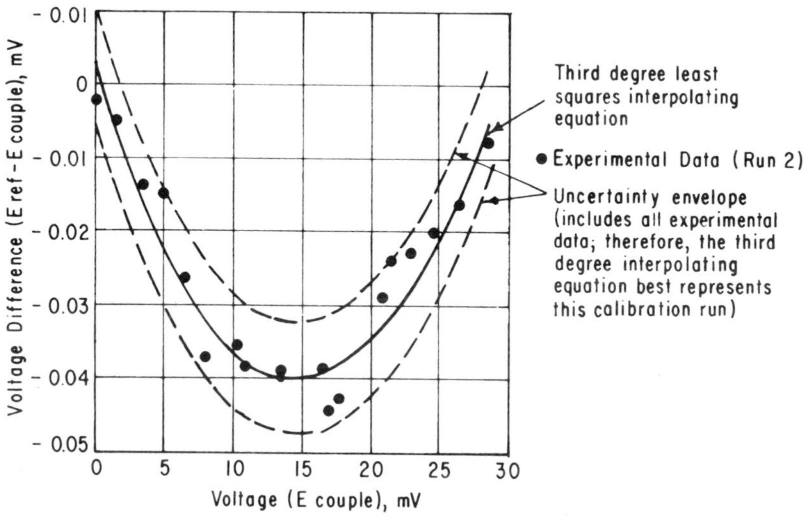

FIG. 8.6—*Uncertainty envelope method for determining degree of least squares interpolating equation for a single calibration run (cubic).*

measure the junction temperatures because the difference in emf changes insignificantly even for large changes in temperature at the junctions. Actually, the need for the accurate control of the measuring and reference junction temperatures is not eliminated, but merely is shifted to an accurate laboratory method which calibrates the secondary standard against the standard platinum [*20*].

8.5.1 Test Specimen

The test specimen is a length of wire, rod, ribbon, or strip of the coil or spool of the thermoelement material to be evaluated. The length is adequate to prevent the transfer of heat from the measuring junction to the reference junction during the period of test. A length of 0.6 to 1.2 m (2 to 4 ft), depending on the length of the testing medium and the transverse size of the thermoelement, is generally satisfactory. The transverse size of the specimen is limited only by the size of the test medium, the relative convenience of handling the specimens and reference thermoelement, and the maintenance of an isothermal test temperature junction.

8.5.2 Reference Thermoelement

A reference thermoelement is used which has emf-temperature properties similar to that of the test specimen and which previously has been standard-

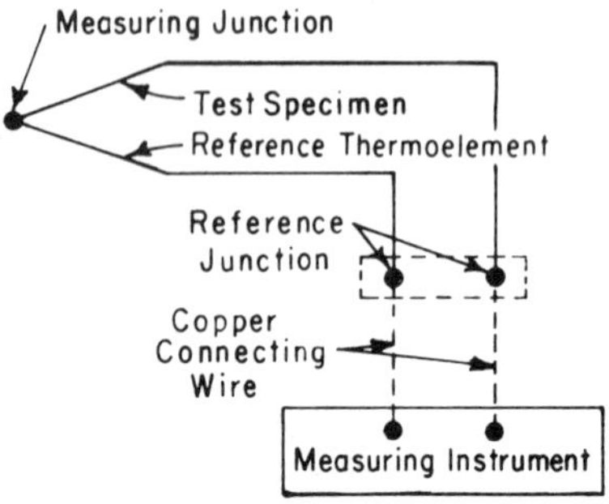

FIG. 8.7—*Circuit diagram for thermal emf test.*

ized thermoelectrically with respect to National Bureau of Standards Platinum 67 (Pt-67). If a large amount of testing is anticipated, a coil or spool of reference thermoelement material is reserved. This material is selected on the basis of thermoelectric uniformity, and a minimum of three samples taken from the center and the ends of the coil is calibrated against the standard platinum. The reference thermoelement and the test specimen may differ in diameter, but it is convenient for their lengths to be about equal.

8.5.3 Reference Junction

The reference junction temperature of the test specimen and of the reference thermoelement is controlled during the period of test by maintaining it at the ice point (0°C). The ice-water junction is used because it is recognized as a convenient means for maintaining a constant reference temperature. The description and maintenance of the ice point are given in Chapter 7. A reference temperature other than the ice point may be used. However, the reference thermoelement should be calibrated against standard platinum using the alternate reference temperature, or the calibration data adjusted to correspond to the alternate reference temperature.

8.5.4 Measuring Junction

The measuring junction consists of a welded union of the test specimen and reference thermoelement. The weldment may be prepared by any method providing a good electrical connection which can be immersed fully in the uniform temperature region of the testing medium. Any number of test specimens and reference thermoelements may be welded together provided the resulting assembly does not introduce heat losses which prevent maintaining a uniform temperature region. If separation of the reference thermoelement and test specimen cannot be maintained during the test (except where they must make contact at the measuring junction), it is necessary that they be insulated from each other.

8.5.5 Test Temperature Medium

For temperatures up to 300°C (570°F), appropriate liquid baths may be used. For temperatures above 150°C (300°F), electrically heated tube or muffle type furnaces are recommended for comparison testing of base metal or noble metal thermoelement materials of similar emf-temperature properties. For convenience, a separate furnace may be controlled and available for each test temperature. This eliminates lost time in changing furnace temperatures when a large volume of testing is to be done. The length of each furnace is at least 381 mm (15 in.) so as to provide a minimum depth of immersion of 178 mm (7 in.) for the test thermocouple assembly. A constant immersion depth is maintained, whether single or multiple furnaces are used. The inside diameter of the furnace tube is approximately 25 to 76 mm (1 to 3 in)., the specific diameter depending upon the size and the number of the specimens to be included in the test thermocouple assembly. The furnace provides a uniform temperature zone extending at least 76 mm (3 in.) back from the measuring junction, or further if required to contain any inhomogeneity in the test thermocouple assembly. The temperature of each furnace is controlled manually or automatically to within ±10°C (±18°F) of the desired value which is ample for comparison testing of thermoelements having compositions similar to that of the reference thermoelement.

8.5.6 Emf Indicator

The emf generated by the thermocouple, consisting of the test specimen and the reference thermoelement is measured with instrumentation sensitive to ±0.001 mV with an accuracy over a 2 mV range of ±1 percent of the reading plus 0.003 mV. Any millivoltmeter, with circuitry errors taken into account, or potentiometer with a galvanometer or null indicator, providing measurements within these tolerances is acceptable. An indicator with a bidirectional scale (zero center) is convenient, but a unidirectional instrument may be used if a polarity switch is provided in the copper connecting circuit, or if the copper connecting wires to the instrument are exchanged whenever the polarity between the reference thermoelement and the test specimen is reversed.

8.5.7 Procedure

For a furnace medium the test thermocouple assembly is inserted into the furnace so that the measuring junction extends at least 76 mm (3 in.) into the uniform temperature zone taking care there is no contact between the wires and the furnace wall. The free ends of the reference thermoelement and the test specimens are bent as required so they may be inserted into the glass tubes of the reference junction bath. Care is exercised to minimize distorting the wires prior to testing because of the effect of cold work on emf output. After bringing the test temperature to the specified value, sufficient time is provided for the test assembly to reach steady state conditions before recording the emf generated between the test specimen and the reference thermoelement.

In a similar manner the emf generated between all other test specimens in the assembly is measured with respect to the reference thermoelement. Then the test temperature is raised to the next higher specified value, or the test assembly is advanced to the next furnace or bath having the next higher specified temperature if multiple furnaces or baths are used. A second set of readings is taken at the new temperature, and the procedure is repeated with readings taken at all specified temperatures. In all cases the readings are taken in sequence from the lowest to the highest temperature to minimize test variations between producer and consumer if any of the alloys are affected by differences in short time heating cycles. A base-metal reference thermoelement is used for one series of temperature changes only. However, if a portion of it considerably exceeding the region previously exposed to the uniform heating zone is discarded, the remainder of the reference thermoelement may be used for another test assembly. For noble metals and their alloys, reuse depends on the known stability of the material involved.

The polarity of the test specimen with respect to the reference thermoelement is determined as follows:

1. If the test specimen is connected to the positive (+) terminal of an unidirectional potentiometer and balance can be achieved, the specimen is positive to the reference thermoelement.
2. If the connections must be reversed to achieve balance, that is, the reference thermoelement must be connected to the positive terminal, the test specimen is negative to the reference thermoelement.
3. If an indicating potentiometer with a bidirectional scale is used, the test specimen to the positive (+) terminal and the reference thermoelement to the negative (−) terminal are connected. The polarity of the test specimen with respect to the reference thermoelement then will be indicated by the direction of balance of the instrument scale.

The emf of the test specimen with respect to Pt-67 is then reported for each test temperature after algebraically adding the measured emf of the test specimen versus the reference thermoelement to the known emf of the reference thermoelement versus Pt-67.

8.6 References

[1] "The International Practical Temperature Scale of 1968," *Metrologia*, Vol. 5, No. 2, April 1969, pp. 34–49.

[2] Benedict, R. P., *Technical Journal*, Leeds & Northrup Co., Vol. 6, Summer Issue, 1969.
[3] "The International Practical Temperature Scale of 1968 Amended Edition of 1975," *Metrologia*, Vol. 12, No. 1, March 1976, pp. 7–17.
[4] Stimson, H. F., "International Practical Temperature Scale of 1948, Text Revision of 1960," NBS Monograph 37, National Bureau of Standards, 1961.
[5] ASTM Method E 1-71, "Standard Specifications for ASTM Thermometers," *1979 Annual Book of ASTM Standards*, Part 44, p. 505.
[6] Kostkowski, H. J. and Lee, R. D., "Theory and Methods of Optical Pyrometry," NBS Monograph 41, National Bureau of Standards, 1962.
[7] Burns, G. W. and Gallagher, J. S., "Reference Tables for the Pt-30% Rh Versus Pt-6% Rh Thermocouple," *Journal of Research*, National Bureau of Standards, Vol. 70C, April–June 1966.
[8] Corruccini, R. J., "Annealing of Platinum for Thermometry," *Journal of Research*, National Bureau of Standards, Vol. 47, No. 94, 1951, RP2232.
[9] Diesselhorst, H., "Thermokraftfreier Kompensationapparat mit funf Dekadeu und kunstanter kleinem Widerstand," *Zeitschrift für Instrumentenkunde*, Vol. 28, No. 1, 1908.
[10] White, W. P., "Thermokfraftfreie Kompensationapparat mit kleinem Widerstand und kustanter Galvanometereempfindlicheit," *Zeitschrift für Instrumenteokunde*, Vol. 27, No. 210, 1907.
[11] Behr, L., "The Wenner Potentiometer," *Review of Scientific Instruments*, Vol. 3, No. 108, 1932.
[12] Roeser, W. F. and Lomberger, S. T., "Methods of Testing Thermocouple Materials," NBS Circular 590, National Bureau of Standards, 1958.
[13] Finch, D. I., "General Principles of Thermoelectric Thermometry," *Temperature, Its Measurement and Control in Science and Industry*, Vol. 3, Part 2, Reinhold, New York, 1962, p. 3.
[14] McLaren, E. H., "The Freezing Points of High Purity Metals as Precision Temperature Standards," *Temperature, Its Measurement and Control in Science and Industry*, Vol. 3, Part 1, Reinhold, New York, 1962, p. 185.
[15] Trabold, W. G., *Temperature, Its Measurement and Control in Science and Industry*, Vol. 3, Part 2, Reinhold, New York, p. 45.
[16] Fairchild, C. O., Hoover, W. H., and Peters, M. F., "A New Determination of the Melting Point of Palladium," *Journal of Research*, National Bureau of Standards, Vol. 2, No. 931, 1929, RP 65.
[17] Barber, C. R., "The Calibration of the Platinum/13% Rhodium-Platinum Thermocouple Over the Liquid Steel Temperature Range," *Journal of the Iron and Steel Institute*, Vol. 147, No. 205, 1943.
[18] Bedford, R. E., "Reference Tables for Platinum-20% Rhodium/Platinum-5% Rhodium Thermocouples," *Review of Scientific Instruments*, Vol. 35, No. 1177, 1964.
[19] Quigley, H. C., "Resume of Thermocouple Checking Procedures," *Instruments*, Vol. 25, No. 616, 1952.
[20] ASTM Method E 220, "Standard Method for Calibration of Thermocouples by Comparison Techniques," *1979 Annual Book of ASTM Standards*, Part 44, p. 605.
[21] "Automatic Thermocouple Comparator," *Technical News Bulletin*, National Bureau of Standards, Vol. 47, No. 5, May 1963, p. 82.
[22] Toenshoff, D. A., "Automatic Calibration of Thermocouples," *Technical Bulletin*, Englehard Industries, Vol. 2, No. 3, Dec. 1961, p. 88.
[23] ASTM Method E 77, "Standard Method for Verification and Calibration of Liquid-in-Glass Thermometers," *1979 Annual Book of ASTM Standards*, Part 44, p. 566.
[24] Dahl, A. I., "Stability of Base-Metal Thermocouples in Air from 800 to 2200°F," *Journal of Research*, National Bureau of Standards, Vol. 24, No. 205, 1940, RP 1278.
[25] Benedict, R. P. and Ashby, F. H., "Empirical Determination of Thermocouple Characteristics," *Transactions*, American Society of Mechanical Engineers, *Journal of Engineering for Power*, Jan. 1963, p. 9.
[26] Benedict, R. P., "Engineering Analysis of Experimental Data," *Transactions*, American Society of Mechanical Engineers, *Journal of Engineering for Power*, Jan. 1969, p. 21.
[27] Benedict, R. P. and Godett, T. M., "A Note on Obtaining Temperatures From Thermocouple EMF Measurements," *Transactions*, American Society of Mechanical Engineers, *Journal of Engineering for Power*, Oct. 1975, p. 516.

[28] McLaren, E. H., and Murdock, E. G., "New Considerations on the Preparation, Properties and Limitations of the Standard Thermocouple for Thermometry," *Temperature*, Instrument Society of America, Vol. 4, Part 3, 1972, p. 1543.

Bibliography

Temperature, Its Measurement and Control in Science and Industry, Vol. 1, Reinhold, New York, 1941, Vol. 3, Parts 1 and 2, 1962, Vol. 4, Parts 1, 2, and 3, 1972.

ASME Performance Test Codes, Instruments and Apparatus Supplement, Part 3, Temperature Measurement, American Society of Mechanical Engineers, New York, 1973.

Benedict, R. P., *Fundamentals of Temperature, Pressure, and Flow Measurement*, Second Edition, Wiley, New York, 1977.

ASTM E 563, Preparation and Use of Freezing Point Reference Baths, American Society for Testing and Materials.

Chapter 9—Installation Effects

9.1 Temperature Measurement in Fluids

Fluids are divided readily into two types, compressible and incompressible, or more simply into gases and liquids. However, many concepts involved in the measurement of temperatures in fluids are common to both types, and these are discussed first.

9.1.1 Response

No instrument responds instantly to a change in its environment. Thus in a region where temperature is changing, a thermocouple will not be at the temperature of its environment and, hence, cannot indicate the true temperature. The simplified temperature changes considered here are the step change and the ramp change. In the step change, the temperature of the environment shifts instantaneously from T_1 to T_2. In the ramp change, the environment temperature shifts linearly with time from T_1 to T_2.

It is common practice to characterize the response of a temperature sensor by a first order thermal time constant τ which is defined as:

$$\tau = \frac{\rho \mathbf{V} c}{hA} \tag{1}$$

where ρ is density, $\mathbf{V}$ is volume, and c is specific heat, all of the sensor; while h is the heat transfer coefficient, and A is the area of the fluid film surrounding the sensor.

A solution of the first order, first degree, linear, differential equation [*1,2*][1] resulting from a heat balance between the fluid film surrounding the sensor and the sensor itself is

$$T = Ce^{-t/\tau} + \frac{1}{\tau} e^{-t/\tau} \int_0^t T_e e^{t/\tau} \, dt \tag{2}$$

where T is the sensor temperature, and T_e is the environment temperature, both at time t, and C is a constant of integration.

[1] The italic numbers in brackets refer to the list of references appended to this chapter.

For a ramp change in temperature (as is found in a furnace being heated at a uniform rate) Eq 2 reduces to

$$(T_e - T) = R\tau \qquad (3)$$

Equation 3 states that if an element is immersed for a long time in an environment whose temperature is rising at a constant rate $R = dT_e/dt$, then τ is the interval between the time when the environment reaches a given temperature and the time when the element indicates this temperature.

For a step change in temperature (as when a thermocouple is plunged into a constant temperature bath), Eq 2 reduces to

$$(T_e - T) = (T_e - T_1)e^{-t/\tau} \qquad (4)$$

Equation 4 states that if an element is plunged into a constant-temperature environment, τ is the time required for the temperature difference between the environment and the element to be reduced to $1/e$ of the initial difference. Note that for practical purposes the sensor will reach the new temperature after approximately 5 time constants. See Fig. 9.1 for a graphical presentation of these equations.

Below a Mach number of 0.4, the time constant hardly is affected by the fluid velocity. The size of the temperature change affects τ because physical properties are not necessarily linear functions of temperature. Wormser [3]

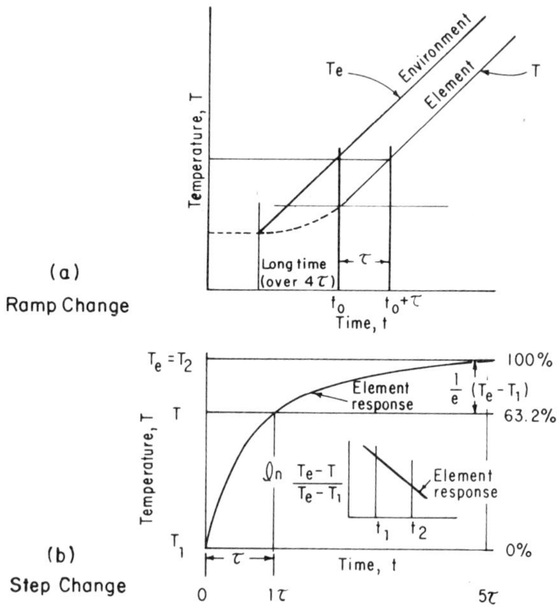

FIG. 9.1—*Graphical presentation of ramp and step changes.*

considers these effects in greater detail. Scadron and Warshawsky [*4*] present convenient nomographs for determining the time constant in the presence of heat transfer. Fluid turbulence tends to reduce the time constant by increasing the film coefficient.

If the sensor is contained in a thermowell the response time is increased because of the extra mass and the additional heat transfer coefficient involved. It may be necessary to consider the response as a second order function in which case a dead time is observed before the sensor responds to a step temperature change. However, Coon and Looney [*5,6*] state that the response time usually is represented adequately by the first order time constant τ.

To achieve the best response time, the measuring junction should be in intimate contact with the well tip. Spring loading is sometimes used to accomplish this.

The response time usually is measured in liquids by plunging the thermocouple assembly into a well-stirred bath held at constant temperature. If the response time is of the same order as the immersion time, the velocity of immersion or depth of immersion or both become important parameters in the measurement. Reference *7* indicates a method for standardizing the liquid baths used in this determination.

For the response time in gases, the literature should be consulted [*8,9*]. Manufacturers sometimes can supply this information for simple conditions.

Multijunctions [*10*] and electrical networks [*11*] have been used successfully to improve the response of a thermocouple.

9.1.2 Recovery

Whenever a gas moves with an appreciable velocity, in addition to the thermal energy in the form of random translational kinetic energy of the molecules, some of its thermal energy is in the form of directed kinetic energy of fluid flow. The static temperature is a measure of the random kinetic energy, while the dynamic temperature is a measure of the directed kinetic energy. The total temperature is a concept (not a measurement) which sums the static and the dynamic temperatures. Such a total temperature would be sensed by an adiabatic probe which completely stagnates an ideal gas. Thus

$$T_{\text{adi, ideal}} = T_t = T + \frac{V^2}{2Jg_c c_p} = T + T_V \qquad (5)$$

where

T_t = total temperature,
T = static temperature,
T_V = dynamic temperature,

V = directed fluid velocity,
J = mechanical equivalent of heat,
g_c = standard acceleration of gravity, and
c_p = specific heat at constant pressure.

In real fluids, an adiabatic recovery factor (α) generally is defined such that

$$T_{\text{adi}} = T + \alpha T_V \tag{6}$$

where α may be more or less than one depending on the relative importance of thermal conductance and thermal capacitance in the boundary layers surrounding the sensor. Since the Prandtl number is the ratio of these two effects, it is common to express recoveries in terms of the Prandtl number. See Refs *12, 13, 14,* and *15* for more information on the recovery factor.

A real sensor immersed in a real fluid tends to radiate to its surroundings. Also there is a tendency for a conductive heat transfer along the probe stem. These two effects are balanced by a convective heat transfer between the probe and the fluid. In addition, real probes do not always stagnate a moving fluid effectively. To account for these realities in temperature measurement in moving fluids, a dynamic correction factor (K) is defined as

$$T_{\text{probe}} = T + KT_V \tag{7}$$

where K corrects for impact, viscosity, and conductivity effects in the fluid, and radiation and conduction effects in the probe. K may take on any value depending on the relative importance of these effects. Variations of K between ± 35 cannot be ruled out. Therefore, the factor KT_V can far outweigh all other factors such as calibration deviations.

9.1.3 Thermowells

Protecting tubes or thermowells (see Chapter 4) or both often are used to separate the measuring junction of a thermocouple from the fluid whose temperature is of interest. Such devices are used to avoid contamination of the thermoelements, to provide safety in case of high pressure installations, to provide strength in the case of significant fluid bending forces in the thermocouple, etc. Thermocouples installed in wells can be withdrawn for inspection, calibration, and replacement. A thermocouple in a well responds more slowly to changes in fluid temperature. Typical wells and their strength requirements are defined in Ref *16.*

The depth of immersion in the fluid is an important consideration. One method of checking for adequacy of immersion is to increase the depth of immersion of the thermocouple well assembly in a constant temperature bath until the thermocouple output becomes constant. A minimum immersion depth of ten times the well outside diameter is a rule of thumb often used.

9.1.4 Thermal Analysis of an Installation

A thermocouple installation may give an indication which differs from the fluid temperature which is to be measured because:

1. The boundary walls are at a temperature different from that of the fluid.
2. There may be a temperature gradient along the well.
3. The fluid may be flowing with an appreciable velocity.
4. The thermocouple may be improperly calibrated.

Item 4 has been covered in Chapter 8 and will not be considered further here.

Basically, the thermocouple temperature is the result of a heat balance between the various modes of heat transfer.

$$q_c = q_r + q_k \tag{8}$$

where q indicates rate of heat transfer and the subscripts c, r, and k signify, respectively, convection, radiation, and conduction. An equation has been developed to describe this heat balance mathematically [*17*] as

$$\frac{d^2T}{dx^2} + a_1(x)\frac{dT_x}{dx} - a_2(x, y)\,T_x = -\,a_2a_3(x, y) \tag{9}$$

where $a_1(x) = dA_k/A_kdx$ which indicates the effect of a change in cross-sectional area of the well; $a_2(x, y) = dA_c(h_r + h_c)/kA_kdx$ which indicates the effect of radiation coefficient (h_r), convection coefficient (h_c), conductivity (k), surface area for convection (A_c) and cross-sectional area for conduction (A_k); $a_3(x, y) = (h_cT_{adi} + h_rT_w)/(h_c + h_r)$ which relates the heat transfer coefficients to the adiabatic fluid temperature (T_{adi}) and the surrounding wall temperature (T_w).

Various solutions are possible for Eq 9 depending on the assumptions one is willing to make. Three simplified solutions are:

1. *Overall Linearization*—When the radiation coefficient is based on an average well temperature, the result is

$$\frac{T_x - a_3}{T_w - a_3} = \frac{e^{mx}}{1 + e^{2mL}} + \frac{e^{-mx}}{1 + e^{-2mL}} \tag{10}$$

where

$$m = \left(\frac{4D(h_r + h_c)}{k(D^2 - d^2)}\right)^{1/2}$$

Typical values for h_r, h_c, and k are given in Ref *17*. This approach leads to quick, approximate answers whenever the fluid can be considered transparent to radiation.

2. *Tip Solution*—When conduction effects are neglected along the well or protecting tube, Eq 9 reduces to

$$T_{\text{tip}} = a_3 \tag{11}$$

which can be solved at once since h_r and h_c are available in the literature (see Ref *17*). This approximation normally would give tip temperatures which are too high since conduction tends to reduce T_{tip}.

3. *Stepwise Linearization*—This is the usual solution to Eq 9. Detailed equations are beyond the scope of this manual, but briefly one divides the well, lengthwise, into a number of elements. The temperature at the center of each element is taken to represent the temperature of that entire element. The heat balance equation is applied successively to one element after another until a match between tip and base temperature is achieved. Each installation is different. Each must be evaluated carefully to determine if the installation is capable of yielding temperatures within the allowable uncertainties.

9.2 Surface Temperature Measurement

9.2.1 General Remarks

There is no easy method of attaching a thermocouple to a surface so that it can be guaranteed to indicate the true surface temperature. To do this, it would be necessary to mount the measuring junction so that it could attain but not affect the surface temperature. In most cases, the presence of the thermocouple (or any alternative transducer) will cause a perturbation of the temperature distribution at the point of attachment, and thus it only will indicate the perturbed temperature.

9.2.1.1 Measurement Error—In many cases, a significant difference will exist between the indicated temperature and the "true" surface temperature, that is, the temperature that the surface would reach if no thermocouple were present. This difference is normally termed a "measurement error," but it should not be confused with calibration or extension wire errors which are common to all thermocouple measurements. The relationship between the indicated and true surface temperature is often defined by the equation

$$Z = \frac{T_s - T_i}{T_s - T_a} \tag{12}$$

where

Z = installation factor,
T_s = true surface temperature,
T_i = indicated surface temperature, and
T_a = temperature of the surroundings or coolant.

Equation 12 expresses the measurement error $T_s - T_i$ as a fraction of the difference between the surface and ambient temperatures.

The value of Z for a particular installation may be calculated or found by experiment; however, as several simplifying assumptions normally are made in any theoretical derivation, experimental verification is necessary if an accurate value of Z is required.

9.2.1.2 Installation Types—There are two basic types of surface thermocouple installation: the permanent, which is used to give a continuous history of the surface temperature, and the temporary, normally made with a sensing probe in mechanical contact with the surface to obtain spot readings. The basic principles for accurate measurement are similar for both types, but the probe type of sensors are more susceptible to measurement errors and generally have a lower accuracy.

9.2.2 Installation Methods

The method of attaching the thermocouple to the surface is governed by considerations of the metallurgical and thermal properties of the materials, their relative sizes, and the modes of heat transfer at the surface. Common methods are shown in Fig. 9.2.

9.2.2.1 Permanent Installations—For thin materials, the thermocouple junction is attached either directly to the surface (Fig. 9.2*a*) or is mounted in a heat collecting pad (Fig. 9.2*b*). It may be welded, brazed, cemented, or clamped to the surface. Good mechanical support of the leads is necessary so that no stresses are applied to the junction.

For thicker materials, the thermocouple junction may be peened into the surface or installed in a groove (Fig. 9.2*c*). The groove may be filled so that the surface is restored to its original profile. A thermocouple in a groove normally will have its junction below the surface and will indicate the subsurface temperature. A similar technique used with tubes is shown in Fig. 9.2*d*.

The configuration shown in Fig. 9.2*e* may be used where rapid response is required, as the junction can be made very thin by electroplating or mechanical polishing techniques [*18–22*].

Several installation methods are illustrated in the literature cited, particularly in Refs *23* and *24*.

Metal sheathed thermocouples are suited particularly to surface measurements, especially for severe environments. They combine good strength and small size, and the measuring junction may be reduced in diameter or flattened to achieve good response with small errors.

9.2.2.2 Measuring Junctions—The measuring junction may be formed in several ways, each having its own advantages and disadvantages.

The bead junction commonly is used. The temperature indicated is a function of the temperatures where the wires leave the bead [*25,26*] so that the bead should be small, and the wires should leave the bead as close to the sur-

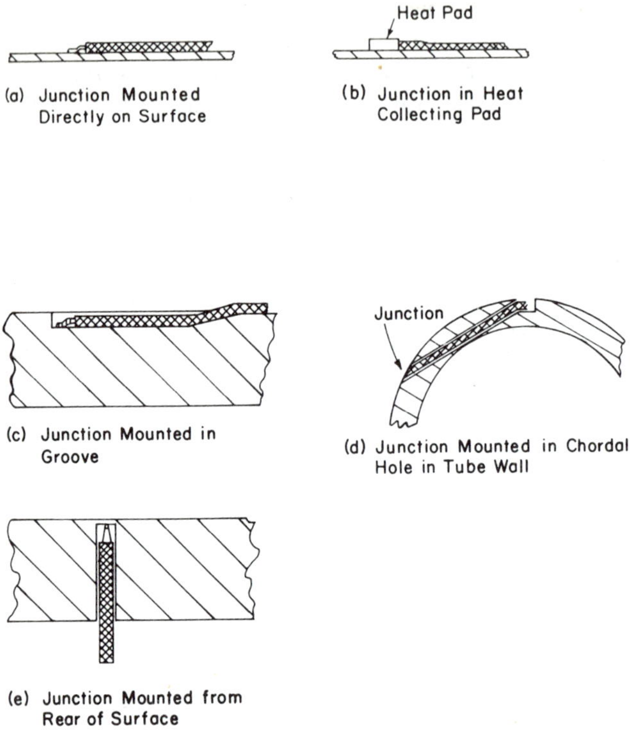

FIG. 9.2—*Common attachment methods.*

face as possible. This may be accomplished by using a flattened bead. Good thermal contact between the bead and the surface is essential, especially if there are temperature gradients. If the surface is a material of poor thermal conductivity, it may be advantageous to mount the measuring junction in a heat collecting pad, or button, which has a good conductivity [*27*].

The simplest junction is shown in Fig. 9.3 in which a single wire is brought to the surface which acts as the second thermoelement. The circuit is completed with a wire of the same material as the surface. This technique usually involves calibration of the wire/surface-material thermocouple. This calibration may not be very reproducible since the surface material is probably not an alloy with controlled thermoelectric properties. It, however, does provide a junction exactly at the surface, and the perturbation errors (see Section 9.2.3) can be reduced to a very low value [*18,28,29*].

A common variation is the separated junction in which each wire is joined separately to the surface (which must be an electrical conductor). This type, which is really two series junctions, has the advantage that the two junctions

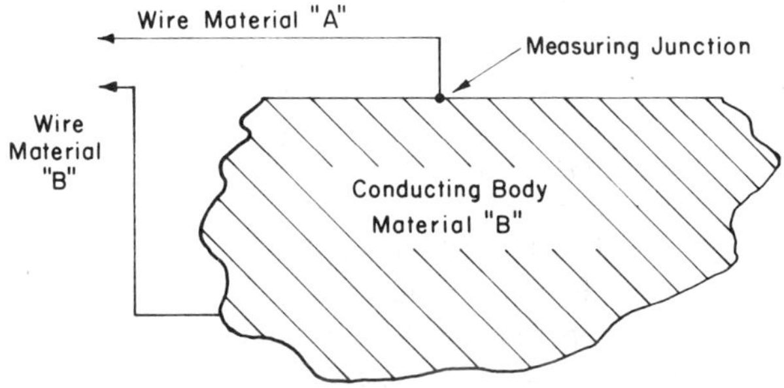

FIG. 9.3—"*Single wire*" *thermocouple.*

form a part of the surface. The output of such a thermocouple is a weighted mean of the two individual junction temperatures, of the form

$$e_0 = e_m + (b_1 - b_2)\frac{(T_1 - T_2)}{2} \tag{13}$$

where

T_1 and T_2 = junction temperatures,
e_0 = measured output,
e_m = output which would be measured if both junctions were at the mean temperature $(T_1 + T_2)/2$, and
b_1 and b_2 = Seebeck coefficients of the two thermocouple wires versus the surface material.

Moffat [*30*] gives a graphical method of analysis.

Thus, the output will be greater or less than the mean depending upon which wire is at the higher temperature. The output generally cannot be calculated, as neither T_1, T_2, nor the relationship between each wire and the surface will be known. There will be, therefore, an uncertainty in the measured temperature if temperature gradients exist. This error is minimized if the wires are bonded to the surface as close together as possible, to reduce $(T_1 - T_2)$. This type of junction has been shown to be more accurate than a bead junction [*31,32*].

9.2.2.3 Probes—It is often desirable to know the temperature distribution over a surface or to make a spot check at one particular point. These measurements are made with a probe containing a thermocouple junction which is held in mechanical contact with the surface. The configuration of the junction is based on the intended application, and several types are commercially available. The probes are in most cases held normal to the surface, and for ease in use should be spring-loaded, which also reduces the error.

Probes are subject to the same errors as permanent installations, but the designer has no control over the conditions of use, and so the errors associated with this type of measurement may be significant. Since the probe provides a heat conducting path from the surface, thermal resistance due to oxide or dirt causes an additional error.

Correction factors [*33*] for several types of junctions range from 0.013 to 0.168, but in general must be determined for specific conditions.

The size of the junction should be as small as possible. Several types of junction are illustrated in Fig. 9.4. The junction types in order of decreasing measurement error are: grounded, exposed, button, and separated.

In order to reduce measurement errors, probes with an auxiliary heater have been used [*34,35*]. The probe thermocouple is heated to the surface temperature so that no temperature gradients are set up when the probe is applied to the surface. One form of probe uses two thermocouples (Fig. 9.5). Equality of the auxiliary and surface junction temperatures indicates that no heat is being transferred along the probe and that the surface junction is at the surface temperature. With this type of probe, the two junctions must be very close, and the response to a change in heater power must be fast or an error can occur in transient measurements. A probe which is controlled automatically has been described recently [*36,37*]. It is claimed to have an accuracy of 3/8 percent and can be used to 760°C (1400°F) on a variety of materials, with a measurement time of less than one second. Sasaki and Kamanda [*38*] used a different approach and eliminated the auxiliary junction. The surface junction was arranged to contact the surface at two second intervals only. The heater input was modulated over a twenty second period and adjusted so that the maxima and minima were above and below the surface temperature. At contact the surface junction temperature changed due to heat exchange unless the two temperatures were equal. With this method the surface temperature of glass bulbs was determined with an accuracy of 0.3°C (0.5°F).

9.2.2.4 Moving Surfaces—Surface temperatures of moving bodies are measured by several methods. A junction mounted in a probe may be held against the body [*39*], but this method results in errors caused by friction. In metal cutting investigations, the metal body and the cutting tool are used as the thermocouple materials. The output of this type of junction has been investigated extensively and has been analyzed by Shu et al [*40*]. Slip-rings to

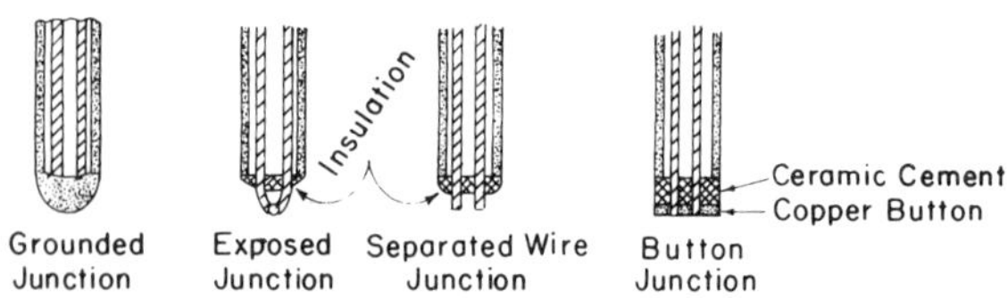

FIG. 9.4—*Types of junction using metal sheathed thermocouples.*

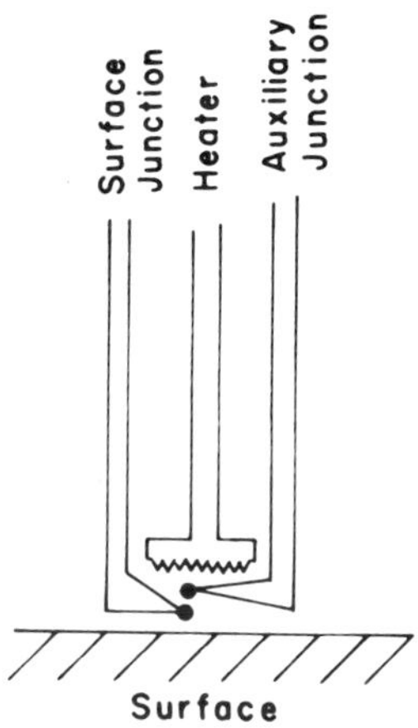

FIG. 9.5—*Thermocouple probe with auxiliary heater, diagramatic arrangement.*

rotating members, intermittent and sliding contacts have been used also. A general review of such installations is given in Ref *24*.

9.2.2.5 Current Carrying Surfaces—A technique of eliminating errors caused by voltage drop in surfaces heated by the passage of d-c current has been described by Dutton and Lee [*41*]. A three-wire thermocouple forming two junctions is used (Fig. 9.6), and the emfs due to the voltage drop in the surface are balanced out during successive reversals of current. When the balance is correct, the thermocouple output is constant regardless of the direction of the heating current.

This technique is also useful for surfaces carrying large alternating currents. In other cases, filters will suffice to attenuate the a-c component. Self-balancing potentiometers usually are affected adversely if the a-c pickup level is high. Galvanometric instruments are normally insensitive to ac and present no problem unless the current is high enough to damage the coils. If the thermocouple junction is not isolated from the surface, voltages appearing between the surface and the instrument ground (common-mode voltages) cause an error with some instruments.

9.2.3 Sources of Error

When a thermocouple is attached to a surface, its presence alters the heat transfer characteristics of the surface and normally will change the temperature distribution. This causes an error which will be referred to as perturbation error.

9.2.3.1—The causes of perturbation error can be broken down into the following:

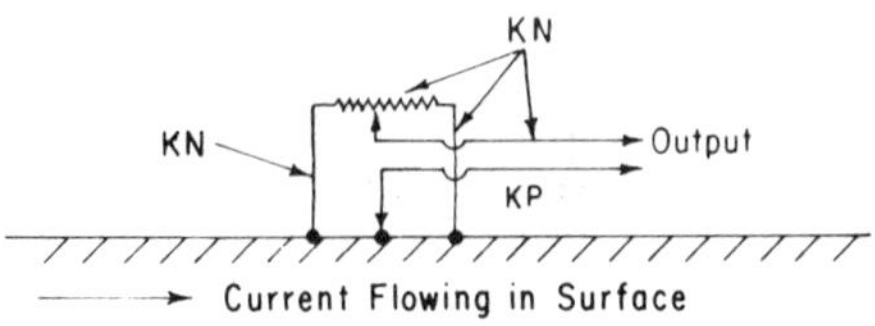

FIG. 9.6—*Three wire Type K thermocouple to compensate for voltage drop induced by surface current. (Other materials may be used.)*

The heat transfer characteristics of the surface are changed by the installation, that is, the surface emissivity or effective thermal conductivity will be altered or the wires may act as fins providing additional heat transfer paths.

Thermal contact resistance between the junction and the surface will cause a temperature gradient which will prevent the measuring junction from attaining the correct temperature if heat is being exchanged between the surface and its surrounding.

If temperature gradients exist, there will be an error due to uncertainty in the exact position of the junction or junctions relative to the surface [*25,26*].

The response time of the thermocouple installation introduces errors during transient conditions [*42,43*] (see also Section 9.1.1). The response time will be that of the surface installation and not of the thermocouple alone.

Although not discussed in this chapter, the errors associated with any thermocouple measurement, such as deviations from standard emf and lead wire errors, must be taken also into consideration [*44,45*].

9.2.4 Error Determination

The perturbation error must be determined if the surface temperature must be known with a high degree of accuracy. The installation should be designed to minimize this error, but in many cases materials and size make compromises necessary, so that an ideal design cannot be achieved. The error, and hence the correction factor, can be determined by the following methods.

9.2.4.1 Steady State Conditions—The direct calculation of the error involves solving the heat flow equation for the measuring junction and surface geometry. Normally simplifying assumptions are made, and the results must be interpreted in relation to them. The calculations show clearly the major sources of error and indicate means of reducing errors to a minimum [*46–50*].

Analog methods of solving the heat transfer equations using resistance or resistance-capacitance networks indicate the overall temperature distribution and show the perturbation effect of the thermocouple clearly. They are, however, difficult to make flexible, and so a number of analog models are required if a range of heat transfer conditions is to be studied [*49–52*].

Relaxation methods of solving the heat transfer equations have been used to calculate the temperature distribution; this method is attractive if a computer is available to perform the considerable amount of arithmetic required.

Direct experimental measurement on the installation is often the only satisfactory way to accurately determine the error. Care must be taken to simulate the service conditions exactly as a change in a variable can significantly affect the error. The major problem is to determine the true surface temperature. This is discussed extensively in the literature [*19,52–57*].

9.2.4.2 Transient Conditions—If surface temperatures are changing, the response of the thermocouple attachment may cause a significant error. The response time of the thermocouple alone will have little significance for surface measurements if the heat transfer path between the surface and the measuring junction is poor or adds thermal mass.

The time required to change the measuring junction temperature causes the thermocouple output to lag the surface temperature in time and decreases its amplitude [*32,42,43*]. The response time may be determined experimentally from the response to step or ramp functions of surface temperature change.

A. Insulated Thermocouple Normal to an Electrically Heated Surface—For surfaces with changing temperatures a bead thermocouple attached normal to an electrically heated surface and insulated from the ambient has been analyzed by Quant and Fink [*59*] and Green and Hunt [*60*].

The analysis showed that in order to obtain a rapid response with a small, steady-state error, it is necessary to use a small junction bead with good surface contact, small diameter wires, and good insulation between the wires and the surroundings.

B. Surface Heated by Radiation—Thermocouples mounted on a surface subject to radiant heating at temperature-rise rates up to 17°C/s were investigated by White [*31*]. His results showed that a separated-junction thermocouple produced the least error.

The thermocouple errors increased with increasing plate thickness and with increasing rates of temperature rise. Furthermore, the amount of bare thermocouple wire between the junction and the insulation should be a minimum.

Kovacs and Mesler [*60*] investigated the response of very fast surface thermocouples subject to radiant heating as a function of size and type of junction. Junctions were formed by electroplating or mechanical abrasion. Very thin junctions were subject to an overshoot error for high rates of temperature rise, as heat could not be transmitted back through the thermocouple to subsurface layers fast enough. On the other hand, too thick a junction corresponded to a junction beneath the surface. The junction thickness should be of the same order of magnitude as the distance between the two thermocouple wires to avoid overshooting.

C. Surface Subject to Aerodynamic Heating—An analysis and design of a thermocouple installation which can be mounted in a thin-metal skin subjected to aerodynamic heating is given in Ref *61*. A finite difference calculation of the distorted temperature field indicated that errors due to insulation resistance at high temperatures were of the same order of magnitude as those due to the uncertainty of the exact junction location.

9.2.5 Procedures for Minimizing Error

The examples quoted in the preceding section have been treated separately. It is generally impossible to extrapolate from one set of conditions to another unless the installation is identical. The analyses show procedures that should be followed to reduce measurement errors. These are:

A. Use the smallest possible installation to avoid perturbation errors.

B. Bring the thermocouple wires away from the junction along an isotherm for at least 20 wire diameters to reduce conduction errors. The use of thermocouple materials with low thermal conductivity also will reduce this error.

C. Locate the measuring junction as close to the surface as possible rather than above or below it.

D. Design the installation so that it causes a minimum disturbance of any fluid flow or change in the emissivity of the surface, to avoid changes in convective or radiative heat transfer.

E. Design the installation so that the total response is fast enough to cause negligible lag for the transients expected in service.

F. Reduce the thermal resistance between the measuring junction and the surface to as low a value as possible. If the surface has a low thermal conductivity, a heat collecting pad may be used.

9.2.6 Commercial Surface Thermocouples

Many surface installations are custom engineered, but industry does offer several standard surface thermocouples intended for specific applications [*39,62*].

9.2.6.1 Surface Types—Figure 9.7 shows several types which are mounted on the surface. Type *a* is a gasket thermocouple which normally is mounted on a stud and Type *b* is a rivet head. The clamp attachment, Type *c*, is used on pipelines and will be reasonably accurate if the pipe is lagged thermally. The weldable pad attachment, Type *d*, is used on boiler or superheater tubes and uses a metal sheathed thermocouple with a grounded junction. The sheath and pad materials are chosen to be compatible with the boiler

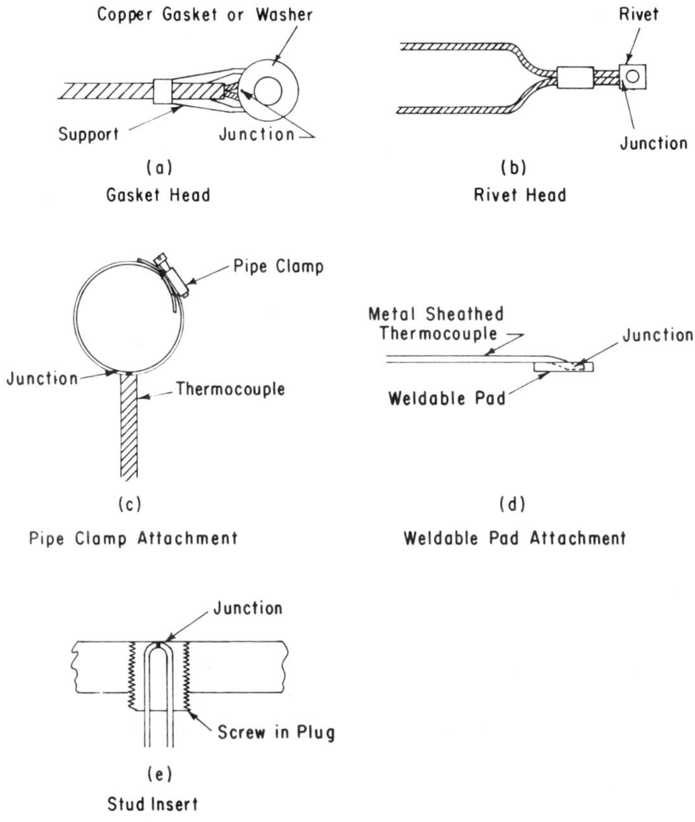

FIG. 9.7—*Commercially available types of surface thermocouples.*

environment. An accuracy of ±2 percent is claimed for these thermocouples (±1 percent if a correction factor supplied by the manufacturer is used).

For installations where it is possible to mount the thermocouples in the surface, stud- or rivet-mounted plugs similar to Type *e* are offered by several manufacturers. Variations of this type have been used extensively for heat transfer measurements [*46,63*] and for applications requiring very rapid response such as the measurement of surface temperatures in gun barrels and rocket exhaust chambers. The material of the plug must match the material of the surface, otherwise significant errors can be introduced [*64*], especially for materials with low thermal conductivity.

9.2.6.2 Probe Types—Probe type thermocouples for temporary or spot readings are offered usually as a complete package consisting of a thermocouple head which contains the measuring junction, a hand probe, and an indicating milliammeter calibrated in degrees. The measuring junctions are

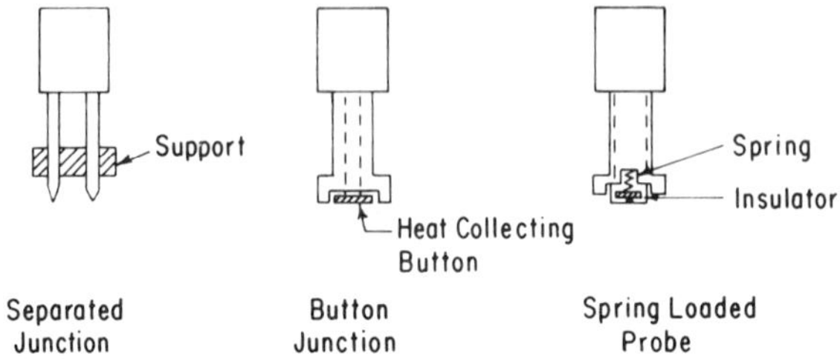

FIG. 9.8—*Commercial probe thermocouple junctions.*

normally interchangeable so that one instrument can be used with a variety of heads. The type of head will depend on the surface characteristics. Common types are shown in Fig. 9.8.

The separated-junction probe is used on electrically conducting surfaces only. Dirt or oxide layers will introduce a thermal resistance error or even prevent the completion of the circuit. For greatest convenience, the two wires should be separately spring-loaded against the surface. The button type of junction must be held carefully as any deviation from the normal will cause a change in the height of the junction above the surface, and the readings will be inconsistent. The spring-loaded type of junction is available in several forms and has been adapted for measurements on moving surfaces [*39*].

The accuracy obtainable with these probes is not high. However, the errors can be reduced to 2 or 3 percent for good conducting surfaces in still air cooled by natural convection. If the surface is a poor thermal conductor or the rate of heat transfer is high, the error will be considerably higher than this.

For rotating or moving surfaces, probe instruments utilizing a junction spring-loaded against the surface are used. Heat generated by friction causes an error which can be significant. (Bowden and Ridler [*65*] have shown that the temperatures may reach the melting point of one of the metals.)

A heated thermocouple probe instrument for measuring the temperature of wires or filaments is described by Bensen and Horne [*66*], and a wire temperature meter [*67*] has been marketed recently.

9.3 References

[*1*] Harper, D. R., "Thermometric Lag," NBS Scientific Paper 185, National Bureau of Standards, March 1912.

[*2*] Benedict, R. P., *Fundamentals of Temperature, Pressure, and Flow Measurements*, Second Edition, Wiley, New York, 1977.

[*3*] Wormser, A. F., "Experimental Determination of Thermocouple Time Constants," National AERO Meeting Paper 158D, Society of Automotive Engineers, April 1960.

[*4*] Scadron, M. D. and Warshawsky, I., "Experimental Determination of Time Constants and Nusselt Numbers for Bare-Wire Thermocouples," NACA TN 2599, National Advisory Committee for Aeronautics, Jan. 1952.

[*5*] Coon, G. A., "Response of Temperature-Sensing-Element Analyses," *Transactions*, American Society of Mechanical Engineers, Nov. 1957, p. 1857.

[*6*] Looney, R., "Method for Presenting the Response of Temperature-Measuring Systems," *Transactions*, American Society of Mechanical Engineers, Nov. 1957, p. 1851.

[*7*] Murdock, J. W., Foltz, C. J., and Gregory, C., "A Practical Method of Determining Response Time of Thermometers in Liquid Baths," *Transactions*, American Society of Mechanical Engineers, *Journal of Engineering for Power*, Jan. 1963, p. 27.

[*8*] Caldwell, F. R., Olsen, L. O., and Freeze, P. D., "Intercomparison of Thermocouple Response Data," SAE Paper 158F, Society of Automotive Engineers, April, 1960.

[*9*] Hornfeck, A. J., "Response Characteristics of Thermometer Elements," *Transactions*, American Society of Mechanical Engineers, Feb. 1949, p. 121.

[*10*] Benedict, R. P., *Fundamentals of Temperature, Pressure, and Flow Measurements*, Second Edition, Wiley, New York, 1977, p. 280.

[*11*] Shepard, C. E. and Warshawsky, I., "Electrical Techniques for Compensation of Thermal Time Lag of Thermocouples and Resistance Thermometer Elements," NACA TN 2703, National Advisory Committee for Aeronautics, Jan. 1952.

[*12*] Benedict, R. P., *Fundamentals of Temperature, Pressure, and Flow Measurements*, Second Edition, Wiley, New York, 1977, p. 213.

[*13*] Roughton, J. E., "Design of Thermometer Pockets for Steam Mains," *Proceedings 1965–1966*, Institute of Mechanical Engineers, Vol. 180, Part 1, No. 39.

[*14*] Faul, J. C., "Thermocouple Performance in Gas Streams," *Instruments and Control Systems*, Dec. 1962.

[*15*] Werner, F. D., "Total Temperature Measurement," ASME Paper 58-AU-17, American Society of Mechanical Engineers.

[*16*] "Temperature Measurement," ASME PTC 19.3, American Society of Mechanical Engineers, 1974.

[*17*] Benedict, R. P. and Murdock, J. W., "Steady-State Thermal Analysis of a Thermometer Well," *Transactions*, American Society of Mechanical Engineers, *Journal of Engineering for Power*, July 1963, p. 235.

[*18*] Bendersby, D., "A Special Thermocouple for Measuring Transient Temperatures," *Mechanical Engineering*, Vol. 75, No. 2, Feb. 1953, p. 117.

[*19*] Powell, W. B. and Price, T. W., "A Method for the Determination of Local Heat Flux from Transient Temperature Measurements," *Transactions*, Instrument Society of America, Vol. 3, No. 3, July 1964, p. 246.

[*20*] Moeller, C. E., "Thermocouples for the Measurement of Transient Surface Temperatures," *Temperature, Its Measurement and Control in Science and Industry*, Vol. 3, Part 2, Reinhold, New York, 1962, p. 617.

[*21*] Ongkiehong, L. and Van Duijn, J., "Construction of a Thermocouple for Measuring Surface Temperatures," *Journal of Scientific Instruments*, Vol. 37, June 1960, p. 221.

[*22*] Vigor, C. W. and Hornaday, J. R., "A Thermocouple for a Measurement of Temperature Transients in Forging Dies," *Temperature, Its Measurement and Control in Science and Industry*, Vol. 3, Part 2, Reinhold, New York, 1962, p. 625.

[*23*] Baker, H. D., Ryder, E. A., and Baker, N. H., *Temperature Measurement in Engineering*, Vol. 1, Wiley, New York, 1953, Chapters 8, 11, 12.

[*24*] Baker, H. D., Ryder, E. A., and Baker, N. H., *Temperature Measurement in Engineering*, Vol. 2, Wiley, New York, 1961.

[*25*] Bailey, N. P., "The Response of Thermocouples," *Mechanical Engineering*, Vol. 53, No. 11, Nov. 1931, p. 797.

[*26*] McCann, J. A., "Temperature Measurement Theory," KAPL-2067-2, 1962, Office of Technical Service, Department of Commerce, Washington, D.C.

[*27*] Chapman, A. J., *Heat Transfer*, McMillan, New York, 1960.

[*28*] Bailey, N. P., "The Measurement of Surface Temperatures, Accuracies Obtainable with Thermocouples," *Mechanical Engineering*, Vol. 54, Aug. 1932, p. 553.

[*29*] Baker, H. D., Ryder, E. A., and Baker N. H., *Temperature Measurement in Engineering*, Vol. 1, Wiley, New York, 1961, p. 90.

[*30*] Moffat, R. J., "The Gradient Approach to Thermocouple Circuitry," *Temperature, Its Measurement and Control in Science and Industry*, Vol. 3, Part 2, Reinhold, New York, 1962, pp. 33–38.

[31] White, F. J., "Accuracy of Thermocouples in Radiant Heat Testing," *Experimental Mechanics*, Vol. 2, July 1962, p. 204.
[32] Moen, W. K., "Surface Temperature Measurements," *Instruments and Control Systems*, Vol. 33, Jan. 1960, p. 71.
[33] Otter, A. J., "Thermocouples and Surface Temperature Measurement," AECL-3062, March 1968, SDDO, AECL, Chalk River, Ont., Canada.
[34] Roeser, W. F. and Mueller, E. F., "Measurement of Surface Temperatures," *Bureau of Standards Journal of Research*, Vol. 5, No. 4, Oct. 1930, p. 793.
[35] Sasaki, N., "A New Method for Surface Temperature Measurement," *Review of Scientific Instruments*, Vol. 21, No. 1, Jan. 1950, p. 1.
[36] Robertson, D. and Sterbutzel, G. A., "An Accurate Surface Temperature Measuring System," *Proceedings of IEEE Industrial Heating Conference of Philadelphia*, April 1969. See also *Leeds and Northrup Technical Journal*, Issue 7, Summer 1969.
[37] "A Probe for the Instantaneous Measurement of Surface Temperature," RTD-TDR-63-4015, OTS, Department of Commerce, Washington, D.C.
[38] Sasaki, N. and Kamanda, A., "A Recording Device for Surface Temperature Measurements," *Review of Scientific Instruments*, Vol. 23, No. 6, June 1952, p. 261.
[39] "Unusual Thermocouples and Accessories," *Instruments and Control Systems*, Vol. 36, No. 8, Aug. 1963, p. 133.
[40] Shu, H. H., Gaylard, E. W., and Hughes, W. F., "The Relation Between the Rubbing Interface Temperature Distribution and Dynamic Thermocouple Temperature," *Transactions ASME Journal of Basic Engineering*, Vol. 86, Series D, No. 3, Sept. 1964, p. 417.
[41] Dutton, R. and Lee, E. C., "Surface-Temperature Measurement of Current-Carrying Objects," *Journal*, Instrument Society of America, Vol. 6, No. 12, Dec. 1959, p. 49.
[42] Jakob, M., *Heat Transfer*, Vol. 2, Wiley, New York, 1957, p. 183.
[43] Caldwell, W. I., Coon, G. A., and Zoss, L. M., *Frequency Response for Process Control*, McGraw-Hill, New York, 1959, p. 295.
[44] "Thermocouples and Thermocouple Extension Wires," *Recommended Practice RPI-7*, Instrument Society of America, 7 July 1959.
[45] Finch, D. I., "General Principles of Thermoelectric Thermometry," *Temperature, Its Measurement and Control in Science and Industry*, Vol. 3, Part 2, Reinhold, New York, p. 3.
[46] Jakob, M., *Heat Transfer*, Vol. 2, Wiley, New York, 1957, p. 153.
[47] Chapman, A. J., *Heat Transfer*, McMillan, New York, 1960.
[48] Boelter, L. M. K. et al, "An Investigation of Aircraft Heaters XXVIII-Equations for Steady State Temperature Distribution Caused by Thermal Sources in Flat Plates Applied to Calculation of Thermocouple Errors, Heat Meter Corrections, and Heat Transfer by Pin-Fin Plates," Technical Note 1452, National Advisory Committee for Aeronautics, 1944.
[49] Schneider, P. J., *Conduction Heat Transfer*, Addison-Wesley, Cambridge, Mass., 1955, p. 176.
[50] Fitts, R. L., Flemons, R. S., and Rogers, J. T., "Study of Temperature Distribution in a Finned Nuclear Fuel Sheath by Electrical Analogue and Mathematic Analysis (Revision)," R65CAP1, 19 Jan. 1965, Canadian General Electric, Peterborough, Ont.
[51] Chan, K. S. and Rushton, K. R., "The Simulation of Boundary Conditions in Heat Conduction Problems in a Resistance-Capacitance Electrical Analogue," *Journal of Scientific Instruments*, Vol. 41, No. 9, Sept. 1964, p. 535.
[52] Brindley, J. H., "Calibration of Surface-Attached Thermocouples on a Flat-Plate Fuel Element by Electrical Analogue and Analytical Techniques," *Transactions of the American Nuclear Society*, Vol. 6, No. 2, Nov. 1963, p. 333.
[53] Boelter, L. M. K. and Lockhart, R. W., "Thermocouple Conduction Error Observed in Measuring Surface Temperatures," Technical Note 2427, National Advisory Committee for Aeronautics, 1946.
[54] Stoll, A. M. and Hardy, J. D., "Direct Experimental Comparison of Several Surface Temperature Measuring Devices," *Review of Scientific Instruments*, Vol. 20, No. 9, Sept. 1949, p. 678.
[55] Oetken, E. R., "Evaluation of Surface-Attached Thermocouples During Forced Convection Heat Transfer," IDO-16889, OTS, Department of Commerce, Washington, D.C.

[56] Sudar, S., "Calibration of OMRE Fuel-Element Surface Thermocouple Assembly," NAA-SR-Memo 3671, 1959.
[57] Browning, W. E. and Hemphill, H. L., "Thermocouples for Measurement of the Surface Temperature of Nuclear Fuel Elements," *Temperature, Its Measurement and Control in Science and Industry*, Vol. 3, Part 2, Reinhold, New York, p. 723.
[58] Quant, E. R. and Fink, E. W., "Experimental and Theoretical Analysis of the Transient Response of Surface Bonded Thermocouples," *Bettis Technical Review*, WAPD-BT-19, Reactor Technology, June 1960, p. 31.
[59] Green, S. J. and Hunt, T. W., "Accuracy and Response of Thermocouples for Surface and Fluid Temperature Measurements," *Temperature, Its Measurement and Control in Science and Industry*, Vol. 3, Part 2, Reinhold, New York, p. 695.
[60] Kovacs, A. and Mesler, R. B., "Making and Testing Small Surface Thermocouples for Fast Response," *Review of Scientific Instruments*, Vol. 35, No. 4, April 1964, p. 485.
[61] "Design and Construction of a Unit for Measuring Metal Skin Temperatures, Phase I, Theoretical Analysis and Design," SC-4461 (RR), Dec. 1960, OTS, Department of Commerce, Washington, D.C.
[62] See Thermocouple Manufacturers' Catalogs.
[63] Sellers, J. P., "Thermocouple Probes for Evaluating Local Heat Transfer Coefficients in Rocket Motors," *Temperature, Its Measurement and Control in Science and Industry*, Vol. 3, Part 2, Reinhold, New York, p. 673.
[64] Nanigan, J., "Thermal Properties of Thermocouples," *Instruments and Control Systems*, Vol. 36, No. 10, Oct. 1963, p. 87.
[65] Bowden, F. P. and Ridler, K. E. W., "Physical Properties of Surfaces III. The Surface Temperature of Sliding Metals," *Proceedings*, Royal Society, London A, Vol. 154, 1936, p. 644.
[66] Bensen, J. and Horne, R., "Surface Temperature of Filaments and Thin Sheets," *Instruments and Control Systems*, Vol. 35, No. 10, Part 1, Oct. 1962, p. 115.
[67] Report No. 152, British Iron and Steel Research Organization.
[68] Lovuola, V. J., "Preventing Noise in Grounded Thermocouple Measurements," *Instruments and Control Systems*, Jan. 1980, p. 31.
[69] Dean, R. C. et al, *Aerodynamic Measurements*, M.I.T. Press, Cambridge, Mass., 1953, Eighth printing, 1963.

Chapter 10—Reference Tables for Thermocouples

The practical use of thermocouples in industrial and laboratory applications requires that the thermocouple conform to an established temperature—electromotive-force relationship within acceptable limits of error. Since the thermocouple in a thermoelectric thermometer system is usually expendable, conformance to established temperature-emf relationships is necessary in order to permit interchangeability.

Section 10.2 consists of reference tables that give temperature-emf relationships for the thermocouple types most commonly used in industry. These are identified as thermocouple Types B, E, J, K, R, S, and T, as defined in ANSI Standard MC96.1.

Data in these tables are based upon absolute electrical units and the International Practical Temperature Scale of 1968. All temperature emf data contained in Section 10.2 have been extracted from ASTM Standard E 230.

Reference tables giving temperature emf relationships for single leg thermoelements referenced to platinum (NBS Pt 67) are not included in this manual but are contained in ASTM Standard E 230.

10.1 Thermocouple Types and Limits of Error

10.1.1 Thermocouple Types

The letter symbols identifying each reference table are those defined in ANSI Standard MC96.1. These symbols which are used in common throughout industry, identify the following thermocouple calibrations:

Type B—Platinum-30 percent rhodium (+) versus platinum 6 percent rhodium (−).
Type E—Nickel-10 percent chromium (+) versus constantan (−).
Type J—Iron (+) versus constantan (−).
Type K—Nickel-10 percent chromium (+) versus nickel-5 percent aluminum and silicon (−).[1]
Type R—Platinum-13 percent rhodium (+) versus platinum (−).
Type S—Platinum-10 percent rhodium (+) versus platinum (−).
Type T—Copper (+) versus constantan (−).

[1]Silicon, or aluminum and silicon, may be present in combination with other elements.

Detailed information covering the advantages and limitations of each of these thermocouple types, their recommended temperature ranges, and detailed physical properties of the thermoelements comprising them are contained in Section 3.1 of this manual.

10.1.2 Limits of Error

The limits of error for the common letter designated thermocouple types, as listed in Table 10.1 are taken from ASTM E 230. Most manufacturers supply thermocouples and thermocouple wire to these limits of error or better.

10.2 Thermocouples Reference Tables

Following is a list of the reference tables included in this section:

Table Number	Thermocouple Type	Temperature Range
10.2	B	0 to + 3308°F
10.3	B	0 to + 1820°C
10.4	E	−454 to + 1832°F
10.5	E	−270 to + 1000°C
10.6	J	−350 to + 2192°F
10.7	J	−210 to + 1200°C
10.8	K	−454 to + 2500°F
10.9	K	−270 to + 1372°C
10.10	R	−58 to + 3214°F
10.11	R	−50 to + 1768°C
10.12	S	−58 to + 3214°F
10.13	S	−50 to + 1768°C
10.14	T	−454 to + 752°F
10.15	T	−270 to + 400°C

10.3 Generation of Smooth Temperature-Emf Relationships

10.3.1 Need for Smooth Temperature-Emf Relationship

A table of reference values for use with thermocouples should be capable of easy and unique generation to facilitate its use in computer and similar applications. Furthermore, the reference values should agree closely with the characteristics of the thermocouple type under study, so that differences will change only slowly and smoothly with temperature level.

Reference Tables 10.2 to 10.15 give values to three decimal places only, and this roundoff in table values results in inherent discontinuities in the temperature-emf relationship as represented by the tables.

This section provides means for generating smooth, continuous tem-

TABLE 10.1—*Limits of error for thermocouples.*

Type	Temperature Range		Limits of Error: Standard		Limits of Error: Special	
	°F	°C	°F	°C (Note 2)	°F	°C (Note 2)
B	1600 to 3100	871 to 1705	± 1/2 percent			
E	32 to 600	0 to 316	± 3°F		± 2 1/4 °F	
	600 to 1600	316 to 871	± 1/2 percent		± 3/8 percent	
J	32 to 530	0 to 277	± 4°F		± 2°F	
	530 to 1400	277 to 760	± 3/4 percent		± 3/8 percent	
K	32 to 530	0 to 277	± 4°F		± 2°F	
	530 to 2300	277 to 1260	± 3/4 percent		± 3/8 percent	
R or S	32 to 1000	0 to 538	± 2 1/2 °F			
	1000 to 2700	538 to 1482	± 1/4 percent			
T	−300 to −150	−184 to −101			± 1 percent	
	−150 to −75	−101 to −59	± 2 percent		± 1 percent	
	−75 to 200	−59 to 93	± 1 1/2 °F		± 3/4 °F	
	200 to 700	93 to 371	± 3/4 percent		± 3/8 percent	

NOTE 1—In this table the limits of error for each type of thermocouple apply only over the temperature range for which the wire size in question is recommended (see Table 3.5). These limits of error should be applied only to standard wire sizes. The same limits may not be obtainable in special sizes.

NOTE 2—Where limits of error are given in percent, the percentage applies to the temperature being measured when expressed in degrees Fahrenheit. To determine the limit of error in degrees Celsius, multiply the limit of error in degrees Fahrenheit by 5/9.

NOTE 3—Limits of error apply to thermocouples as supplied by the manufacturer. The calibration of a thermocouple may change during use. The magnitude of the change depends upon such factors as temperature, the length of time, and the conditions under which it was used.

NOTE 4—Type T wire cannot be expected to meet the limits of error at temperatures below the ice point unless so specified at time of purchase. Selection is usually required.

TABLE 10.2—*Type B thermocouples (deg F-millivolts).*

Temperature in Degrees Fahrenheit[a]

EMF in Absolute Millivolts **Reference Junctions at 32 F**

DEG F	0	1	2	3	4	5	6	7	8	9	10	DEG F
	THERMOELECTRIC VOLTAGE IN ABSOLUTE MILLIVOLTS											
0	0.006	0.006	0.006	0.006	0.005	0.005	0.005	0.005	0.004	0.004	0.004	0
10	0.004	0.004	0.003	0.003	0.003	0.003	0.003	0.002	0.002	0.002	0.002	10
20	0.002	0.002	0.002	0.001	0.001	0.001	0.001	0.001	0.001	0.000	0.000	20
30	0.000	0.000	0.000	-0.000	-0.000	-0.000	-0.001	-0.001	-0.001	-0.001	-0.001	30
40	-0.001	-0.001	-0.001	-0.001	-0.001	-0.001	-0.002	-0.002	-0.002	-0.002	-0.002	40
50	-0.002	-0.002	-0.002	-0.002	-0.002	-0.002	-0.002	-0.002	-0.002	-0.002	-0.002	50
60	-0.002	-0.002	-0.002	-0.003	-0.003	-0.003	-0.003	-0.003	-0.003	-0.003	-0.003	60
70	-0.003	-0.003	-0.003	-0.003	-0.003	-0.003	-0.003	-0.002	-0.002	-0.002	-0.002	70
80	-0.002	-0.002	-0.002	-0.002	-0.002	-0.002	-0.002	-0.002	-0.002	-0.002	-0.002	80
90	-0.002	-0.002	-0.002	-0.002	-0.002	-0.001	-0.001	-0.001	-0.001	-0.001	-0.001	90
100	-0.001	-0.001	-0.001	-0.001	-0.000	-0.000	-0.000	-0.000	0.000	0.000	0.000	100
110	0.000	0.000	0.001	0.001	0.001	0.001	0.001	0.001	0.002	0.002	0.002	110
120	0.002	0.002	0.002	0.002	0.003	0.003	0.003	0.003	0.003	0.004	0.004	120
130	0.004	0.004	0.004	0.005	0.005	0.005	0.005	0.005	0.006	0.006	0.006	130
140	0.006	0.006	0.007	0.007	0.007	0.007	0.008	0.008	0.008	0.009	0.009	140
150	0.009	0.009	0.009	0.010	0.010	0.010	0.011	0.011	0.011	0.012	0.012	150
160	0.012	0.012	0.013	0.013	0.013	0.014	0.014	0.014	0.015	0.015	0.015	160
170	0.015	0.016	0.016	0.016	0.017	0.017	0.017	0.018	0.018	0.019	0.019	170
180	0.019	0.019	0.020	0.020	0.021	0.021	0.021	0.022	0.022	0.023	0.023	180
190	0.023	0.023	0.024	0.024	0.025	0.025	0.026	0.026	0.027	0.027	0.027	190
200	0.027	0.028	0.028	0.029	0.029	0.030	0.030	0.031	0.031	0.032	0.032	200
210	0.032	0.033	0.033	0.034	0.034	0.035	0.035	0.036	0.036	0.037	0.037	210
220	0.037	0.038	0.038	0.039	0.039	0.040	0.041	0.041	0.042	0.042	0.043	220
230	0.043	0.043	0.044	0.044	0.045	0.046	0.046	0.047	0.047	0.048	0.049	230
240	0.049	0.049	0.050	0.050	0.051	0.052	0.052	0.053	0.053	0.054	0.055	240
250	0.055	0.055	0.056	0.057	0.057	0.058	0.058	0.059	0.060	0.060	0.061	250
260	0.061	0.062	0.062	0.063	0.064	0.064	0.065	0.066	0.067	0.067	0.068	260
270	0.068	0.069	0.069	0.070	0.071	0.071	0.072	0.073	0.074	0.074	0.075	270
280	0.075	0.076	0.077	0.077	0.078	0.079	0.080	0.080	0.081	0.082	0.083	280
290	0.083	0.083	0.084	0.085	0.086	0.086	0.087	0.088	0.089	0.090	0.090	290
300	0.090	0.091	0.092	0.093	0.094	0.094	0.095	0.096	0.097	0.098	0.099	300
310	0.099	0.099	0.100	0.101	0.102	0.103	0.104	0.104	0.105	0.106	0.107	310
320	0.107	0.108	0.109	0.110	0.111	0.111	0.112	0.113	0.114	0.115	0.116	320
330	0.116	0.117	0.118	0.119	0.120	0.120	0.121	0.122	0.123	0.124	0.125	330
340	0.125	0.126	0.127	0.128	0.129	0.130	0.131	0.132	0.133	0.134	0.135	340
350	0.135	0.136	0.137	0.138	0.138	0.139	0.140	0.141	0.142	0.143	0.144	350
360	0.144	0.145	0.146	0.147	0.148	0.149	0.151	0.152	0.153	0.154	0.155	360
370	0.155	0.156	0.157	0.158	0.159	0.160	0.161	0.162	0.163	0.164	0.165	370
380	0.165	0.166	0.167	0.168	0.169	0.171	0.172	0.173	0.174	0.175	0.176	380
390	0.176	0.177	0.178	0.179	0.180	0.182	0.183	0.184	0.185	0.186	0.187	390
400	0.187	0.188	0.189	0.191	0.192	0.193	0.194	0.195	0.196	0.197	0.199	400
410	0.199	0.200	0.201	0.202	0.203	0.205	0.206	0.207	0.208	0.209	0.210	410
420	0.210	0.212	0.213	0.214	0.215	0.217	0.218	0.219	0.220	0.221	0.223	420
430	0.223	0.224	0.225	0.226	0.228	0.229	0.230	0.231	0.233	0.234	0.235	430
440	0.235	0.236	0.238	0.239	0.240	0.242	0.243	0.244	0.245	0.247	0.248	440
450	0.248	0.249	0.251	0.252	0.253	0.254	0.256	0.257	0.258	0.260	0.261	450
460	0.261	0.262	0.264	0.265	0.266	0.268	0.269	0.271	0.272	0.273	0.275	460
470	0.275	0.276	0.277	0.279	0.280	0.281	0.283	0.284	0.286	0.287	0.288	470
480	0.288	0.290	0.291	0.293	0.294	0.295	0.297	0.298	0.300	0.301	0.303	480
490	0.303	0.304	0.305	0.307	0.308	0.310	0.311	0.313	0.314	0.315	0.317	490
500	0.317	0.318	0.320	0.321	0.323	0.324	0.326	0.327	0.329	0.330	0.332	500
510	0.332	0.333	0.335	0.336	0.338	0.339	0.341	0.342	0.344	0.345	0.347	510
520	0.347	0.348	0.350	0.351	0.353	0.355	0.356	0.358	0.359	0.361	0.362	520
530	0.362	0.364	0.365	0.367	0.369	0.370	0.372	0.373	0.375	0.376	0.378	530
540	0.378	0.380	0.381	0.383	0.384	0.386	0.388	0.389	0.391	0.392	0.394	540
550	0.394	0.396	0.397	0.399	0.401	0.402	0.404	0.405	0.407	0.409	0.410	550
560	0.410	0.412	0.414	0.415	0.417	0.419	0.420	0.422	0.424	0.425	0.427	560
570	0.427	0.429	0.431	0.432	0.434	0.436	0.437	0.439	0.441	0.442	0.444	570
580	0.444	0.446	0.448	0.449	0.451	0.453	0.455	0.456	0.458	0.460	0.462	580
590	0.462	0.463	0.465	0.467	0.469	0.470	0.472	0.474	0.476	0.477	0.479	590
600	0.479	0.481	0.483	0.485	0.486	0.488	0.490	0.492	0.494	0.495	0.497	600
610	0.497	0.499	0.501	0.503	0.504	0.506	0.508	0.510	0.512	0.514	0.515	610
620	0.515	0.517	0.519	0.521	0.523	0.525	0.527	0.528	0.530	0.532	0.534	620
630	0.534	0.536	0.538	0.540	0.542	0.544	0.545	0.547	0.549	0.551	0.553	630
640	0.553	0.555	0.557	0.559	0.561	0.563	0.565	0.566	0.568	0.570	0.572	640
DEG F	0	1	2	3	4	5	6	7	8	9	10	DEG F

[a] **Converted from degrees Celsius (IPTS 1968).**

TABLE 10.2—*Type B thermocouples (continued).*

Temperature in Degrees Fahrenheit[a]

EMF in Absolute Millivolts | **Reference Junctions at 32 F**

DEG F	0	1	2	3	4	5	6	7	8	9	10	DEG F
					THERMOELECTRIC VOLTAGE IN ABSOLUTE MILLIVOLTS							
650	0.572	0.574	0.576	0.578	0.580	0.582	0.584	0.586	0.588	0.590	0.592	650
660	0.592	0.594	0.596	0.598	0.600	0.602	0.604	0.606	0.608	0.610	0.612	660
670	0.612	0.614	0.616	0.618	0.620	0.622	0.624	0.626	0.628	0.630	0.632	670
680	0.632	0.634	0.636	0.638	0.640	0.642	0.644	0.646	0.648	0.650	0.652	680
690	0.652	0.654	0.656	0.659	0.661	0.663	0.665	0.667	0.669	0.671	0.673	690
700	0.673	0.675	0.677	0.679	0.682	0.684	0.686	0.688	0.690	0.692	0.694	700
710	0.694	0.696	0.699	0.701	0.703	0.705	0.707	0.709	0.711	0.714	0.716	710
720	0.716	0.718	0.720	0.722	0.724	0.727	0.729	0.731	0.733	0.735	0.737	720
730	0.737	0.740	0.742	0.744	0.746	0.748	0.751	0.753	0.755	0.757	0.759	730
740	0.759	0.762	0.764	0.766	0.768	0.771	0.773	0.775	0.777	0.780	0.782	740
750	0.782	0.784	0.786	0.789	0.791	0.793	0.795	0.798	0.800	0.802	0.804	750
760	0.804	0.807	0.809	0.811	0.814	0.816	0.818	0.821	0.823	0.825	0.827	760
770	0.827	0.830	0.832	0.834	0.837	0.839	0.841	0.844	0.846	0.848	0.851	770
780	0.851	0.853	0.855	0.858	0.860	0.862	0.865	0.867	0.870	0.872	0.874	780
790	0.874	0.877	0.879	0.881	0.884	0.886	0.889	0.891	0.893	0.896	0.898	790
800	0.898	0.901	0.903	0.905	0.908	0.910	0.913	0.915	0.918	0.920	0.922	800
810	0.922	0.925	0.927	0.930	0.932	0.935	0.937	0.939	0.942	0.944	0.947	810
820	0.947	0.949	0.952	0.954	0.957	0.959	0.962	0.964	0.967	0.969	0.972	820
830	0.972	0.974	0.977	0.979	0.982	0.984	0.987	0.989	0.992	0.994	0.997	830
840	0.997	0.999	1.002	1.004	1.007	1.009	1.012	1.014	1.017	1.020	1.022	840
850	1.022	1.025	1.027	1.030	1.032	1.035	1.037	1.040	1.043	1.045	1.048	850
860	1.048	1.050	1.053	1.056	1.058	1.061	1.063	1.066	1.069	1.071	1.074	860
870	1.074	1.076	1.079	1.082	1.084	1.087	1.090	1.092	1.095	1.097	1.100	870
880	1.100	1.103	1.105	1.108	1.111	1.113	1.116	1.119	1.121	1.124	1.127	880
890	1.127	1.129	1.132	1.135	1.137	1.140	1.143	1.145	1.148	1.151	1.153	890
900	1.153	1.156	1.159	1.162	1.164	1.167	1.170	1.172	1.175	1.178	1.181	900
910	1.181	1.183	1.186	1.189	1.192	1.194	1.197	1.200	1.203	1.205	1.208	910
920	1.208	1.211	1.214	1.216	1.219	1.222	1.225	1.228	1.230	1.233	1.236	920
930	1.236	1.239	1.241	1.244	1.247	1.250	1.253	1.255	1.258	1.261	1.264	930
940	1.264	1.267	1.270	1.272	1.275	1.278	1.281	1.284	1.287	1.289	1.292	940
950	1.292	1.295	1.298	1.301	1.304	1.307	1.309	1.312	1.315	1.318	1.321	950
960	1.321	1.324	1.327	1.330	1.332	1.335	1.338	1.341	1.344	1.347	1.350	960
970	1.350	1.353	1.356	1.359	1.361	1.364	1.367	1.370	1.373	1.376	1.379	970
980	1.379	1.382	1.385	1.388	1.391	1.394	1.397	1.400	1.403	1.406	1.409	980
990	1.409	1.411	1.414	1.417	1.420	1.423	1.426	1.429	1.432	1.435	1.438	990
1,000	1.438	1.441	1.444	1.447	1.450	1.453	1.456	1.459	1.462	1.465	1.468	1,000
1,010	1.468	1.471	1.474	1.477	1.480	1.483	1.487	1.490	1.493	1.496	1.499	1,010
1,020	1.499	1.502	1.505	1.508	1.511	1.514	1.517	1.520	1.523	1.526	1.529	1,020
1,030	1.529	1.532	1.536	1.539	1.542	1.545	1.548	1.551	1.554	1.557	1.560	1,030
1,040	1.560	1.563	1.566	1.570	1.573	1.576	1.579	1.582	1.585	1.588	1.591	1,040
1,050	1.591	1.595	1.598	1.601	1.604	1.607	1.610	1.613	1.617	1.620	1.623	1,050
1,060	1.623	1.626	1.629	1.632	1.636	1.639	1.642	1.645	1.648	1.652	1.655	1,060
1,070	1.655	1.658	1.661	1.664	1.668	1.671	1.674	1.677	1.680	1.684	1.687	1,070
1,080	1.687	1.690	1.693	1.696	1.700	1.703	1.706	1.709	1.713	1.716	1.719	1,080
1,090	1.719	1.722	1.726	1.729	1.732	1.735	1.739	1.742	1.745	1.748	1.752	1,090
1,100	1.752	1.755	1.758	1.762	1.765	1.768	1.771	1.775	1.778	1.781	1.785	1,100
1,110	1.785	1.788	1.791	1.795	1.798	1.801	1.804	1.808	1.811	1.814	1.818	1,110
1,120	1.818	1.821	1.824	1.828	1.831	1.834	1.838	1.841	1.844	1.848	1.851	1,120
1,130	1.851	1.855	1.858	1.861	1.865	1.868	1.871	1.875	1.878	1.882	1.885	1,130
1,140	1.885	1.888	1.892	1.895	1.898	1.902	1.905	1.909	1.912	1.915	1.919	1,140
1,150	1.919	1.922	1.926	1.929	1.933	1.936	1.939	1.943	1.946	1.950	1.953	1,150
1,160	1.953	1.957	1.960	1.963	1.967	1.970	1.974	1.977	1.981	1.984	1.988	1,160
1,170	1.988	1.991	1.995	1.998	2.002	2.005	2.009	2.012	2.015	2.019	2.022	1,170
1,180	2.022	2.026	2.029	2.033	2.036	2.040	2.043	2.047	2.051	2.054	2.058	1,180
1,190	2.058	2.061	2.065	2.068	2.072	2.075	2.079	2.082	2.086	2.089	2.093	1,190
1,200	2.093	2.096	2.100	2.104	2.107	2.111	2.114	2.118	2.121	2.125	2.128	1,200
1,210	2.128	2.132	2.136	2.139	2.143	2.146	2.150	2.154	2.157	2.161	2.164	1,210
1,220	2.164	2.168	2.172	2.175	2.179	2.182	2.186	2.190	2.193	2.197	2.201	1,220
1,230	2.201	2.204	2.208	2.211	2.215	2.219	2.222	2.226	2.230	2.233	2.237	1,230
1,240	2.237	2.241	2.244	2.248	2.252	2.255	2.259	2.263	2.266	2.270	2.274	1,240
1,250	2.274	2.277	2.281	2.285	2.288	2.292	2.296	2.299	2.303	2.307	2.311	1,250
1,260	2.311	2.314	2.318	2.322	2.325	2.329	2.333	2.337	2.340	2.344	2.348	1,260
1,270	2.348	2.351	2.355	2.359	2.363	2.366	2.370	2.374	2.378	2.381	2.385	1,270
1,280	2.385	2.389	2.393	2.396	2.400	2.404	2.408	2.412	2.415	2.419	2.423	1,280
1,290	2.423	2.427	2.430	2.434	2.438	2.442	2.446	2.449	2.453	2.457	2.461	1,290
DEG F	0	1	2	3	4	5	6	7	8	9	10	DEG F

[a] Converted from degrees Celsius (IPTS 1968).

TABLE 10.2—*Type B thermocouples (continued).*

Temperature in Degrees Fahrenheit*

EMF in Absolute Millivolts | **Reference Junctions at 32 F**

DEG F	0	1	2	3	4	5	6	7	8	9	10	DEG F
THERMOELECTRIC VOLTAGE IN ABSOLUTE MILLIVOLTS												
1,300	2.461	2.465	2.469	2.472	2.476	2.480	2.484	2.488	2.491	2.495	2.499	1,300
1,310	2.499	2.503	2.507	2.511	2.515	2.518	2.522	2.526	2.530	2.534	2.538	1,310
1,320	2.538	2.542	2.545	2.549	2.553	2.557	2.561	2.565	2.569	2.573	2.576	1,320
1,330	2.576	2.580	2.584	2.588	2.592	2.596	2.600	2.604	2.608	2.612	2.615	1,330
1,340	2.615	2.619	2.623	2.627	2.631	2.635	2.639	2.643	2.647	2.651	2.655	1,340
1,350	2.655	2.659	2.663	2.667	2.670	2.674	2.678	2.682	2.686	2.690	2.694	1,350
1,360	2.694	2.698	2.702	2.706	2.710	2.714	2.718	2.722	2.726	2.730	2.734	1,360
1,370	2.734	2.738	2.742	2.746	2.750	2.754	2.758	2.762	2.766	2.770	2.774	1,370
1,380	2.774	2.778	2.782	2.786	2.790	2.794	2.798	2.802	2.806	2.810	2.814	1,380
1,390	2.814	2.818	2.822	2.826	2.830	2.835	2.839	2.843	2.847	2.851	2.855	1,390
1,400	2.855	2.859	2.863	2.867	2.871	2.875	2.879	2.883	2.887	2.892	2.896	1,400
1,410	2.896	2.900	2.904	2.908	2.912	2.916	2.920	2.924	2.928	2.933	2.937	1,410
1,420	2.937	2.941	2.945	2.949	2.953	2.957	2.961	2.966	2.970	2.974	2.978	1,420
1,430	2.978	2.982	2.986	2.990	2.995	2.999	3.003	3.007	3.011	3.015	3.019	1,430
1,440	3.019	3.024	3.028	3.032	3.036	3.040	3.045	3.049	3.053	3.057	3.061	1,440
1,450	3.061	3.065	3.070	3.074	3.078	3.082	3.086	3.091	3.095	3.099	3.103	1,450
1,460	3.103	3.107	3.112	3.116	3.120	3.124	3.129	3.133	3.137	3.141	3.146	1,460
1,470	3.146	3.150	3.154	3.158	3.163	3.167	3.171	3.175	3.180	3.184	3.188	1,470
1,480	3.188	3.192	3.197	3.201	3.205	3.209	3.214	3.218	3.222	3.227	3.231	1,480
1,490	3.231	3.235	3.239	3.244	3.248	3.252	3.257	3.261	3.265	3.269	3.274	1,490
1,500	3.274	3.278	3.282	3.287	3.291	3.295	3.300	3.304	3.308	3.313	3.317	1,500
1,510	3.317	3.321	3.326	3.330	3.334	3.339	3.343	3.347	3.352	3.356	3.361	1,510
1,520	3.361	3.365	3.369	3.374	3.378	3.382	3.387	3.391	3.395	3.400	3.404	1,520
1,530	3.404	3.409	3.413	3.417	3.422	3.426	3.431	3.435	3.439	3.444	3.448	1,530
1,540	3.448	3.453	3.457	3.461	3.466	3.470	3.475	3.479	3.484	3.488	3.492	1,540
1,550	3.492	3.497	3.501	3.506	3.510	3.515	3.519	3.523	3.528	3.532	3.537	1,550
1,560	3.537	3.541	3.546	3.550	3.555	3.559	3.564	3.568	3.573	3.577	3.581	1,560
1,570	3.581	3.586	3.590	3.595	3.599	3.604	3.608	3.613	3.617	3.622	3.626	1,570
1,580	3.626	3.631	3.635	3.640	3.644	3.649	3.653	3.658	3.662	3.667	3.672	1,580
1,590	3.672	3.676	3.681	3.685	3.690	3.694	3.699	3.703	3.708	3.712	3.717	1,590
1,600	3.717	3.721	3.726	3.731	3.735	3.740	3.744	3.749	3.753	3.758	3.762	1,600
1,610	3.762	3.767	3.772	3.776	3.781	3.785	3.790	3.795	3.799	3.804	3.808	1,610
1,620	3.808	3.813	3.818	3.822	3.827	3.831	3.836	3.841	3.845	3.850	3.854	1,620
1,630	3.854	3.859	3.864	3.868	3.873	3.877	3.882	3.887	3.891	3.896	3.901	1,630
1,640	3.901	3.905	3.910	3.915	3.919	3.924	3.929	3.933	3.938	3.943	3.947	1,640
1,650	3.947	3.952	3.957	3.961	3.966	3.971	3.975	3.980	3.985	3.989	3.994	1,650
1,660	3.994	3.999	4.003	4.008	4.013	4.017	4.022	4.027	4.031	4.036	4.041	1,660
1,670	4.041	4.046	4.050	4.055	4.060	4.064	4.069	4.074	4.079	4.083	4.088	1,670
1,680	4.088	4.093	4.098	4.102	4.107	4.112	4.117	4.121	4.126	4.131	4.136	1,680
1,690	4.136	4.140	4.145	4.150	4.155	4.159	4.164	4.169	4.174	4.178	4.183	1,690
1,700	4.183	4.188	4.193	4.198	4.202	4.207	4.212	4.217	4.221	4.226	4.231	1,700
1,710	4.231	4.236	4.241	4.245	4.250	4.255	4.260	4.265	4.269	4.274	4.279	1,710
1,720	4.279	4.284	4.289	4.294	4.298	4.303	4.308	4.313	4.318	4.323	4.327	1,720
1,730	4.327	4.332	4.337	4.342	4.347	4.352	4.357	4.361	4.366	4.371	4.376	1,730
1,740	4.376	4.381	4.386	4.391	4.395	4.400	4.405	4.410	4.415	4.420	4.425	1,740
1,750	4.425	4.430	4.435	4.439	4.444	4.449	4.454	4.459	4.464	4.469	4.474	1,750
1,760	4.474	4.479	4.484	4.488	4.493	4.498	4.503	4.508	4.513	4.518	4.523	1,760
1,770	4.523	4.528	4.533	4.538	4.543	4.548	4.552	4.557	4.562	4.567	4.572	1,770
1,780	4.572	4.577	4.582	4.587	4.592	4.597	4.602	4.607	4.612	4.617	4.622	1,780
1,790	4.622	4.627	4.632	4.637	4.642	4.647	4.652	4.657	4.662	4.667	4.672	1,790
1,800	4.672	4.677	4.682	4.687	4.692	4.697	4.702	4.707	4.712	4.717	4.722	1,800
1,810	4.722	4.727	4.732	4.737	4.742	4.747	4.752	4.757	4.762	4.767	4.772	1,810
1,820	4.772	4.777	4.782	4.787	4.792	4.797	4.802	4.807	4.812	4.817	4.823	1,820
1,830	4.823	4.828	4.833	4.838	4.843	4.848	4.853	4.858	4.863	4.868	4.873	1,830
1,840	4.873	4.878	4.883	4.888	4.894	4.899	4.904	4.909	4.914	4.919	4.924	1,840
1,850	4.924	4.929	4.934	4.939	4.945	4.950	4.955	4.960	4.965	4.970	4.975	1,850
1,860	4.975	4.980	4.985	4.991	4.996	5.001	5.006	5.011	5.016	5.021	5.027	1,860
1,870	5.027	5.032	5.037	5.042	5.047	5.052	5.057	5.063	5.068	5.073	5.078	1,870
1,880	5.078	5.083	5.088	5.094	5.099	5.104	5.109	5.114	5.119	5.125	5.130	1,880
1,890	5.130	5.135	5.140	5.145	5.150	5.156	5.161	5.166	5.171	5.176	5.182	1,890
1,900	5.182	5.187	5.192	5.197	5.202	5.208	5.213	5.218	5.223	5.229	5.234	1,900
1,910	5.234	5.239	5.244	5.249	5.255	5.260	5.265	5.270	5.276	5.281	5.286	1,910
1,920	5.286	5.291	5.297	5.302	5.307	5.312	5.318	5.323	5.328	5.333	5.339	1,920
1,930	5.339	5.344	5.349	5.354	5.360	5.365	5.370	5.376	5.381	5.386	5.391	1,930
1,940	5.391	5.397	5.402	5.407	5.413	5.418	5.423	5.428	5.434	5.439	5.444	1,940
DEG F	0	1	2	3	4	5	6	7	8	9	10	DEG F

*Converted from degrees Celsius (IPTS 1968).

TABLE 10.2—*Type B thermocouples (continued).*

Temperature in Degrees Fahrenheit[a]

EMF in Absolute Millivolts — Reference Junctions at 32 F

DEG F	0	1	2	3	4	5	6	7	8	9	10	DEG F
					THERMOELECTRIC VOLTAGE IN ABSOLUTE MILLIVOLTS							
1,950	5.444	5.450	5.455	5.460	5.466	5.471	5.476	5.482	5.487	5.492	5.497	1,950
1,960	5.497	5.503	5.508	5.513	5.519	5.524	5.529	5.535	5.540	5.545	5.551	1,960
1,970	5.551	5.556	5.561	5.567	5.572	5.578	5.583	5.588	5.594	5.599	5.604	1,970
1,980	5.604	5.610	5.615	5.620	5.626	5.631	5.637	5.642	5.647	5.653	5.658	1,980
1,990	5.658	5.663	5.669	5.674	5.680	5.685	5.690	5.696	5.701	5.707	5.712	1,990
2,000	5.712	5.717	5.723	5.728	5.734	5.739	5.744	5.750	5.755	5.761	5.766	2,000
2,010	5.766	5.771	5.777	5.782	5.788	5.793	5.799	5.804	5.810	5.815	5.820	2,010
2,020	5.820	5.826	5.831	5.837	5.842	5.848	5.853	5.859	5.864	5.869	5.875	2,020
2,030	5.875	5.880	5.886	5.891	5.897	5.902	5.908	5.913	5.919	5.924	5.930	2,030
2,040	5.930	5.935	5.941	5.946	5.951	5.957	5.962	5.968	5.973	5.979	5.984	2,040
2,050	5.984	5.990	5.995	6.001	6.006	6.012	6.017	6.023	6.028	6.034	6.039	2,050
2,060	6.039	6.045	6.051	6.056	6.062	6.067	6.073	6.078	6.084	6.089	6.095	2,060
2,070	6.095	6.100	6.106	6.111	6.117	6.122	6.128	6.134	6.139	6.145	6.150	2,070
2,080	6.150	6.156	6.161	6.167	6.172	6.178	6.184	6.189	6.195	6.200	6.206	2,080
2,090	6.206	6.211	6.217	6.223	6.228	6.234	6.239	6.245	6.250	6.256	6.262	2,090
2,100	6.262	6.267	6.273	6.278	6.284	6.290	6.295	6.301	6.306	6.312	6.318	2,100
2,110	6.318	6.323	6.329	6.334	6.340	6.346	6.351	6.357	6.362	6.368	6.374	2,110
2,120	6.374	6.379	6.385	6.391	6.396	6.402	6.408	6.413	6.419	6.424	6.430	2,120
2,130	6.430	6.436	6.441	6.447	6.453	6.458	6.464	6.470	6.475	6.481	6.487	2,130
2,140	6.487	6.492	6.498	6.504	6.509	6.515	6.521	6.526	6.532	6.538	6.543	2,140
2,150	6.543	6.549	6.555	6.560	6.566	6.572	6.577	6.583	6.589	6.594	6.600	2,150
2,160	6.600	6.606	6.612	6.617	6.623	6.629	6.634	6.640	6.646	6.651	6.657	2,160
2,170	6.657	6.663	6.669	6.674	6.680	6.686	6.692	6.697	6.703	6.709	6.714	2,170
2,180	6.714	6.720	6.726	6.732	6.737	6.743	6.749	6.755	6.760	6.766	6.772	2,180
2,190	6.772	6.778	6.783	6.789	6.795	6.801	6.806	6.812	6.818	6.824	6.829	2,190
2,200	6.829	6.835	6.841	6.847	6.852	6.858	6.864	6.870	6.876	6.881	6.887	2,200
2,210	6.887	6.893	6.899	6.904	6.910	6.916	6.922	6.928	6.933	6.939	6.945	2,210
2,220	6.945	6.951	6.957	6.962	6.968	6.974	6.980	6.986	6.991	6.997	7.003	2,220
2,230	7.003	7.009	7.015	7.021	7.026	7.032	7.038	7.044	7.050	7.055	7.061	2,230
2,240	7.061	7.067	7.073	7.079	7.085	7.090	7.096	7.102	7.108	7.114	7.120	2,240
2,250	7.120	7.126	7.131	7.137	7.143	7.149	7.155	7.161	7.167	7.172	7.178	2,250
2,260	7.178	7.184	7.190	7.196	7.202	7.208	7.213	7.219	7.225	7.231	7.237	2,260
2,270	7.237	7.243	7.249	7.255	7.260	7.266	7.272	7.278	7.284	7.290	7.296	2,270
2,280	7.296	7.302	7.308	7.314	7.319	7.325	7.331	7.337	7.343	7.349	7.355	2,280
2,290	7.355	7.361	7.367	7.373	7.378	7.384	7.390	7.396	7.402	7.408	7.414	2,290
2,300	7.414	7.420	7.426	7.432	7.438	7.444	7.450	7.456	7.461	7.467	7.473	2,300
2,310	7.473	7.479	7.485	7.491	7.497	7.503	7.509	7.515	7.521	7.527	7.533	2,310
2,320	7.533	7.539	7.545	7.551	7.557	7.563	7.569	7.575	7.581	7.587	7.592	2,320
2,330	7.592	7.598	7.604	7.610	7.616	7.622	7.628	7.634	7.640	7.646	7.652	2,330
2,340	7.652	7.658	7.664	7.670	7.676	7.682	7.688	7.694	7.700	7.706	7.712	2,340
2,350	7.712	7.718	7.724	7.730	7.736	7.742	7.748	7.754	7.760	7.766	7.772	2,350
2,360	7.772	7.778	7.784	7.790	7.796	7.802	7.808	7.814	7.820	7.827	7.833	2,360
2,370	7.833	7.839	7.845	7.851	7.857	7.863	7.869	7.875	7.881	7.887	7.893	2,370
2,380	7.893	7.899	7.905	7.911	7.917	7.923	7.929	7.935	7.941	7.947	7.953	2,380
2,390	7.953	7.959	7.966	7.972	7.978	7.984	7.990	7.996	8.002	8.008	8.014	2,390
2,400	8.014	8.020	8.026	8.032	8.038	8.044	8.051	8.057	8.063	8.069	8.075	2,400
2,410	8.075	8.081	8.087	8.093	8.099	8.105	8.111	8.118	8.124	8.130	8.136	2,410
2,420	8.136	8.142	8.148	8.154	8.160	8.166	8.172	8.179	8.185	8.191	8.197	2,420
2,430	8.197	8.203	8.209	8.215	8.221	8.227	8.234	8.240	8.246	8.252	8.258	2,430
2,440	8.258	8.264	8.270	8.276	8.283	8.289	8.295	8.301	8.307	8.313	8.319	2,440
2,450	8.319	8.326	8.332	8.338	8.344	8.350	8.356	8.362	8.369	8.375	8.381	2,450
2,460	8.381	8.387	8.393	8.399	8.405	8.412	8.418	8.424	8.430	8.436	8.442	2,460
2,470	8.442	8.449	8.455	8.461	8.467	8.473	8.479	8.486	8.492	8.498	8.504	2,470
2,480	8.504	8.510	8.516	8.523	8.529	8.535	8.541	8.547	8.554	8.560	8.566	2,480
2,490	8.566	8.572	8.578	8.585	8.591	8.597	8.603	8.609	8.616	8.622	8.628	2,490
2,500	8.628	8.634	8.640	8.647	8.653	8.659	8.665	8.671	8.678	8.684	8.690	2,500
2,510	8.690	8.696	8.702	8.709	8.715	8.721	8.727	8.733	8.740	8.746	8.752	2,510
2,520	8.752	8.758	8.765	8.771	8.777	8.783	8.790	8.796	8.802	8.808	8.814	2,520
2,530	8.814	8.821	8.827	8.833	8.839	8.846	8.852	8.858	8.864	8.871	8.877	2,530
2,540	8.877	8.883	8.889	8.896	8.902	8.908	8.914	8.921	8.927	8.933	8.939	2,540
2,550	8.939	8.946	8.952	8.958	8.964	8.971	8.977	8.983	8.989	8.996	9.002	2,550
2,560	9.002	9.008	9.015	9.021	9.027	9.033	9.040	9.046	9.052	9.058	9.065	2,560
2,570	9.065	9.071	9.077	9.084	9.090	9.096	9.102	9.109	9.115	9.121	9.128	2,570
2,580	9.128	9.134	9.140	9.146	9.153	9.159	9.165	9.172	9.178	9.184	9.191	2,580
2,590	9.191	9.197	9.203	9.209	9.216	9.222	9.228	9.235	9.241	9.247	9.254	2,590
DEG F	0	1	2	3	4	5	6	7	8	9	10	DEG F

[a] Converted from degrees Celsius (IPTS 1968).

TABLE 10.2—*Type B thermocouples (continued).*

EMF in Absolute Millivolts — Temperature in Degrees Fahrenheit[a] — Reference Junctions at 32 F

DEG F	0	1	2	3	4	5	6	7	8	9	10	DEG F
	THERMOELECTRIC VOLTAGE IN ABSOLUTE MILLIVOLTS											
2,600	9.254	9.260	9.266	9.273	9.279	9.285	9.291	9.298	9.304	9.310	9.317	2,600
2,610	9.317	9.323	9.329	9.336	9.342	9.348	9.355	9.361	9.367	9.374	9.380	2,610
2,620	9.380	9.386	9.393	9.399	9.405	9.412	9.418	9.424	9.431	9.437	9.443	2,620
2,630	9.443	9.450	9.456	9.462	9.469	9.475	9.481	9.488	9.494	9.500	9.507	2,630
2,640	9.507	9.513	9.519	9.526	9.532	9.538	9.545	9.551	9.558	9.564	9.570	2,640
2,650	9.570	9.577	9.583	9.589	9.596	9.602	9.608	9.615	9.621	9.627	9.634	2,650
2,660	9.634	9.640	9.647	9.653	9.659	9.666	9.672	9.678	9.685	9.691	9.697	2,660
2,670	9.697	9.704	9.710	9.717	9.723	9.729	9.736	9.742	9.748	9.755	9.761	2,670
2,680	9.761	9.768	9.774	9.780	9.787	9.793	9.800	9.806	9.812	9.819	9.825	2,680
2,690	9.825	9.831	9.838	9.844	9.851	9.857	9.863	9.870	9.876	9.883	9.889	2,690
2,700	9.889	9.895	9.902	9.908	9.915	9.921	9.927	9.934	9.940	9.947	9.953	2,700
2,710	9.953	9.959	9.966	9.972	9.979	9.985	9.991	9.998	10.004	10.011	10.017	2,710
2,720	10.017	10.023	10.030	10.036	10.043	10.049	10.056	10.062	10.068	10.075	10.081	2,720
2,730	10.081	10.088	10.094	10.100	10.107	10.113	10.120	10.126	10.133	10.139	10.145	2,730
2,740	10.145	10.152	10.158	10.165	10.171	10.178	10.184	10.190	10.197	10.203	10.210	2,740
2,750	10.210	10.216	10.223	10.229	10.235	10.242	10.248	10.255	10.261	10.268	10.274	2,750
2,760	10.274	10.280	10.287	10.293	10.300	10.306	10.313	10.319	10.325	10.332	10.338	2,760
2,770	10.338	10.345	10.351	10.358	10.364	10.371	10.377	10.383	10.390	10.396	10.403	2,770
2,780	10.403	10.409	10.416	10.422	10.429	10.435	10.441	10.448	10.454	10.461	10.467	2,780
2,790	10.467	10.474	10.480	10.487	10.493	10.500	10.506	10.512	10.519	10.525	10.532	2,790
2,800	10.532	10.538	10.545	10.551	10.558	10.564	10.571	10.577	10.584	10.590	10.596	2,800
2,810	10.596	10.603	10.609	10.616	10.622	10.629	10.635	10.642	10.648	10.655	10.661	2,810
2,820	10.661	10.668	10.674	10.680	10.687	10.693	10.700	10.706	10.713	10.719	10.726	2,820
2,830	10.726	10.732	10.739	10.745	10.752	10.758	10.765	10.771	10.778	10.784	10.790	2,830
2,840	10.790	10.797	10.803	10.810	10.816	10.823	10.829	10.836	10.842	10.849	10.855	2,840
2,850	10.855	10.862	10.868	10.875	10.881	10.888	10.894	10.901	10.907	10.914	10.920	2,850
2,860	10.920	10.926	10.933	10.939	10.946	10.952	10.959	10.965	10.972	10.978	10.985	2,860
2,870	10.985	10.991	10.998	11.004	11.011	11.017	11.024	11.030	11.037	11.043	11.050	2,870
2,880	11.050	11.056	11.063	11.069	11.076	11.082	11.089	11.095	11.102	11.108	11.115	2,880
2,890	11.115	11.121	11.128	11.134	11.141	11.147	11.154	11.160	11.166	11.173	11.179	2,890
2,900	11.179	11.186	11.192	11.199	11.205	11.212	11.218	11.225	11.231	11.238	11.244	2,900
2,910	11.244	11.251	11.257	11.264	11.270	11.277	11.283	11.290	11.296	11.303	11.309	2,910
2,920	11.309	11.316	11.322	11.329	11.335	11.342	11.348	11.355	11.361	11.368	11.374	2,920
2,930	11.374	11.381	11.387	11.394	11.400	11.407	11.413	11.420	11.426	11.433	11.439	2,930
2,940	11.439	11.446	11.452	11.459	11.465	11.472	11.478	11.485	11.491	11.498	11.504	2,940
2,950	11.504	11.511	11.517	11.524	11.530	11.537	11.543	11.550	11.556	11.563	11.569	2,950
2,960	11.569	11.576	11.582	11.589	11.595	11.602	11.608	11.615	11.621	11.628	11.634	2,960
2,970	11.634	11.641	11.647	11.654	11.660	11.667	11.673	11.680	11.686	11.693	11.699	2,970
2,980	11.699	11.706	11.712	11.719	11.725	11.732	11.738	11.745	11.751	11.758	11.764	2,980
2,990	11.764	11.771	11.777	11.784	11.790	11.797	11.803	11.810	11.816	11.823	11.829	2,990
3,000	11.829	11.836	11.842	11.849	11.855	11.862	11.868	11.875	11.881	11.888	11.894	3,000
3,010	11.894	11.901	11.907	11.914	11.920	11.927	11.933	11.940	11.946	11.953	11.959	3,010
3,020	11.959	11.966	11.972	11.979	11.985	11.992	11.998	12.005	12.011	12.018	12.024	3,020
3,030	12.024	12.031	12.037	12.044	12.050	12.057	12.063	12.070	12.076	12.083	12.089	3,030
3,040	12.089	12.096	12.102	12.109	12.115	12.121	12.128	12.134	12.141	12.147	12.154	3,040
3,050	12.154	12.160	12.167	12.173	12.180	12.186	12.193	12.199	12.206	12.212	12.219	3,050
3,060	12.219	12.225	12.232	12.238	12.245	12.251	12.258	12.264	12.271	12.277	12.284	3,060
3,070	12.284	12.290	12.297	12.303	12.310	12.316	12.323	12.329	12.336	12.342	12.349	3,070
3,080	12.349	12.355	12.362	12.368	12.374	12.381	12.387	12.394	12.400	12.407	12.413	3,080
3,090	12.413	12.420	12.426	12.433	12.439	12.446	12.452	12.459	12.465	12.472	12.478	3,090
3,100	12.478	12.485	12.491	12.498	12.504	12.511	12.517	12.523	12.530	12.536	12.543	3,100
3,110	12.543	12.549	12.556	12.562	12.569	12.575	12.582	12.588	12.595	12.601	12.608	3,110
3,120	12.608	12.614	12.621	12.627	12.633	12.640	12.646	12.653	12.659	12.666	12.672	3,120
3,130	12.672	12.679	12.685	12.692	12.698	12.705	12.711	12.718	12.724	12.730	12.737	3,130
3,140	12.737	12.743	12.750	12.756	12.763	12.769	12.776	12.782	12.789	12.795	12.801	3,140
3,150	12.801	12.808	12.814	12.821	12.827	12.834	12.840	12.847	12.853	12.860	12.866	3,150
3,160	12.866	12.872	12.879	12.885	12.892	12.898	12.905	12.911	12.918	12.924	12.930	3,160
3,170	12.930	12.937	12.943	12.950	12.956	12.963	12.969	12.976	12.982	12.988	12.995	3,170
3,180	12.995	13.001	13.008	13.014	13.021	13.027	13.034	13.040	13.046	13.053	13.059	3,180
3,190	13.059	13.066	13.072	13.079	13.085	13.091	13.098	13.104	13.111	13.117	13.124	3,190
3,200	13.124	13.130	13.136	13.143	13.149	13.156	13.162	13.169	13.175	13.181	13.188	3,200
3,210	13.188	13.194	13.201	13.207	13.213	13.220	13.226	13.233	13.239	13.246	13.252	3,210
3,220	13.252	13.258	13.265	13.271	13.278	13.284	13.290	13.297	13.303	13.310	13.316	3,220
3,230	13.316	13.322	13.329	13.335	13.342	13.348	13.354	13.361	13.367	13.374	13.380	3,230
3,240	13.380	13.387	13.393	13.399	13.406	13.412	13.418	13.425	13.431	13.438	13.444	3,240
DEG F	0	1	2	3	4	5	6	7	8	9	10	DEG F

[a] Converted from degrees Celsius (IPTS 1968).

TABLE 10.2—*Type B thermocouples (continued).*

Temperature in Degrees Fahrenheit[a]

EMF in Absolute Millivolts | Reference Junctions at 32 F

DEG F	0	1	2	3	4	5	6	7	8	9	10	DEG F
THERMOELECTRIC VOLTAGE IN ABSOLUTE MILLIVOLTS												
3,250	13.444	13.450	13.457	13.463	13.470	13.476	13.482	13.489	13.495	13.502	13.508	3,250
3,260	13.508	13.514	13.521	13.527	13.533	13.540	13.546	13.553	13.559	13.565	13.572	3,260
3,270	13.572	13.578	13.585	13.591	13.597	13.604	13.610	13.616	13.623	13.629	13.635	3,270
3,280	13.635	13.642	13.648	13.655	13.661	13.667	13.674	13.680	13.686	13.693	13.699	3,280
3,290	13.699	13.706	13.712	13.718	13.725	13.731	13.737	13.744	13.750	13.756	13.763	3,290
3,300	13.763	13.769	13.775	13.782	13.788	13.794	13.801	13.807	13.814			3,300
DEG F	0	1	2	3	4	5	6	7	8	9	10	DEG F

[a] Converted from degrees Celsius (IPTS 1968).

TABLE 10.3—*Type B thermocouples (deg C-millivolts).*

Temperature in Degrees Celsius (IPTS 1968)

EMF in Absolute Millivolts — Reference Junctions at 0 C

DEG C	0	1	2	3	4	5	6	7	8	9	10	DEG C
						THERMOELECTRIC VOLTAGE IN ABSOLUTE MILLIVOLTS						
0	0.000	-0.000	-0.000	-0.001	-0.001	-0.001	-0.001	-0.001	-0.002	-0.002	-0.002	0
10	-0.002	-0.002	-0.002	-0.002	-0.002	-0.002	-0.002	-0.002	-0.003	-0.003	-0.003	10
20	-0.003	-0.003	-0.003	-0.003	-0.003	-0.002	-0.002	-0.002	-0.002	-0.002	-0.002	20
30	-0.002	-0.002	-0.002	-0.002	-0.002	-0.001	-0.001	-0.001	-0.001	-0.001	-0.000	30
40	-0.000	-0.000	-0.000	0.000	0.000	0.001	0.001	0.001	0.002	0.002	0.002	40
50	0.002	0.003	0.003	0.003	0.004	0.004	0.004	0.005	0.005	0.006	0.006	50
60	0.006	0.007	0.007	0.008	0.008	0.009	0.009	0.010	0.010	0.011	0.011	60
70	0.011	0.012	0.012	0.013	0.014	0.014	0.015	0.015	0.016	0.017	0.017	70
80	0.017	0.018	0.019	0.020	0.020	0.021	0.022	0.022	0.023	0.024	0.025	80
90	0.025	0.026	0.026	0.027	0.028	0.029	0.030	0.031	0.031	0.032	0.033	90
100	0.033	0.034	0.035	0.036	0.037	0.038	0.039	0.040	0.041	0.042	0.043	100
110	0.043	0.044	0.045	0.046	0.047	0.048	0.049	0.050	0.051	0.052	0.053	110
120	0.053	0.055	0.056	0.057	0.058	0.059	0.060	0.062	0.063	0.064	0.065	120
130	0.065	0.066	0.068	0.069	0.070	0.071	0.073	0.074	0.075	0.077	0.078	130
140	0.078	0.079	0.081	0.082	0.083	0.085	0.086	0.088	0.089	0.091	0.092	140
150	0.092	0.093	0.095	0.096	0.098	0.099	0.101	0.102	0.104	0.106	0.107	150
160	0.107	0.109	0.110	0.112	0.113	0.115	0.117	0.118	0.120	0.122	0.123	160
170	0.123	0.125	0.127	0.128	0.130	0.132	0.133	0.135	0.137	0.139	0.140	170
180	0.140	0.142	0.144	0.146	0.148	0.149	0.151	0.153	0.155	0.157	0.159	180
190	0.159	0.161	0.163	0.164	0.166	0.168	0.170	0.172	0.174	0.176	0.178	190
200	0.178	0.180	0.182	0.184	0.186	0.188	0.190	0.192	0.194	0.197	0.199	200
210	0.199	0.201	0.203	0.205	0.207	0.209	0.211	0.214	0.216	0.218	0.220	210
220	0.220	0.222	0.225	0.227	0.229	0.231	0.234	0.236	0.238	0.240	0.243	220
230	0.243	0.245	0.247	0.250	0.252	0.254	0.257	0.259	0.262	0.264	0.266	230
240	0.266	0.269	0.271	0.274	0.276	0.279	0.281	0.284	0.286	0.289	0.291	240
250	0.291	0.294	0.296	0.299	0.301	0.304	0.307	0.309	0.312	0.314	0.317	250
260	0.317	0.320	0.322	0.325	0.328	0.330	0.333	0.336	0.338	0.341	0.344	260
260	0.317	0.320	0.322	0.325	0.328	0.330	0.333	0.336	0.338	0.341	0.344	260
270	0.344	0.347	0.349	0.352	0.355	0.358	0.360	0.363	0.366	0.369	0.372	270
280	0.372	0.375	0.377	0.380	0.383	0.386	0.389	0.392	0.395	0.398	0.401	280
290	0.401	0.404	0.406	0.409	0.412	0.415	0.418	0.421	0.424	0.427	0.431	290
300	0.431	0.434	0.437	0.440	0.443	0.446	0.449	0.452	0.455	0.458	0.462	300
310	0.462	0.465	0.468	0.471	0.474	0.477	0.481	0.484	0.487	0.490	0.494	310
320	0.494	0.497	0.500	0.503	0.507	0.510	0.513	0.517	0.520	0.523	0.527	320
330	0.527	0.530	0.533	0.537	0.540	0.544	0.547	0.550	0.554	0.557	0.561	330
340	0.561	0.564	0.568	0.571	0.575	0.578	0.582	0.585	0.589	0.592	0.596	340
350	0.596	0.599	0.603	0.606	0.610	0.614	0.617	0.621	0.625	0.628	0.632	350
360	0.632	0.636	0.639	0.643	0.647	0.650	0.654	0.658	0.661	0.665	0.669	360
370	0.669	0.673	0.677	0.680	0.684	0.688	0.692	0.696	0.699	0.703	0.707	370
380	0.707	0.711	0.715	0.719	0.723	0.727	0.730	0.734	0.738	0.742	0.746	380
390	0.746	0.750	0.754	0.758	0.762	0.766	0.770	0.774	0.778	0.782	0.786	390
400	0.786	0.790	0.794	0.799	0.803	0.807	0.811	0.815	0.819	0.823	0.827	400
410	0.827	0.832	0.836	0.840	0.844	0.848	0.853	0.857	0.861	0.865	0.870	410
420	0.870	0.874	0.878	0.882	0.887	0.891	0.895	0.900	0.904	0.908	0.913	420
430	0.913	0.917	0.921	0.926	0.930	0.935	0.939	0.943	0.948	0.952	0.957	430
440	0.957	0.961	0.966	0.970	0.975	0.979	0.984	0.988	0.993	0.997	1.002	440
450	1.002	1.006	1.011	1.015	1.020	1.025	1.029	1.034	1.039	1.043	1.048	450
460	1.048	1.052	1.057	1.062	1.066	1.071	1.076	1.081	1.085	1.090	1.095	460
470	1.095	1.100	1.104	1.109	1.114	1.119	1.123	1.128	1.133	1.138	1.143	470
480	1.143	1.148	1.152	1.157	1.162	1.167	1.172	1.177	1.182	1.187	1.192	480
490	1.192	1.197	1.202	1.206	1.211	1.216	1.221	1.226	1.231	1.236	1.241	490
500	1.241	1.246	1.252	1.257	1.262	1.267	1.272	1.277	1.282	1.287	1.292	500
510	1.292	1.297	1.303	1.308	1.313	1.318	1.323	1.328	1.334	1.339	1.344	510
520	1.344	1.349	1.354	1.360	1.365	1.370	1.375	1.381	1.386	1.391	1.397	520
530	1.397	1.402	1.407	1.413	1.418	1.423	1.429	1.434	1.439	1.445	1.450	530
540	1.450	1.456	1.461	1.467	1.472	1.477	1.483	1.488	1.494	1.499	1.505	540
550	1.505	1.510	1.516	1.521	1.527	1.532	1.538	1.544	1.549	1.555	1.560	550
560	1.560	1.566	1.571	1.577	1.583	1.588	1.594	1.600	1.605	1.611	1.617	560
570	1.617	1.622	1.628	1.634	1.639	1.645	1.651	1.657	1.662	1.668	1.674	570
580	1.674	1.680	1.685	1.691	1.697	1.703	1.709	1.715	1.720	1.726	1.732	580
590	1.732	1.738	1.744	1.750	1.756	1.762	1.767	1.773	1.779	1.785	1.791	590
600	1.791	1.797	1.803	1.809	1.815	1.821	1.827	1.833	1.839	1.845	1.851	600
610	1.851	1.857	1.863	1.869	1.875	1.882	1.888	1.894	1.900	1.906	1.912	610
620	1.912	1.918	1.924	1.931	1.937	1.943	1.949	1.955	1.961	1.968	1.974	620
630	1.974	1.980	1.986	1.993	1.999	2.005	2.011	2.018	2.024	2.030	2.036	630
640	2.036	2.043	2.049	2.055	2.062	2.068	2.074	2.081	2.087	2.094	2.100	640
DEG C	0	1	2	3	4	5	6	7	8	9	10	DEG C

TABLE 10.3—*Type B thermocouples (continued).*

Temperature in Degrees Celsius (IPTS 1968)

EMF in Absolute Millivolts — Reference Junctions at 0 C

DEG C	0	1	2	3	4	5	6	7	8	9	10	DEG C
	THERMOELECTRIC VOLTAGE IN ABSOLUTE MILLIVOLTS											
650	2.100	2.106	2.113	2.119	2.126	2.132	2.139	2.145	2.151	2.158	2.164	650
660	2.164	2.171	2.177	2.184	2.190	2.197	2.203	2.210	2.216	2.223	2.230	660
670	2.230	2.236	2.243	2.249	2.256	2.263	2.269	2.276	2.282	2.289	2.296	670
680	2.296	2.302	2.309	2.316	2.322	2.329	2.336	2.343	2.349	2.356	2.363	680
690	2.363	2.369	2.376	2.383	2.390	2.396	2.403	2.410	2.417	2.424	2.430	690
700	2.430	2.437	2.444	2.451	2.458	2.465	2.472	2.478	2.485	2.492	2.499	700
710	2.499	2.506	2.513	2.520	2.527	2.534	2.541	2.548	2.555	2.562	2.569	710
720	2.569	2.576	2.583	2.590	2.597	2.604	2.611	2.618	2.625	2.632	2.639	720
730	2.639	2.646	2.653	2.660	2.667	2.674	2.682	2.689	2.696	2.703	2.710	730
740	2.710	2.717	2.724	2.732	2.739	2.746	2.753	2.760	2.768	2.775	2.782	740
750	2.782	2.789	2.797	2.804	2.811	2.818	2.826	2.833	2.840	2.848	2.855	750
760	2.855	2.862	2.869	2.877	2.884	2.892	2.899	2.906	2.914	2.921	2.928	760
770	2.928	2.936	2.943	2.951	2.958	2.966	2.973	2.980	2.988	2.995	3.003	770
780	3.003	3.010	3.018	3.025	3.033	3.040	3.048	3.055	3.063	3.070	3.078	780
790	3.073	3.086	3.093	3.101	3.108	3.116	3.124	3.131	3.139	3.146	3.154	790
800	3.154	3.162	3.169	3.177	3.185	3.192	3.200	3.208	3.215	3.223	3.231	800
810	3.231	3.239	3.246	3.254	3.262	3.269	3.277	3.285	3.293	3.301	3.308	810
820	3.308	3.316	3.324	3.332	3.340	3.347	3.355	3.363	3.371	3.379	3.387	820
830	3.387	3.395	3.402	3.410	3.418	3.426	3.434	3.442	3.450	3.458	3.466	830
840	3.466	3.474	3.482	3.490	3.498	3.506	3.514	3.522	3.530	3.538	3.546	840
850	3.546	3.554	3.562	3.570	3.578	3.586	3.594	3.602	3.610	3.618	3.626	850
860	3.626	3.634	3.643	3.651	3.659	3.667	3.675	3.683	3.691	3.700	3.708	860
870	3.708	3.716	3.724	3.732	3.741	3.749	3.757	3.765	3.773	3.782	3.790	870
880	3.790	3.798	3.806	3.815	3.823	3.831	3.840	3.848	3.856	3.865	3.873	880
890	3.873	3.881	3.890	3.898	3.906	3.915	3.923	3.931	3.940	3.948	3.957	890
900	3.957	3.965	3.973	3.982	3.990	3.999	4.007	4.016	4.024	4.032	4.041	900
910	4.041	4.049	4.058	4.066	4.075	4.083	4.092	4.100	4.109	4.117	4.126	910
920	4.126	4.135	4.143	4.152	4.160	4.169	4.177	4.186	4.195	4.203	4.212	920
930	4.212	4.220	4.229	4.238	4.246	4.255	4.264	4.272	4.281	4.290	4.298	930
940	4.298	4.307	4.316	4.325	4.333	4.342	4.351	4.359	4.368	4.377	4.386	940
950	4.386	4.394	4.403	4.412	4.421	4.430	4.438	4.447	4.456	4.465	4.474	950
960	4.474	4.483	4.491	4.500	4.509	4.518	4.527	4.536	4.545	4.553	4.562	960
970	4.562	4.571	4.580	4.589	4.598	4.607	4.616	4.625	4.634	4.643	4.652	970
980	4.652	4.661	4.670	4.679	4.688	4.697	4.706	4.715	4.724	4.733	4.742	980
990	4.742	4.751	4.760	4.769	4.778	4.787	4.796	4.805	4.814	4.824	4.833	990
1,000	4.833	4.842	4.851	4.860	4.869	4.878	4.887	4.897	4.906	4.915	4.924	1,000
1,010	4.924	4.933	4.942	4.952	4.961	4.970	4.979	4.989	4.998	5.007	5.016	1,010
1,020	5.016	5.025	5.035	5.044	5.053	5.063	5.072	5.081	5.090	5.100	5.109	1,020
1,030	5.109	5.118	5.128	5.137	5.146	5.156	5.165	5.174	5.184	5.193	5.202	1,030
1,040	5.202	5.212	5.221	5.231	5.240	5.249	5.259	5.268	5.278	5.287	5.297	1,040
1,050	5.297	5.306	5.316	5.325	5.334	5.344	5.353	5.363	5.372	5.382	5.391	1,050
1,060	5.391	5.401	5.410	5.420	5.429	5.439	5.449	5.458	5.468	5.477	5.487	1,060
1,070	5.487	5.496	5.506	5.516	5.525	5.535	5.544	5.554	5.564	5.573	5.583	1,070
1,080	5.583	5.593	5.602	5.612	5.621	5.631	5.641	5.651	5.660	5.670	5.680	1,080
1,090	5.680	5.689	5.699	5.709	5.718	5.728	5.738	5.748	5.757	5.767	5.777	1,090
1,100	5.777	5.787	5.796	5.806	5.816	5.826	5.836	5.845	5.855	5.865	5.875	1,100
1,110	5.875	5.885	5.895	5.904	5.914	5.924	5.934	5.944	5.954	5.964	5.973	1,110
1,120	5.973	5.983	5.993	6.003	6.013	6.023	6.033	6.043	6.053	6.063	6.073	1,120
1,130	6.073	6.083	6.093	6.102	6.112	6.122	6.132	6.142	6.152	6.162	6.172	1,130
1,140	6.172	6.182	6.192	6.202	6.212	6.223	6.233	6.243	6.253	6.263	6.273	1,140
1,150	6.273	6.283	6.293	6.303	6.313	6.323	6.333	6.343	6.353	6.364	6.374	1,150
1,160	6.374	6.384	6.394	6.404	6.414	6.424	6.435	6.445	6.455	6.465	6.475	1,160
1,170	6.475	6.485	6.496	6.506	6.516	6.526	6.536	6.547	6.557	6.567	6.577	1,170
1,180	6.577	6.588	6.598	6.608	6.618	6.629	6.639	6.649	6.659	6.670	6.680	1,180
1,190	6.680	6.690	6.701	6.711	6.721	6.732	6.742	6.752	6.763	6.773	6.783	1,190
1,200	6.783	6.794	6.804	6.814	6.825	6.835	6.846	6.856	6.866	6.877	6.887	1,200
1,210	6.887	6.898	6.908	6.918	6.929	6.939	6.950	6.960	6.971	6.981	6.991	1,210
1,220	6.991	7.002	7.012	7.023	7.033	7.044	7.054	7.065	7.075	7.086	7.096	1,220
1,230	7.096	7.107	7.117	7.128	7.138	7.149	7.159	7.170	7.181	7.191	7.202	1,230
1,240	7.202	7.212	7.223	7.233	7.244	7.255	7.265	7.276	7.286	7.297	7.308	1,240
1,250	7.308	7.318	7.329	7.339	7.350	7.361	7.371	7.382	7.393	7.403	7.414	1,250
1,260	7.414	7.425	7.435	7.446	7.457	7.467	7.478	7.489	7.500	7.510	7.521	1,260
1,270	7.521	7.532	7.542	7.553	7.564	7.575	7.585	7.596	7.607	7.618	7.628	1,270
1,280	7.628	7.639	7.650	7.661	7.671	7.682	7.693	7.704	7.715	7.725	7.736	1,280
1,290	7.736	7.747	7.758	7.769	7.780	7.790	7.801	7.812	7.823	7.834	7.845	1,290
DEG C	0	1	2	3	4	5	6	7	8	9	10	DEG C

TABLE 10.3—*Type B thermocouples (continued).*

Temperature in Degrees Celsius (IPTS 1968)

EMF in Absolute Millivolts | Reference Junctions at 0 C

DEG C	0	1	2	3	4	5	6	7	8	9	10	DEG C
					THERMOELECTRIC VOLTAGE IN ABSOLUTE MILLIVOLTS							
1,300	7.845	7.855	7.866	7.877	7.888	7.899	7.910	7.921	7.932	7.943	7.953	1,300
1,310	7.953	7.964	7.975	7.986	7.997	8.008	8.019	8.030	8.041	8.052	8.063	1,310
1,320	8.063	8.074	8.085	8.096	8.107	8.118	8.128	8.139	8.150	8.161	8.172	1,320
1,330	8.172	8.183	8.194	8.205	8.216	8.227	8.238	8.249	8.261	8.272	8.283	1,330
1,340	8.283	8.294	8.305	8.316	8.327	8.338	8.349	8.360	8.371	8.382	8.393	1,340
1,350	8.393	8.404	8.415	8.426	8.437	8.449	8.460	8.471	8.482	8.493	8.504	1,350
1,360	8.504	8.515	8.526	8.538	8.549	8.560	8.571	8.582	8.593	8.604	8.616	1,360
1,370	8.616	8.627	8.638	8.649	8.660	8.671	8.683	8.694	8.705	8.716	8.727	1,370
1,380	8.727	8.738	8.750	8.761	8.772	8.783	8.795	8.806	8.817	8.828	8.839	1,380
1,390	8.839	8.851	8.862	8.873	8.884	8.896	8.907	8.918	8.929	8.941	8.952	1,390
1,400	8.952	8.963	8.974	8.986	8.997	9.008	9.020	9.031	9.042	9.053	9.065	1,400
1,410	9.065	9.076	9.087	9.099	9.110	9.121	9.133	9.144	9.155	9.167	9.178	1,410
1,420	9.178	9.189	9.201	9.212	9.223	9.235	9.246	9.257	9.269	9.280	9.291	1,420
1,430	9.291	9.303	9.314	9.326	9.337	9.348	9.360	9.371	9.382	9.394	9.405	1,430
1,440	9.405	9.417	9.428	9.439	9.451	9.462	9.474	9.485	9.497	9.508	9.519	1,440
1,450	9.519	9.531	9.542	9.554	9.565	9.577	9.588	9.599	9.611	9.622	9.634	1,450
1,460	9.634	9.645	9.657	9.668	9.680	9.691	9.703	9.714	9.726	9.737	9.748	1,460
1,470	9.748	9.760	9.771	9.783	9.794	9.806	9.817	9.829	9.840	9.852	9.863	1,470
1,480	9.863	9.875	9.886	9.898	9.909	9.921	9.933	9.944	9.956	9.967	9.979	1,480
1,490	9.979	9.990	10.002	10.013	10.025	10.036	10.048	10.059	10.071	10.082	10.094	1,490
1,500	10.094	10.106	10.117	10.129	10.140	10.152	10.163	10.175	10.187	10.198	10.210	1,500
1,510	10.210	10.221	10.233	10.244	10.256	10.268	10.279	10.291	10.302	10.314	10.325	1,510
1,520	10.325	10.337	10.349	10.360	10.372	10.383	10.395	10.407	10.418	10.430	10.441	1,520
1,530	10.441	10.453	10.465	10.476	10.488	10.500	10.511	10.523	10.534	10.546	10.558	1,530
1,540	10.558	10.569	10.581	10.593	10.604	10.616	10.627	10.639	10.651	10.662	10.674	1,540
1,550	10.674	10.686	10.697	10.709	10.721	10.732	10.744	10.756	10.767	10.779	10.790	1,550
1,560	10.790	10.802	10.814	10.825	10.837	10.849	10.860	10.872	10.884	10.895	10.907	1,560
1,570	10.907	10.919	10.930	10.942	10.954	10.965	10.977	10.989	11.000	11.012	11.024	1,570
1,580	11.024	11.035	11.047	11.059	11.070	11.082	11.094	11.105	11.117	11.129	11.141	1,580
1,590	11.141	11.152	11.164	11.176	11.187	11.199	11.211	11.222	11.234	11.246	11.257	1,590
1,600	11.257	11.269	11.281	11.292	11.304	11.316	11.328	11.339	11.351	11.363	11.374	1,600
1,610	11.374	11.386	11.398	11.409	11.421	11.433	11.444	11.456	11.468	11.480	11.491	1,610
1,620	11.491	11.503	11.515	11.526	11.538	11.550	11.561	11.573	11.585	11.597	11.608	1,620
1,630	11.608	11.620	11.632	11.643	11.655	11.667	11.678	11.690	11.702	11.714	11.725	1,630
1,640	11.725	11.737	11.749	11.760	11.772	11.784	11.795	11.807	11.819	11.830	11.842	1,640
1,650	11.842	11.854	11.866	11.877	11.889	11.901	11.912	11.924	11.936	11.947	11.959	1,650
1,660	11.959	11.971	11.983	11.994	12.006	12.018	12.029	12.041	12.053	12.064	12.076	1,660
1,670	12.076	12.088	12.099	12.111	12.123	12.134	12.146	12.158	12.170	12.181	12.193	1,670
1,680	12.193	12.205	12.216	12.228	12.240	12.251	12.263	12.275	12.286	12.298	12.310	1,680
1,690	12.310	12.321	12.333	12.345	12.356	12.368	12.380	12.391	12.403	12.415	12.426	1,690
1,700	12.426	12.438	12.450	12.461	12.473	12.485	12.496	12.508	12.520	12.531	12.543	1,700
1,710	12.543	12.555	12.566	12.578	12.590	12.601	12.613	12.624	12.636	12.648	12.659	1,710
1,720	12.659	12.671	12.683	12.694	12.706	12.718	12.729	12.741	12.752	12.764	12.776	1,720
1,730	12.776	12.787	12.799	12.811	12.822	12.834	12.845	12.857	12.869	12.880	12.892	1,730
1,740	12.892	12.903	12.915	12.927	12.938	12.950	12.961	12.973	12.985	12.996	13.008	1,740
1,750	13.008	13.019	13.031	13.043	13.054	13.066	13.077	13.089	13.100	13.112	13.124	1,750
1,760	13.124	13.135	13.147	13.158	13.170	13.181	13.193	13.204	13.216	13.228	13.239	1,760
1,770	13.239	13.251	13.262	13.274	13.285	13.297	13.308	13.320	13.331	13.343	13.354	1,770
1,780	13.354	13.366	13.378	13.389	13.401	13.412	13.424	13.435	13.447	13.458	13.470	1,780
1,790	13.470	13.481	13.493	13.504	13.516	13.527	13.539	13.550	13.562	13.573	13.585	1,790
1,800	13.585	13.596	13.607	13.619	13.630	13.642	13.653	13.665	13.676	13.688	13.699	1,800
1,810	13.699	13.711	13.722	13.733	13.745	13.756	13.768	13.779	13.791	13.802	13.814	1,810
1,820	13.814											1,820
DEG C	0	1	2	3	4	5	6	7	8	9	10	DEG C

TABLE 10.4—*Type E thermocouples (deg F-millivolts).*

Temperature in Degrees Fahrenheit[a]

EMF in Absolute Millivolts — Reference Junctions at 32 F

DEG F	0	1	2	3	4	5	6	7	8	9	10	DEG F
					THERMOELECTRIC VOLTAGE IN ABSOLUTE MILLIVOLTS							
-450	-9.830	-9.832	-9.833	-9.834	-9.835							-450
-440	-9.809	-9.812	-9.814	-9.817	-9.819	-9.821	-9.823	-9.825	-9.827	-9.829	-9.830	-440
-430	-9.775	-9.779	-9.783	-9.786	-9.790	-9.793	-9.797	-9.800	-9.803	-9.806	-9.809	-430
-420	-9.729	-9.734	-9.739	-9.744	-9.749	-9.753	-9.758	-9.762	-9.767	-9.771	-9.775	-420
-410	-9.672	-9.678	-9.684	-9.690	-9.696	-9.702	-9.708	-9.713	-9.719	-9.724	-9.729	-410
-400	-9.604	-9.611	-9.619	-9.626	-9.633	-9.639	-9.646	-9.653	-9.659	-9.666	-9.672	-400
-390	-9.526	-9.534	-9.542	-9.550	-9.558	-9.566	-9.574	-9.582	-9.589	-9.597	-9.604	-390
-380	-9.437	-9.446	-9.455	-9.464	-9.473	-9.482	-9.491	-9.500	-9.509	-9.517	-9.526	-380
-370	-9.338	-9.348	-9.358	-9.368	-9.378	-9.388	-9.398	-9.408	-9.418	-9.427	-9.437	-370
-360	-9.229	-9.241	-9.252	-9.263	-9.274	-9.285	-9.296	-9.306	-9.317	-9.327	-9.338	-360
-350	-9.112	-9.124	-9.136	-9.148	-9.160	-9.172	-9.184	-9.195	-9.207	-9.218	-9.229	-350
-340	-8.986	-8.999	-9.012	-9.025	-9.038	-9.050	-9.063	-9.075	-9.088	-9.100	-9.112	-340
-330	-8.852	-8.866	-8.880	-8.893	-8.907	-8.920	-8.934	-8.947	-8.960	-8.973	-8.986	-330
-320	-8.710	-8.725	-8.739	-8.754	-8.768	-8.782	-8.796	-8.810	-8.824	-8.838	-8.852	-320
-310	-8.561	-8.576	-8.591	-8.606	-8.621	-8.636	-8.651	-8.666	-8.681	-8.696	-8.710	-310
-300	-8.404	-8.420	-8.436	-8.452	-8.468	-8.483	-8.499	-8.514	-8.530	-8.545	-8.561	-300
-290	-8.240	-8.257	-8.273	-8.290	-8.306	-8.323	-8.339	-8.355	-8.372	-8.388	-8.404	-290
-280	-8.069	-8.086	-8.104	-8.121	-8.138	-8.155	-8.172	-8.189	-8.206	-8.223	-8.240	-280
-270	-7.891	-7.909	-7.927	-7.945	-7.963	-7.981	-7.999	-8.016	-8.034	-8.051	-8.069	-270
-260	-7.707	-7.726	-7.744	-7.763	-7.781	-7.800	-7.818	-7.837	-7.855	-7.873	-7.891	-260
-250	-7.516	-7.535	-7.555	-7.574	-7.593	-7.612	-7.631	-7.650	-7.669	-7.688	-7.707	-250
-240	-7.319	-7.339	-7.359	-7.379	-7.399	-7.418	-7.438	-7.458	-7.477	-7.497	-7.516	-240
-230	-7.116	-7.137	-7.157	-7.178	-7.198	-7.218	-7.239	-7.259	-7.279	-7.299	-7.319	-230
-220	-6.907	-6.928	-6.949	-6.970	-6.991	-7.012	-7.033	-7.054	-7.075	-7.095	-7.116	-220
-210	-6.692	-6.714	-6.735	-6.757	-6.779	-6.800	-6.822	-6.843	-6.864	-6.886	-6.907	-210
-200	-6.471	-6.494	-6.516	-6.538	-6.560	-6.582	-6.604	-6.626	-6.648	-6.670	-6.692	-200
-190	-6.245	-6.268	-6.291	-6.314	-6.336	-6.359	-6.382	-6.404	-6.427	-6.449	-6.471	-190
-180	-6.013	-6.037	-6.060	-6.084	-6.107	-6.130	-6.153	-6.176	-6.199	-6.222	-6.245	-180
-170	-5.776	-5.800	-5.824	-5.848	-5.872	-5.896	-5.919	-5.943	-5.967	-5.990	-6.013	-170
-160	-5.534	-5.559	-5.583	-5.607	-5.632	-5.656	-5.680	-5.704	-5.728	-5.752	-5.776	-160
-150	-5.287	-5.312	-5.337	-5.362	-5.386	-5.411	-5.436	-5.460	-5.485	-5.510	-5.534	-150
-140	-5.034	-5.060	-5.085	-5.111	-5.136	-5.161	-5.186	-5.212	-5.237	-5.262	-5.287	-140
-130	-4.777	-4.803	-4.829	-4.855	-4.880	-4.906	-4.932	-4.958	-4.983	-5.009	-5.034	-130
-120	-4.515	-4.541	-4.567	-4.594	-4.620	-4.646	-4.672	-4.699	-4.725	-4.751	-4.777	-120
-110	-4.248	-4.274	-4.301	-4.328	-4.355	-4.382	-4.408	-4.435	-4.462	-4.488	-4.515	-110
-100	-3.976	-4.003	-4.031	-4.058	-4.085	-4.112	-4.139	-4.167	-4.194	-4.221	-4.248	-100
-90	-3.700	-3.728	-3.755	-3.783	-3.811	-3.838	-3.866	-3.894	-3.921	-3.949	-3.976	-90
-80	-3.419	-3.447	-3.476	-3.504	-3.532	-3.560	-3.588	-3.616	-3.644	-3.672	-3.700	-80
-70	-3.134	-3.163	-3.192	-3.220	-3.249	-3.277	-3.306	-3.334	-3.363	-3.391	-3.419	-70
-60	-2.845	-2.874	-2.903	-2.932	-2.961	-2.990	-3.019	-3.048	-3.077	-3.106	-3.134	-60
-50	-2.552	-2.581	-2.611	-2.640	-2.670	-2.699	-2.728	-2.758	-2.787	-2.816	-2.845	-50
-40	-2.254	-2.284	-2.314	-2.344	-2.374	-2.404	-2.433	-2.463	-2.493	-2.522	-2.552	-40
-30	-1.953	-1.983	-2.014	-2.044	-2.074	-2.104	-2.134	-2.164	-2.194	-2.224	-2.254	-30
-20	-1.648	-1.678	-1.709	-1.740	-1.770	-1.801	-1.831	-1.862	-1.892	-1.923	-1.953	-20
-10	-1.339	-1.370	-1.401	-1.432	-1.463	-1.494	-1.525	-1.555	-1.586	-1.617	-1.648	-10
- 0	-1.026	-1.057	-1.089	-1.120	-1.151	-1.183	-1.214	-1.245	-1.276	-1.308	-1.339	- 0
0	-1.026	-0.994	-0.963	-0.931	-0.900	-0.868	-0.836	-0.805	-0.773	-0.741	-0.709	0
10	-0.709	-0.677	-0.645	-0.613	-0.581	-0.549	-0.517	-0.485	-0.453	-0.421	-0.389	10
20	-0.389	-0.357	-0.324	-0.292	-0.260	-0.227	-0.195	-0.163	-0.130	-0.098	-0.065	20
30	-0.065	-0.033	0.000	0.033	0.065	0.098	0.131	0.163	0.196	0.229	0.262	30
40	0.262	0.295	0.327	0.360	0.393	0.426	0.459	0.492	0.525	0.558	0.591	40
50	0.591	0.624	0.658	0.691	0.724	0.757	0.790	0.824	0.857	0.890	0.924	50
60	0.924	0.957	0.990	1.024	1.057	1.091	1.124	1.158	1.192	1.225	1.259	60
70	1.259	1.292	1.326	1.360	1.394	1.427	1.461	1.495	1.529	1.563	1.597	70
80	1.597	1.631	1.665	1.699	1.733	1.767	1.801	1.835	1.869	1.903	1.937	80
90	1.937	1.972	2.006	2.040	2.075	2.109	2.143	2.178	2.212	2.247	2.281	90
100	2.281	2.316	2.350	2.385	2.419	2.454	2.489	2.523	2.558	2.593	2.627	100
110	2.627	2.662	2.697	2.732	2.767	2.802	2.837	2.872	2.907	2.942	2.977	110
120	2.977	3.012	3.047	3.082	3.117	3.152	3.187	3.223	3.258	3.293	3.329	120
130	3.329	3.364	3.399	3.435	3.470	3.506	3.541	3.577	3.612	3.648	3.683	130
140	3.683	3.719	3.755	3.790	3.826	3.862	3.898	3.933	3.969	4.005	4.041	140
DEG F	0	1	2	3	4	5	6	7	8	9	10	DEG F

[a] Converted from degrees Celsius (IPTS 1968).

TABLE 10.4—*Type E thermocouples (continued).*

EMF in Absolute Millivolts — Temperature in Degrees Fahrenheit[a] — Reference Junctions at 32 F

DEG F	0	1	2	3	4	5	6	7	8	9	10	DEG F
	THERMOELECTRIC VOLTAGE IN ABSOLUTE MILLIVOLTS											
150	4.041	4.077	4.113	4.149	4.185	4.221	4.257	4.293	4.329	4.365	4.401	150
160	4.401	4.437	4.474	4.510	4.546	4.582	4.619	4.655	4.691	4.728	4.764	160
170	4.764	4.801	4.837	4.874	4.910	4.947	4.983	5.020	5.056	5.093	5.130	170
180	5.130	5.166	5.203	5.240	5.277	5.314	5.350	5.387	5.424	5.461	5.498	180
190	5.498	5.535	5.572	5.609	5.646	5.683	5.720	5.757	5.794	5.832	5.869	190
200	5.869	5.906	5.943	5.981	6.018	6.055	6.092	6.130	6.167	6.205	6.242	200
210	6.242	6.280	6.317	6.355	6.392	6.430	6.467	6.505	6.543	6.580	6.618	210
220	6.618	6.656	6.693	6.731	6.769	6.807	6.845	6.882	6.920	6.958	6.996	220
230	6.996	7.034	7.072	7.110	7.148	7.186	7.224	7.262	7.300	7.339	7.377	230
240	7.377	7.415	7.453	7.491	7.530	7.568	7.606	7.645	7.683	7.721	7.760	240
250	7.760	7.798	7.837	7.875	7.914	7.952	7.991	8.029	8.068	8.106	8.145	250
260	8.145	8.184	8.222	8.261	8.300	8.338	8.377	8.416	8.455	8.494	8.532	260
270	8.532	8.571	8.610	8.649	8.688	8.727	8.766	8.805	8.844	8.883	8.922	270
280	8.922	8.961	9.000	9.039	9.078	9.118	9.157	9.196	9.235	9.274	9.314	280
290	9.314	9.353	9.392	9.432	9.471	9.510	9.550	9.589	9.629	9.668	9.708	290
300	9.708	9.747	9.787	9.826	9.866	9.905	9.945	9.984	10.024	10.064	10.103	300
310	10.103	10.143	10.183	10.223	10.262	10.302	10.342	10.382	10.421	10.461	10.501	310
320	10.501	10.541	10.581	10.621	10.661	10.701	10.741	10.781	10.821	10.861	10.901	320
330	10.901	10.941	10.981	11.021	11.061	11.101	11.142	11.182	11.222	11.262	11.302	330
340	11.302	11.343	11.383	11.423	11.464	11.504	11.544	11.585	11.625	11.665	11.706	340
350	11.706	11.746	11.787	11.827	11.868	11.908	11.949	11.989	12.030	12.070	12.111	350
360	12.111	12.152	12.192	12.233	12.273	12.314	12.355	12.396	12.436	12.477	12.518	360
370	12.518	12.559	12.599	12.640	12.681	12.722	12.763	12.804	12.844	12.885	12.926	370
380	12.926	12.967	13.008	13.049	13.090	13.131	13.172	13.213	13.254	13.295	13.336	380
390	13.336	13.378	13.419	13.460	13.501	13.542	13.583	13.624	13.666	13.707	13.748	390
400	13.748	13.789	13.831	13.872	13.913	13.955	13.996	14.037	14.079	14.120	14.161	400
410	14.161	14.203	14.244	14.286	14.327	14.368	14.410	14.451	14.493	14.534	14.576	410
420	14.576	14.618	14.659	14.701	14.742	14.784	14.826	14.867	14.909	14.950	14.992	420
430	14.992	15.034	15.076	15.117	15.159	15.201	15.243	15.284	15.326	15.368	15.410	430
440	15.410	15.451	15.493	15.535	15.577	15.619	15.661	15.703	15.745	15.787	15.829	440
450	15.829	15.871	15.912	15.954	15.996	16.038	16.080	16.123	16.165	16.207	16.249	450
460	16.249	16.291	16.333	16.375	16.417	16.459	16.501	16.544	16.586	16.628	16.670	460
470	16.670	16.712	16.755	16.797	16.839	16.881	16.924	16.966	17.008	17.051	17.093	470
480	17.093	17.135	17.178	17.220	17.262	17.305	17.347	17.389	17.432	17.474	17.517	480
490	17.517	17.559	17.602	17.644	17.687	17.729	17.772	17.814	17.857	17.899	17.942	490
500	17.942	17.984	18.027	18.070	18.112	18.155	18.197	18.240	18.283	18.325	18.368	500
510	18.368	18.411	18.453	18.496	18.539	18.581	18.624	18.667	18.710	18.752	18.795	510
520	18.795	18.838	18.881	18.924	18.966	19.009	19.052	19.095	19.138	19.181	19.223	520
530	19.223	19.266	19.309	19.352	19.395	19.438	19.481	19.524	19.567	19.610	19.653	530
540	19.653	19.696	19.739	19.782	19.825	19.868	19.911	19.954	19.997	20.040	20.083	540
550	20.083	20.126	20.169	20.212	20.256	20.299	20.342	20.385	20.428	20.471	20.514	550
560	20.514	20.558	20.601	20.644	20.687	20.730	20.774	20.817	20.860	20.903	20.947	560
570	20.947	20.990	21.033	21.076	21.120	21.163	21.206	21.250	21.293	21.336	21.380	570
580	21.380	21.423	21.466	21.510	21.553	21.597	21.640	21.683	21.727	21.770	21.814	580
590	21.814	21.857	21.901	21.944	21.987	22.031	22.074	22.118	22.161	22.205	22.248	590
600	22.248	22.292	22.336	22.379	22.423	22.466	22.510	22.553	22.597	22.640	22.684	600
610	22.684	22.728	22.771	22.815	22.859	22.902	22.946	22.989	23.033	23.077	23.120	610
620	23.120	23.164	23.208	23.252	23.295	23.339	23.383	23.426	23.470	23.514	23.558	620
630	23.558	23.601	23.645	23.689	23.733	23.777	23.820	23.864	23.908	23.952	23.996	630
640	23.996	24.039	24.083	24.127	24.171	24.215	24.259	24.302	24.346	24.390	24.434	640
650	24.434	24.478	24.522	24.566	24.610	24.654	24.698	24.742	24.786	24.829	24.873	650
660	24.873	24.917	24.961	25.005	25.049	25.093	25.137	25.181	25.225	25.269	25.313	660
670	25.313	25.357	25.401	25.445	25.490	25.534	25.578	25.622	25.666	25.710	25.754	670
680	25.754	25.798	25.842	25.886	25.930	25.974	26.019	26.063	26.107	26.151	26.195	680
690	26.195	26.239	26.283	26.328	26.372	26.416	26.460	26.504	26.549	26.593	26.637	690
700	26.637	26.681	26.725	26.770	26.814	26.858	26.902	26.947	26.991	27.035	27.079	700
710	27.079	27.124	27.168	27.212	27.256	27.301	27.345	27.389	27.434	27.478	27.522	710
720	27.522	27.566	27.611	27.655	27.699	27.744	27.788	27.832	27.877	27.921	27.966	720
730	27.966	28.010	28.054	28.099	28.143	28.187	28.232	28.276	28.321	28.365	28.409	730
740	28.409	28.454	28.498	28.543	28.587	28.632	28.676	28.720	28.765	28.809	28.854	740
750	28.854	28.898	28.943	28.987	29.032	29.076	29.121	29.165	29.210	29.254	29.299	750
760	29.299	29.343	29.388	29.432	29.477	29.521	29.566	29.610	29.655	29.699	29.744	760
770	29.744	29.788	29.833	29.878	29.922	29.967	30.011	30.056	30.100	30.145	30.190	770
780	30.190	30.234	30.279	30.323	30.368	30.412	30.457	30.502	30.546	30.591	30.636	780
790	30.636	30.680	30.725	30.769	30.814	30.859	30.903	30.948	30.993	31.037	31.082	790
DEG F	0	1	2	3	4	5	6	7	8	9	10	DEG F

[a] Converted from degrees Celsius (IPTS 1968).

TABLE 10.4—*Type E thermocouples (continued).*

Temperature in Degrees Fahrenheit[a]

EMF in Absolute Millivolts — Reference Junctions at 32 F

DEG F	0	1	2	3	4	5	6	7	8	9	10	DEG F
					THERMOELECTRIC VOLTAGE IN ABSOLUTE MILLIVOLTS							
800	31.082	31.127	31.171	31.216	31.261	31.305	31.350	31.395	31.439	31.484	31.529	800
810	31.529	31.573	31.618	31.663	31.707	31.752	31.797	31.842	31.886	31.931	31.976	810
820	31.976	32.020	32.065	32.110	32.155	32.199	32.244	32.289	32.334	32.378	32.423	820
830	32.423	32.468	32.513	32.557	32.602	32.647	32.692	32.736	32.781	32.826	32.871	830
840	32.871	32.916	32.960	33.005	33.050	33.095	33.140	33.184	33.229	33.274	33.319	840
850	33.319	33.364	33.408	33.453	33.498	33.543	33.588	33.632	33.677	33.722	33.767	850
860	33.767	33.812	33.857	33.901	33.946	33.991	34.036	34.081	34.126	34.170	34.215	860
870	34.215	34.260	34.305	34.350	34.395	34.440	34.484	34.529	34.574	34.619	34.664	870
880	34.664	34.709	34.754	34.798	34.843	34.888	34.933	34.978	35.023	35.068	35.113	880
890	35.113	35.157	35.202	35.247	35.292	35.337	35.382	35.427	35.472	35.517	35.562	890
900	35.562	35.606	35.651	35.696	35.741	35.786	35.831	35.876	35.921	35.966	36.011	900
910	36.011	36.056	36.100	36.145	36.190	36.235	36.280	36.325	36.370	36.415	36.460	910
920	36.460	36.505	36.550	36.595	36.640	36.684	36.729	36.774	36.819	36.864	36.909	920
930	36.909	36.954	36.999	37.044	37.089	37.134	37.179	37.224	37.269	37.314	37.358	930
940	37.358	37.403	37.448	37.493	37.538	37.583	37.628	37.673	37.718	37.763	37.808	940
950	37.808	37.853	37.898	37.943	37.988	38.033	38.078	38.123	38.168	38.213	38.257	950
960	38.257	38.302	38.347	38.392	38.437	38.482	38.527	38.572	38.617	38.662	38.707	960
970	38.707	38.752	38.797	38.842	38.887	38.932	38.977	39.022	39.067	39.112	39.157	970
980	39.157	39.202	39.247	39.291	39.336	39.381	39.426	39.471	39.516	39.561	39.606	980
990	39.606	39.651	39.696	39.741	39.786	39.831	39.876	39.921	39.966	40.011	40.056	990
1,000	40.056	40.101	40.146	40.191	40.236	40.280	40.325	40.370	40.415	40.460	40.505	1,000
1,010	40.505	40.550	40.595	40.640	40.685	40.730	40.775	40.820	40.865	40.910	40.955	1,010
1,020	40.955	41.000	41.045	41.090	41.134	41.179	41.224	41.269	41.314	41.359	41.404	1,020
1,030	41.404	41.449	41.494	41.539	41.584	41.629	41.674	41.719	41.764	41.808	41.853	1,030
1,040	41.853	41.898	41.943	41.988	42.033	42.078	42.123	42.168	42.213	42.258	42.303	1,040
1,050	42.303	42.348	42.392	42.437	42.482	42.527	42.572	42.617	42.662	42.707	42.752	1,050
1,060	42.752	42.797	42.842	42.886	42.931	42.976	43.021	43.066	43.111	43.156	43.201	1,060
1,070	43.201	43.246	43.290	43.335	43.380	43.425	43.470	43.515	43.560	43.605	43.650	1,070
1,080	43.650	43.694	43.739	43.784	43.829	43.874	43.919	43.964	44.008	44.053	44.098	1,080
1,090	44.098	44.143	44.188	44.233	44.278	44.322	44.367	44.412	44.457	44.502	44.547	1,090
1,100	44.547	44.592	44.636	44.681	44.726	44.771	44.816	44.861	44.905	44.950	44.995	1,100
1,110	44.995	45.040	45.085	45.130	45.174	45.219	45.264	45.309	45.354	45.398	45.443	1,110
1,120	45.443	45.488	45.533	45.578	45.622	45.667	45.712	45.757	45.802	45.846	45.891	1,120
1,130	45.891	45.936	45.981	46.025	46.070	46.115	46.160	46.205	46.249	46.294	46.339	1,130
1,140	46.339	46.384	46.428	46.473	46.518	46.563	46.607	46.652	46.697	46.742	46.786	1,140
1,150	46.786	46.831	46.876	46.921	46.965	47.010	47.055	47.099	47.144	47.189	47.234	1,150
1,160	47.234	47.278	47.323	47.368	47.412	47.457	47.502	47.546	47.591	47.636	47.681	1,160
1,170	47.681	47.725	47.770	47.815	47.859	47.904	47.949	47.993	48.038	48.083	48.127	1,170
1,180	48.127	48.172	48.217	48.261	48.306	48.351	48.395	48.440	48.484	48.529	48.574	1,180
1,190	48.574	48.618	48.663	48.708	48.752	48.797	48.842	48.886	48.931	48.975	49.020	1,190
1,200	49.020	49.065	49.109	49.154	49.198	49.243	49.288	49.332	49.377	49.421	49.466	1,200
1,210	49.466	49.510	49.555	49.600	49.644	49.689	49.733	49.778	49.822	49.867	49.911	1,210
1,220	49.911	49.956	50.001	50.045	50.090	50.134	50.179	50.223	50.268	50.312	50.357	1,220
1,230	50.357	50.401	50.446	50.490	50.535	50.579	50.624	50.668	50.713	50.757	50.802	1,230
1,240	50.802	50.846	50.891	50.935	50.980	51.024	51.069	51.113	51.157	51.202	51.246	1,240
1,250	51.246	51.291	51.335	51.380	51.424	51.469	51.513	51.557	51.602	51.646	51.691	1,250
1,260	51.691	51.735	51.780	51.824	51.868	51.913	51.957	52.002	52.046	52.090	52.135	1,260
1,270	52.135	52.179	52.223	52.268	52.312	52.357	52.401	52.445	52.490	52.534	52.578	1,270
1,280	52.578	52.623	52.667	52.711	52.756	52.800	52.844	52.889	52.933	52.977	53.022	1,280
1,290	53.022	53.066	53.110	53.155	53.199	53.243	53.288	53.332	53.376	53.420	53.465	1,290
1,300	53.465	53.509	53.553	53.597	53.642	53.686	53.730	53.774	53.819	53.863	53.907	1,300
1,310	53.907	53.951	53.996	54.040	54.084	54.128	54.173	54.217	54.261	54.305	54.349	1,310
1,320	54.349	54.394	54.438	54.482	54.526	54.570	54.615	54.659	54.703	54.747	54.791	1,320
1,330	54.791	54.835	54.880	54.924	54.968	55.012	55.056	55.100	55.145	55.189	55.233	1,330
1,340	55.233	55.277	55.321	55.365	55.409	55.453	55.498	55.542	55.586	55.630	55.674	1,340
1,350	55.674	55.718	55.762	55.806	55.850	55.894	55.938	55.982	56.026	56.071	56.115	1,350
1,360	56.115	56.159	56.203	56.247	56.291	56.335	56.379	56.423	56.467	56.511	56.555	1,360
1,370	56.555	56.599	56.643	56.687	56.731	56.775	56.819	56.863	56.907	56.951	56.995	1,370
1,380	56.995	57.039	57.083	57.127	57.171	57.215	57.259	57.303	57.346	57.390	57.434	1,380
1,390	57.434	57.478	57.522	57.566	57.610	57.654	57.698	57.742	57.786	57.830	57.873	1,390
1,400	57.873	57.917	57.961	58.005	58.049	58.093	58.137	58.181	58.224	58.268	58.312	1,400
1,410	58.312	58.356	58.400	58.444	58.487	58.531	58.575	58.619	58.663	58.707	58.750	1,410
1,420	58.750	58.794	58.838	58.882	58.926	58.969	59.013	59.057	59.101	59.144	59.188	1,420
1,430	59.188	59.232	59.276	59.319	59.363	59.407	59.451	59.494	59.538	59.582	59.626	1,430
1,440	59.626	59.669	59.713	59.757	59.800	59.844	59.888	59.932	59.975	60.019	60.063	1,440
DEG F	0	1	2	3	4	5	6	7	8	9	10	DEG F

[a] Converted from degrees Celsius (IPTS 1968).

TABLE 10.4—*Type E thermocouples (continued).*

Temperature in Degrees Fahrenheit[a]

EMF in Absolute Millivolts — Reference Junctions at 32 F

DEG F	0	1	2	3	4	5	6	7	8	9	10	DEG F
					THERMOELECTRIC VOLTAGE IN ABSOLUTE MILLIVOLTS							
1,450	60.063	60.106	60.150	60.194	60.237	60.281	60.325	60.368	60.412	60.455	60.499	1,450
1,460	60.499	60.543	60.586	60.630	60.674	60.717	60.761	60.804	60.848	60.892	60.935	1,460
1,470	60.935	60.979	61.022	61.066	61.109	61.153	61.197	61.240	61.284	61.327	61.371	1,470
1,480	61.371	61.414	61.458	61.501	61.545	61.588	61.632	61.675	61.719	61.762	61.806	1,480
1,490	61.806	61.849	61.893	61.936	61.980	62.023	62.067	62.110	62.154	62.197	62.240	1,490
1,500	62.240	62.284	62.327	62.371	62.414	62.458	62.501	62.544	62.588	62.631	62.675	1,500
1,510	62.675	62.718	62.761	62.805	62.848	62.892	62.935	62.978	63.022	63.065	63.108	1,510
1,520	63.108	63.152	63.195	63.238	63.282	63.325	63.368	63.412	63.455	63.498	63.542	1,520
1,530	63.542	63.585	63.628	63.671	63.715	63.758	63.801	63.844	63.888	63.931	63.974	1,530
1,540	63.974	64.017	64.061	64.104	64.147	64.190	64.234	64.277	64.320	64.363	64.406	1,540
1,550	64.406	64.450	64.493	64.536	64.579	64.622	64.665	64.709	64.752	64.795	64.838	1,550
1,560	64.838	64.881	64.924	64.967	65.011	65.054	65.097	65.140	65.183	65.226	65.269	1,560
1,570	65.269	65.312	65.355	65.398	65.441	65.484	65.528	65.571	65.614	65.657	65.700	1,570
1,580	65.700	65.743	65.786	65.829	65.872	65.915	65.958	66.001	66.044	66.087	66.130	1,580
1,590	66.130	66.173	66.216	66.259	66.302	66.345	66.387	66.430	66.473	66.516	66.559	1,590
1,600	66.559	66.602	66.645	66.688	66.731	66.774	66.817	66.859	66.902	66.945	66.988	1,600
1,610	66.988	67.031	67.074	67.117	67.159	67.202	67.245	67.288	67.331	67.374	67.416	1,610
1,620	67.416	67.459	67.502	67.545	67.588	67.630	67.673	67.716	67.759	67.801	67.844	1,620
1,630	67.844	67.887	67.930	67.972	68.015	68.058	68.101	68.143	68.186	68.229	68.271	1,630
1,640	68.271	68.314	68.357	68.399	68.442	68.485	68.527	68.570	68.613	68.655	68.698	1,640
1,650	68.698	68.740	68.783	68.826	68.868	68.911	68.953	68.996	69.039	69.081	69.124	1,650
1,660	69.124	69.166	69.209	69.251	69.294	69.337	69.379	69.422	69.464	69.507	69.549	1,660
1,670	69.549	69.592	69.634	69.677	69.719	69.762	69.804	69.847	69.889	69.931	69.974	1,670
1,680	69.974	70.016	70.059	70.101	70.144	70.186	70.228	70.271	70.313	70.356	70.398	1,680
1,690	70.398	70.440	70.483	70.525	70.567	70.610	70.652	70.694	70.737	70.779	70.821	1,690
1,700	70.821	70.864	70.906	70.948	70.991	71.033	71.075	71.118	71.160	71.202	71.244	1,700
1,710	71.244	71.287	71.329	71.371	71.413	71.456	71.498	71.540	71.582	71.624	71.667	1,710
1,720	71.667	71.709	71.751	71.793	71.835	71.878	71.920	71.962	72.004	72.046	72.088	1,720
1,730	72.088	72.130	72.173	72.215	72.257	72.299	72.341	72.383	72.425	72.467	72.509	1,730
1,740	72.509	72.551	72.593	72.635	72.678	72.720	72.762	72.804	72.846	72.888	72.930	1,740
1,750	72.930	72.972	73.014	73.056	73.098	73.140	73.182	73.224	73.266	73.308	73.350	1,750
1,760	73.350	73.392	73.434	73.475	73.517	73.559	73.601	73.643	73.685	73.727	73.769	1,760
1,770	73.769	73.811	73.853	73.895	73.936	73.978	74.020	74.062	74.104	74.146	74.188	1,770
1,780	74.188	74.229	74.271	74.313	74.355	74.397	74.439	74.480	74.522	74.564	74.606	1,780
1,790	74.606	74.648	74.689	74.731	74.773	74.815	74.857	74.898	74.940	74.982	75.024	1,790
1,800	75.024	75.065	75.107	75.149	75.191	75.232	75.274	75.316	75.357	75.399	75.441	1,800
1,810	75.441	75.483	75.524	75.566	75.608	75.649	75.691	75.733	75.774	75.816	75.858	1,810
1,820	75.858	75.899	75.941	75.983	76.024	76.066	76.108	76.149	76.191	76.233	76.274	1,820
1,830	76.274	76.316	76.358									1,830
DEG F	0	1	2	3	4	5	6	7	8	9	10	DEG F

[a] Converted from degrees Celsius (IPTS 1968).

TABLE 10.5—*Type E thermocouples (deg C-millivolts).*

Temperature in Degrees Celsius (IPTS 1968)

EMF in Absolute Millivolts | Reference Junctions at 0 C

DEG C	0	1	2	3	4	5	6	7	8	9	10	DEG C
	THERMOELECTRIC VOLTAGE IN ABSOLUTE MILLIVOLTS											
-270	-9.835											-270
-260	-9.797	-9.802	-9.808	-9.813	-9.817	-9.821	-9.825	-9.828	-9.831	-9.833	-9.835	-260
-250	-9.719	-9.728	-9.737	-9.746	-9.754	-9.762	-9.770	-9.777	-9.784	-9.791	-9.797	-250
-240	-9.604	-9.617	-9.630	-9.642	-9.654	-9.666	-9.677	-9.688	-9.699	-9.709	-9.719	-240
-230	-9.455	-9.472	-9.488	-9.503	-9.519	-9.534	-9.549	-9.563	-9.577	-9.591	-9.604	-230
-220	-9.274	-9.293	-9.313	-9.332	-9.350	-9.368	-9.386	-9.404	-9.421	-9.438	-9.455	-220
-210	-9.063	-9.085	-9.107	-9.129	-9.151	-9.172	-9.193	-9.214	-9.234	-9.254	-9.274	-210
-200	-8.824	-8.850	-8.874	-8.899	-8.923	-8.947	-8.971	-8.994	-9.017	-9.040	-9.063	-200
-190	-8.561	-8.588	-8.615	-8.642	-8.669	-8.696	-8.722	-8.748	-8.774	-8.799	-8.824	-190
-180	-8.273	-8.303	-8.333	-8.362	-8.391	-8.420	-8.449	-8.477	-8.505	-8.533	-8.561	-180
-170	-7.963	-7.995	-8.027	-8.058	-8.090	-8.121	-8.152	-8.183	-8.213	-8.243	-8.273	-170
-160	-7.631	-7.665	-7.699	-7.733	-7.767	-7.800	-7.833	-7.866	-7.898	-7.931	-7.963	-160
-150	-7.279	-7.315	-7.351	-7.387	-7.422	-7.458	-7.493	-7.528	-7.562	-7.597	-7.631	-150
-140	-6.907	-6.945	-6.983	-7.020	-7.058	-7.095	-7.132	-7.169	-7.206	-7.243	-7.279	-140
-130	-6.516	-6.556	-6.596	-6.635	-6.675	-6.714	-6.753	-6.792	-6.830	-6.869	-6.907	-130
-120	-6.107	-6.149	-6.190	-6.231	-6.273	-6.314	-6.354	-6.395	-6.436	-6.476	-6.516	-120
-110	-5.680	-5.724	-5.767	-5.810	-5.853	-5.896	-5.938	-5.981	-6.023	-6.065	-6.107	-110
-100	-5.237	-5.282	-5.327	-5.371	-5.416	-5.460	-5.505	-5.549	-5.593	-5.637	-5.680	-100
-90	-4.777	-4.824	-4.870	-4.916	-4.963	-5.009	-5.055	-5.100	-5.146	-5.191	-5.237	-90
-80	-4.301	-4.350	-4.398	-4.446	-4.493	-4.541	-4.588	-4.636	-4.683	-4.730	-4.777	-80
-70	-3.811	-3.860	-3.910	-3.959	-4.009	-4.058	-4.107	-4.156	-4.204	-4.253	-4.301	-70
-60	-3.306	-3.357	-3.408	-3.459	-3.509	-3.560	-3.610	-3.661	-3.711	-3.761	-3.811	-60
-50	-2.787	-2.839	-2.892	-2.944	-2.996	-3.048	-3.100	-3.152	-3.203	-3.254	-3.306	-50
-40	-2.254	-2.308	-2.362	-2.416	-2.469	-2.522	-2.575	-2.628	-2.681	-2.734	-2.787	-40
-30	-1.709	-1.764	-1.819	-1.874	-1.929	-1.983	-2.038	-2.092	-2.146	-2.200	-2.254	-30
-20	-1.151	-1.208	-1.264	-1.320	-1.376	-1.432	-1.487	-1.543	-1.599	-1.654	-1.709	-20
-10	-0.581	-0.639	-0.696	-0.754	-0.811	-0.868	-0.925	-0.982	-1.038	-1.095	-1.151	-10
- 0	0.000	-0.059	-0.117	-0.176	-0.234	-0.292	-0.350	-0.408	-0.466	-0.524	-0.581	- 0
0	0.000	0.059	0.118	0.176	0.235	0.295	0.354	0.413	0.472	0.532	0.591	0
10	0.591	0.651	0.711	0.770	0.830	0.890	0.950	1.011	1.071	1.131	1.192	10
20	1.192	1.252	1.313	1.373	1.434	1.495	1.556	1.617	1.678	1.739	1.801	20
30	1.801	1.862	1.924	1.985	2.047	2.109	2.171	2.233	2.295	2.357	2.419	30
40	2.419	2.482	2.544	2.607	2.669	2.732	2.795	2.858	2.921	2.984	3.047	40
50	3.047	3.110	3.173	3.237	3.300	3.364	3.428	3.491	3.555	3.619	3.683	50
60	3.683	3.748	3.812	3.876	3.941	4.005	4.070	4.134	4.199	4.264	4.329	60
70	4.329	4.394	4.459	4.524	4.590	4.655	4.720	4.786	4.852	4.917	4.983	70
80	4.983	5.049	5.115	5.181	5.247	5.314	5.380	5.446	5.513	5.579	5.646	80
90	5.646	5.713	5.780	5.846	5.913	5.981	6.048	6.115	6.182	6.250	6.317	90
100	6.317	6.385	6.452	6.520	6.588	6.656	6.724	6.792	6.860	6.928	6.996	100
110	6.996	7.064	7.133	7.201	7.270	7.339	7.407	7.476	7.545	7.614	7.683	110
120	7.683	7.752	7.821	7.890	7.960	8.029	8.099	8.168	8.238	8.307	8.377	120
130	8.377	8.447	8.517	8.587	8.657	8.727	8.797	8.867	8.938	9.008	9.078	130
140	9.078	9.149	9.220	9.290	9.361	9.432	9.503	9.573	9.644	9.715	9.787	140
150	9.787	9.858	9.929	10.000	10.072	10.143	10.215	10.286	10.358	10.429	10.501	150
160	10.501	10.573	10.645	10.717	10.789	10.861	10.933	11.005	11.077	11.150	11.222	160
170	11.222	11.294	11.367	11.439	11.512	11.585	11.657	11.730	11.803	11.876	11.949	170
180	11.949	12.022	12.095	12.168	12.241	12.314	12.387	12.461	12.534	12.608	12.681	180
190	12.681	12.755	12.828	12.902	12.975	13.049	13.123	13.197	13.271	13.345	13.419	190
200	13.419	13.493	13.567	13.641	13.715	13.789	13.864	13.938	14.012	14.087	14.161	200
210	14.161	14.236	14.310	14.385	14.460	14.534	14.609	14.684	14.759	14.834	14.909	210
220	14.909	14.984	15.059	15.134	15.209	15.284	15.359	15.435	15.510	15.585	15.661	220
230	15.661	15.736	15.812	15.887	15.963	16.038	16.114	16.190	16.266	16.341	16.417	230
240	16.417	16.493	16.569	16.645	16.721	16.797	16.873	16.949	17.025	17.101	17.178	240
250	17.178	17.254	17.330	17.406	17.483	17.559	17.636	17.712	17.789	17.865	17.942	250
260	17.942	18.018	18.095	18.172	18.248	18.325	18.402	18.479	18.556	18.633	18.710	260
270	18.710	18.787	18.864	18.941	19.018	19.095	19.172	19.249	19.326	19.404	19.481	270
280	19.481	19.558	19.636	19.713	19.790	19.868	19.945	20.023	20.100	20.178	20.256	280
290	20.256	20.333	20.411	20.488	20.566	20.644	20.722	20.800	20.877	20.955	21.033	290
300	21.033	21.111	21.189	21.267	21.345	21.423	21.501	21.579	21.657	21.735	21.814	300
310	21.814	21.892	21.970	22.048	22.127	22.205	22.283	22.362	22.440	22.518	22.597	310
320	22.597	22.675	22.754	22.832	22.911	22.989	23.068	23.147	23.225	23.304	23.383	320
330	23.383	23.461	23.540	23.619	23.698	23.777	23.855	23.934	24.013	24.092	24.171	330
340	24.171	24.250	24.329	24.408	24.487	24.566	24.645	24.724	24.803	24.882	24.961	340
DEG C	0	1	2	3	4	5	6	7	8	9	10	DEG C

TABLE 10.5—*Type E thermocouples (continued).*

Temperature in Degrees Celsius (IPTS 1968)

EMF in Absolute Millivolts — Reference Junctions at 0 C

DEG C	0	1	2	3	4	5	6	7	8	9	10	DEG C
						THERMOELECTRIC VOLTAGE IN ABSOLUTE MILLIVOLTS						
350	24.961	25.041	25.120	25.199	25.278	25.357	25.437	25.516	25.595	25.675	25.754	350
360	25.754	25.833	25.913	25.992	26.072	26.151	26.230	26.310	26.389	26.469	26.549	360
370	26.549	26.628	26.708	26.787	26.867	26.947	27.026	27.106	27.186	27.265	27.345	370
380	27.345	27.425	27.504	27.584	27.664	27.744	27.824	27.903	27.983	28.063	28.143	380
390	28.143	28.223	28.303	28.383	28.463	28.543	28.623	28.703	28.783	28.863	28.943	390
400	28.943	29.023	29.103	29.183	29.263	29.343	29.423	29.503	29.584	29.664	29.744	400
410	29.744	29.824	29.904	29.984	30.065	30.145	30.225	30.305	30.386	30.466	30.546	410
420	30.546	30.627	30.707	30.787	30.868	30.948	31.028	31.109	31.189	31.270	31.350	420
430	31.350	31.430	31.511	31.591	31.672	31.752	31.833	31.913	31.994	32.074	32.155	430
440	32.155	32.235	32.316	32.396	32.477	32.557	32.638	32.719	32.799	32.880	32.960	440
450	32.960	33.041	33.122	33.202	33.283	33.364	33.444	33.525	33.605	33.686	33.767	450
460	33.767	33.848	33.928	34.009	34.090	34.170	34.251	34.332	34.413	34.493	34.574	460
470	34.574	34.655	34.736	34.816	34.897	34.978	35.059	35.140	35.220	35.301	35.382	470
480	35.382	35.463	35.544	35.624	35.705	35.786	35.867	35.948	36.029	36.109	36.190	480
490	36.190	36.271	36.352	36.433	36.514	36.595	36.675	36.756	36.837	36.918	36.999	490
500	36.999	37.080	37.161	37.242	37.323	37.403	37.484	37.565	37.646	37.727	37.808	500
510	37.808	37.889	37.970	38.051	38.132	38.213	38.293	38.374	38.455	38.536	38.617	510
520	38.617	38.698	38.779	38.860	38.941	39.022	39.103	39.184	39.264	39.345	39.426	520
530	39.426	39.507	39.588	39.669	39.750	39.831	39.912	39.993	40.074	40.155	40.236	530
540	40.236	40.316	40.397	40.478	40.559	40.640	40.721	40.802	40.883	40.964	41.045	540
550	41.045	41.125	41.206	41.287	41.368	41.449	41.530	41.611	41.692	41.773	41.853	550
560	41.853	41.934	42.015	42.096	42.177	42.258	42.339	42.419	42.500	42.581	42.662	560
570	42.662	42.743	42.824	42.904	42.985	43.066	43.147	43.228	43.308	43.389	43.470	570
580	43.470	43.551	43.632	43.712	43.793	43.874	43.955	44.035	44.116	44.197	44.278	580
590	44.278	44.358	44.439	44.520	44.601	44.681	44.762	44.843	44.923	45.004	45.085	590
600	45.085	45.165	45.246	45.327	45.407	45.488	45.569	45.649	45.730	45.811	45.891	600
610	45.891	45.972	46.052	46.133	46.213	46.294	46.375	46.455	46.536	46.616	46.697	610
620	46.697	46.777	46.858	46.938	47.019	47.099	47.180	47.260	47.341	47.421	47.502	620
630	47.502	47.582	47.663	47.743	47.824	47.904	47.984	48.065	48.145	48.226	48.306	630
640	48.306	48.386	48.467	48.547	48.627	48.708	48.788	48.868	48.949	49.029	49.109	640
650	49.109	49.189	49.270	49.350	49.430	49.510	49.591	49.671	49.751	49.831	49.911	650
660	49.911	49.992	50.072	50.152	50.232	50.312	50.392	50.472	50.553	50.633	50.713	660
670	50.713	50.793	50.873	50.953	51.033	51.113	51.193	51.273	51.353	51.433	51.513	670
680	51.513	51.593	51.673	51.753	51.833	51.913	51.993	52.073	52.152	52.232	52.312	680
690	52.312	52.392	52.472	52.552	52.632	52.711	52.791	52.871	52.951	53.031	53.110	690
700	53.110	53.190	53.270	53.350	53.429	53.509	53.589	53.668	53.748	53.828	53.907	700
710	53.907	53.987	54.066	54.146	54.226	54.305	54.385	54.464	54.544	54.623	54.703	710
720	54.703	54.782	54.862	54.941	55.021	55.100	55.180	55.259	55.339	55.418	55.498	720
730	55.498	55.577	55.656	55.736	55.815	55.894	55.974	56.053	56.132	56.212	56.291	730
740	56.291	56.370	56.449	56.529	56.608	56.687	56.766	56.845	56.924	57.004	57.083	740
750	57.083	57.162	57.241	57.320	57.399	57.478	57.557	57.636	57.715	57.794	57.873	750
760	57.873	57.952	58.031	58.110	58.189	58.268	58.347	58.426	58.505	58.584	58.663	760
770	58.663	58.742	58.820	58.899	58.978	59.057	59.136	59.214	59.293	59.372	59.451	770
780	59.451	59.529	59.608	59.687	59.765	59.844	59.923	60.001	60.080	60.159	60.237	780
790	60.237	60.316	60.394	60.473	60.551	60.630	60.708	60.787	60.865	60.944	61.022	790
800	61.022	61.101	61.179	61.258	61.336	61.414	61.493	61.571	61.649	61.728	61.806	800
810	61.806	61.884	61.962	62.041	62.119	62.197	62.275	62.353	62.432	62.510	62.588	810
820	62.588	62.666	62.744	62.822	62.900	62.978	63.056	63.134	63.212	63.290	63.368	820
830	63.368	63.446	63.524	63.602	63.680	63.758	63.836	63.914	63.992	64.069	64.147	830
840	64.147	64.225	64.303	64.380	64.458	64.536	64.614	64.691	64.769	64.847	64.924	840
850	64.924	65.002	65.080	65.157	65.235	65.312	65.390	65.467	65.545	65.622	65.700	850
860	65.700	65.777	65.855	65.932	66.009	66.087	66.164	66.241	66.319	66.396	66.473	860
870	66.473	66.551	66.628	66.705	66.782	66.859	66.937	67.014	67.091	67.168	67.245	870
880	67.245	67.322	67.399	67.476	67.553	67.630	67.707	67.784	67.861	67.938	68.015	880
890	68.015	68.092	68.169	68.246	68.323	68.399	68.476	68.553	68.630	68.706	68.783	890
900	68.783	68.860	68.936	69.013	69.090	69.166	69.243	69.320	69.396	69.473	69.549	900
910	69.549	69.626	69.702	69.779	69.855	69.931	70.008	70.084	70.161	70.237	70.313	910
920	70.313	70.390	70.466	70.542	70.618	70.694	70.771	70.847	70.923	70.999	71.075	920
930	71.075	71.151	71.227	71.304	71.380	71.456	71.532	71.608	71.683	71.759	71.835	930
940	71.835	71.911	71.987	72.063	72.139	72.215	72.290	72.366	72.442	72.518	72.593	940
950	72.593	72.669	72.745	72.820	72.896	72.972	73.047	73.123	73.199	73.274	73.350	950
960	73.350	73.425	73.501	73.576	73.652	73.727	73.802	73.878	73.953	74.029	74.104	960
970	74.104	74.179	74.255	74.330	74.405	74.480	74.556	74.631	74.706	74.781	74.857	970
980	74.857	74.932	75.007	75.082	75.157	75.232	75.307	75.382	75.458	75.533	75.608	980
990	75.608	75.683	75.758	75.833	75.908	75.983	76.058	76.133	76.208	76.283	76.358	990
1,000	76.358											1,000
DEG C	0	1	2	3	4	5	6	7	8	9	10	DEG C

TABLE 10.6—*Type J thermocouples (deg F-millivolts).*

NOTE—The maximum recommended temperature limit for Type J thermocouples is 1400 F (760 C) as specified in Table 2. Extension of the Type J tables beyond 1400 F gives temperature-electromotive force data to 2192 F (1200 C). This extension is a mathematical extrapolation based on limited calibration data and caution should be exercised in its use. The basis for the extended curve is discussed in *NBS Monograph 125*.

It should be noted that limits of error for Type J thermocouples (Table 1) do not apply above 1400 F (760 C).

Temperature in Degrees Fahrenheit[a]

EMF in Absolute Millivolts — Reference Junctions at 32 F

DEG F	0	1	2	3	4	5	6	7	8	9	10	DEG F
						THERMOELECTRIC VOLTAGE IN ABSOLUTE MILLIVOLTS						
-350	-8.137											-350
-340	-8.030	-8.041	-8.052	-8.063	-8.074	-8.085	-8.096	-8.106	-8.117	-8.127	-8.137	-340
-330	-7.915	-7.927	-7.938	-7.950	-7.962	-7.973	-7.985	-7.996	-8.008	-8.019	-8.030	-330
-320	-7.791	-7.803	-7.816	-7.829	-7.841	-7.854	-7.866	-7.878	-7.890	-7.903	-7.915	-320
-310	-7.659	-7.672	-7.686	-7.699	-7.712	-7.726	-7.739	-7.752	-7.765	-7.778	-7.791	-310
-300	-7.519	-7.533	-7.548	-7.562	-7.576	-7.590	-7.604	-7.618	-7.631	-7.645	-7.659	-300
-290	-7.372	-7.387	-7.402	-7.417	-7.432	-7.447	-7.461	-7.476	-7.490	-7.505	-7.519	-290
-280	-7.218	-7.234	-7.250	-7.265	-7.281	-7.296	-7.311	-7.327	-7.342	-7.357	-7.372	-280
-270	-7.057	-7.074	-7.090	-7.106	-7.122	-7.139	-7.155	-7.171	-7.187	-7.202	-7.218	-270
-260	-6.890	-6.907	-6.924	-6.941	-6.958	-6.974	-6.991	-7.008	-7.024	-7.041	-7.057	-260
-250	-6.716	-6.734	-6.751	-6.769	-6.786	-6.804	-6.821	-6.838	-6.856	-6.873	-6.890	-250
-240	-6.536	-6.554	-6.572	-6.591	-6.609	-6.627	-6.645	-6.663	-6.680	-6.698	-6.716	-240
-230	-6.350	-6.369	-6.388	-6.407	-6.425	-6.444	-6.462	-6.481	-6.499	-6.518	-6.536	-230
-220	-6.159	-6.178	-6.198	-6.217	-6.236	-6.255	-6.274	-6.293	-6.312	-6.331	-6.350	-220
-210	-5.962	-5.982	-6.002	-6.022	-6.041	-6.061	-6.081	-6.100	-6.120	-6.139	-6.159	-210
-200	-5.760	-5.780	-5.801	-5.821	-5.841	-5.861	-5.882	-5.902	-5.922	-5.942	-5.962	-200
-190	-5.553	-5.574	-5.594	-5.615	-5.636	-5.657	-5.678	-5.698	-5.719	-5.739	-5.760	-190
-180	-5.341	-5.362	-5.383	-5.405	-5.426	-5.447	-5.468	-5.490	-5.511	-5.532	-5.553	-180
-170	-5.124	-5.146	-5.168	-5.190	-5.211	-5.233	-5.255	-5.276	-5.298	-5.319	-5.341	-170
-160	-4.903	-4.925	-4.948	-4.970	-4.992	-5.014	-5.036	-5.058	-5.080	-5.102	-5.124	-160
-150	-4.678	-4.700	-4.723	-4.746	-4.768	-4.791	-4.813	-4.836	-4.858	-4.881	-4.903	-150
-140	-4.448	-4.471	-4.494	-4.517	-4.540	-4.563	-4.586	-4.609	-4.632	-4.655	-4.678	-140
-130	-4.215	-4.238	-4.262	-4.285	-4.309	-4.332	-4.355	-4.379	-4.402	-4.425	-4.448	-130
-120	-3.978	-4.001	-4.025	-4.049	-4.073	-4.097	-4.120	-4.144	-4.168	-4.191	-4.215	-120
-110	-3.737	-3.761	-3.785	-3.809	-3.833	-3.858	-3.882	-3.906	-3.930	-3.954	-3.978	-110
-100	-3.492	-3.517	-3.541	-3.566	-3.590	-3.615	-3.639	-3.664	-3.688	-3.712	-3.737	-100
-90	-3.245	-3.270	-3.294	-3.319	-3.344	-3.369	-3.394	-3.418	-3.443	-3.468	-3.492	-90
-80	-2.994	-3.019	-3.044	-3.069	-3.094	-3.120	-3.145	-3.170	-3.195	-3.220	-3.245	-80
-70	-2.740	-2.765	-2.791	-2.816	-2.842	-2.867	-2.892	-2.918	-2.943	-2.968	-2.994	-70
-60	-2.483	-2.509	-2.534	-2.560	-2.586	-2.612	-2.637	-2.663	-2.689	-2.714	-2.740	-60
-50	-2.223	-2.249	-2.275	-2.301	-2.327	-2.353	-2.379	-2.405	-2.431	-2.457	-2.483	-50
-40	-1.960	-1.987	-2.013	-2.039	-2.066	-2.092	-2.118	-2.144	-2.171	-2.197	-2.223	-40
-30	-1.695	-1.722	-1.748	-1.775	-1.802	-1.828	-1.855	-1.881	-1.908	-1.934	-1.960	-30
-20	-1.428	-1.455	-1.481	-1.508	-1.535	-1.562	-1.589	-1.615	-1.642	-1.669	-1.695	-20
-10	-1.158	-1.185	-1.212	-1.239	-1.266	-1.293	-1.320	-1.347	-1.374	-1.401	-1.428	-10
- 0	-0.885	-0.913	-0.940	-0.967	-0.995	-1.022	-1.049	-1.076	-1.103	-1.131	-1.158	- 0
0	-0.885	-0.858	-0.831	-0.803	-0.776	-0.748	-0.721	-0.694	-0.666	-0.639	-0.611	0
10	-0.611	-0.583	-0.556	-0.528	-0.501	-0.473	-0.445	-0.418	-0.390	-0.362	-0.334	10
20	-0.334	-0.307	-0.279	-0.251	-0.223	-0.195	-0.168	-0.140	-0.112	-0.084	-0.056	20
30	-0.056	-0.028	0.000	0.028	0.056	0.084	0.112	0.140	0.168	0.196	0.224	30
40	0.224	0.253	0.281	0.309	0.337	0.365	0.394	0.422	0.450	0.478	0.507	40
50	0.507	0.535	0.563	0.592	0.620	0.648	0.677	0.705	0.734	0.762	0.791	50
60	0.791	0.819	0.848	0.876	0.905	0.933	0.962	0.990	1.019	1.048	1.076	60
70	1.076	1.105	1.134	1.162	1.191	1.220	1.248	1.277	1.306	1.335	1.363	70
80	1.363	1.392	1.421	1.450	1.479	1.507	1.536	1.565	1.594	1.623	1.652	80
90	1.652	1.681	1.710	1.739	1.768	1.797	1.826	1.855	1.884	1.913	1.942	90
100	1.942	1.971	2.000	2.029	2.058	2.088	2.117	2.146	2.175	2.204	2.233	100
110	2.233	2.263	2.292	2.321	2.350	2.380	2.409	2.438	2.467	2.497	2.526	110
120	2.526	2.555	2.585	2.614	2.644	2.673	2.702	2.732	2.761	2.791	2.820	120
130	2.820	2.849	2.879	2.908	2.938	2.967	2.997	3.026	3.056	3.085	3.115	130
140	3.115	3.145	3.174	3.204	3.233	3.263	3.293	3.322	3.352	3.381	3.411	140
150	3.411	3.441	3.470	3.500	3.530	3.560	3.589	3.619	3.649	3.678	3.708	150
160	3.708	3.738	3.768	3.798	3.827	3.857	3.887	3.917	3.947	3.976	4.006	160
170	4.006	4.036	4.066	4.096	4.126	4.156	4.186	4.216	4.245	4.275	4.305	170
180	4.305	4.335	4.365	4.395	4.425	4.455	4.485	4.515	4.545	4.575	4.605	180
190	4.605	4.635	4.665	4.695	4.725	4.755	4.786	4.816	4.846	4.876	4.906	190
DEG F	0	1	2	3	4	5	6	7	8	9	10	DEG F

[a] Converted from degrees Celsius (IPTS 1968).

TABLE 10.6—*Type J thermocouples (continued).*

Temperature in Degrees Fahrenheit[a]

EMF in Absolute Millivolts — Reference Junctions at 32 F

DEG F	0	1	2	3	4	5	6	7	8	9	10	DEG F
	THERMOELECTRIC VOLTAGE IN ABSOLUTE MILLIVOLTS											
200	4.906	4.936	4.966	4.996	5.026	5.057	5.087	5.117	5.147	5.177	5.207	200
210	5.207	5.238	5.268	5.298	5.328	5.358	5.389	5.419	5.449	5.479	5.509	210
220	5.509	5.540	5.570	5.600	5.630	5.661	5.691	5.721	5.752	5.782	5.812	220
230	5.812	5.843	5.873	5.903	5.934	5.964	5.994	6.025	6.055	6.085	6.116	230
240	6.116	6.146	6.176	6.207	6.237	6.268	6.298	6.328	6.359	6.389	6.420	240
250	6.420	6.450	6.481	6.511	6.541	6.572	6.602	6.633	6.663	6.694	6.724	250
260	6.724	6.755	6.785	6.816	6.846	6.877	6.907	6.938	6.968	6.999	7.029	260
270	7.029	7.060	7.090	7.121	7.151	7.182	7.212	7.243	7.274	7.304	7.335	270
280	7.335	7.365	7.396	7.426	7.457	7.488	7.518	7.549	7.579	7.610	7.641	280
290	7.641	7.671	7.702	7.732	7.763	7.794	7.824	7.855	7.885	7.916	7.947	290
300	7.947	7.977	8.008	8.039	8.069	8.100	8.131	8.161	8.192	8.223	8.253	300
310	8.253	8.284	8.315	8.345	8.376	8.407	8.437	8.468	8.499	8.530	8.560	310
320	8.560	8.591	8.622	8.652	8.683	8.714	8.745	8.775	8.806	8.837	8.867	320
330	8.867	8.898	8.929	8.960	8.990	9.021	9.052	9.083	9.113	9.144	9.175	330
340	9.175	9.206	9.236	9.267	9.298	9.329	9.359	9.390	9.421	9.452	9.483	340
350	9.483	9.513	9.544	9.575	9.606	9.636	9.667	9.698	9.729	9.760	9.790	350
360	9.790	9.821	9.852	9.883	9.914	9.944	9.975	10.006	10.037	10.068	10.098	360
370	10.098	10.129	10.160	10.191	10.222	10.252	10.283	10.314	10.345	10.376	10.407	370
380	10.407	10.437	10.468	10.499	10.530	10.561	10.592	10.622	10.653	10.684	10.715	380
390	10.715	10.746	10.777	10.807	10.838	10.869	10.900	10.931	10.962	10.992	11.023	390
400	11.023	11.054	11.085	11.116	11.147	11.177	11.208	11.239	11.270	11.301	11.332	400
410	11.332	11.363	11.393	11.424	11.455	11.486	11.517	11.548	11.578	11.609	11.640	410
420	11.640	11.671	11.702	11.733	11.764	11.794	11.825	11.856	11.887	11.918	11.949	420
430	11.949	11.980	12.010	12.041	12.072	12.103	12.134	12.165	12.196	12.226	12.257	430
440	12.257	12.288	12.319	12.350	12.381	12.411	12.442	12.473	12.504	12.535	12.566	440
450	12.566	12.597	12.627	12.658	12.689	12.720	12.751	12.782	12.813	12.843	12.874	450
460	12.874	12.905	12.936	12.967	12.998	13.029	13.059	13.090	13.121	13.152	13.183	460
470	13.183	13.214	13.244	13.275	13.306	13.337	13.368	13.399	13.430	13.460	13.491	470
480	13.491	13.522	13.553	13.584	13.615	13.645	13.676	13.707	13.738	13.769	13.800	480
490	13.800	13.830	13.861	13.892	13.923	13.954	13.985	14.015	14.046	14.077	14.108	490
500	14.108	14.139	14.170	14.200	14.231	14.262	14.293	14.324	14.355	14.385	14.416	500
510	14.416	14.447	14.478	14.509	14.539	14.570	14.601	14.632	14.663	14.694	14.724	510
520	14.724	14.755	14.786	14.817	14.848	14.878	14.909	14.940	14.971	15.002	15.032	520
530	15.032	15.063	15.094	15.125	15.156	15.186	15.217	15.248	15.279	15.310	15.340	530
540	15.340	15.371	15.402	15.433	15.464	15.494	15.525	15.556	15.587	15.617	15.648	540
550	15.648	15.679	15.710	15.741	15.771	15.802	15.833	15.864	15.894	15.925	15.956	550
560	15.956	15.987	16.018	16.048	16.079	16.110	16.141	16.171	16.202	16.233	16.264	560
570	16.264	16.294	16.325	16.356	16.387	16.417	16.448	16.479	16.510	16.540	16.571	570
580	16.571	16.602	16.633	16.663	16.694	16.725	16.756	16.786	16.817	16.848	16.879	580
590	16.879	16.909	16.940	16.971	17.001	17.032	17.063	17.094	17.124	17.155	17.186	590
600	17.186	17.217	17.247	17.278	17.309	17.339	17.370	17.401	17.432	17.462	17.493	600
610	17.493	17.524	17.554	17.585	17.616	17.646	17.677	17.708	17.739	17.769	17.800	610
620	17.800	17.831	17.861	17.892	17.923	17.953	17.984	18.015	18.046	18.076	18.107	620
630	18.107	18.138	18.168	18.199	18.230	18.260	18.291	18.322	18.352	18.383	18.414	630
640	18.414	18.444	18.475	18.506	18.537	18.567	18.598	18.629	18.659	18.690	18.721	640
650	18.721	18.751	18.782	18.813	18.843	18.874	18.905	18.935	18.966	18.997	19.027	650
660	19.027	19.058	19.089	19.119	19.150	19.180	19.211	19.242	19.272	19.303	19.334	660
670	19.334	19.364	19.395	19.426	19.456	19.487	19.518	19.548	19.579	19.610	19.640	670
680	19.640	19.671	19.702	19.732	19.763	19.793	19.824	19.855	19.885	19.916	19.947	680
690	19.947	19.977	20.008	20.039	20.069	20.100	20.131	20.161	20.192	20.222	20.253	690
700	20.253	20.284	20.314	20.345	20.376	20.406	20.437	20.467	20.498	20.529	20.559	700
710	20.559	20.590	20.621	20.651	20.682	20.713	20.743	20.774	20.804	20.835	20.866	710
720	20.866	20.896	20.927	20.958	20.988	21.019	21.049	21.080	21.111	21.141	21.172	720
730	21.172	21.203	21.233	21.264	21.295	21.325	21.356	21.386	21.417	21.448	21.478	730
740	21.478	21.509	21.540	21.570	21.601	21.631	21.662	21.693	21.723	21.754	21.785	740
750	21.785	21.815	21.846	21.877	21.907	21.938	21.968	21.999	22.030	22.060	22.091	750
760	22.091	22.122	22.152	22.183	22.214	22.244	22.275	22.305	22.336	22.367	22.397	760
770	22.397	22.428	22.459	22.489	22.520	22.551	22.581	22.612	22.643	22.673	22.704	770
780	22.704	22.735	22.765	22.796	22.826	22.857	22.888	22.918	22.949	22.980	23.010	780
790	23.010	23.041	23.072	23.102	23.133	23.164	23.194	23.225	23.256	23.286	23.317	790
DEG F	0	1	2	3	4	5	6	7	8	9	10	DEG F

[a] Converted from degrees Celsius (IPTS 1968).

TABLE 10.6—*Type J thermocouples (continued).*

Temperature in Degrees Fahrenheit[a]

EMF in Absolute Millivolts — Reference Junctions at 32 F

DEG F	0	1	2	3	4	5	6	7	8	9	10	DEG F
						THERMOELECTRIC VOLTAGE IN ABSOLUTE MILLIVOLTS						
800	23.317	23.348	23.378	23.409	23.440	23.471	23.501	23.532	23.563	23.593	23.624	800
810	23.624	23.655	23.685	23.716	23.747	23.777	23.808	23.839	23.870	23.900	23.931	810
820	23.931	23.962	23.992	24.023	24.054	24.085	24.115	24.146	24.177	24.207	24.238	820
830	24.238	24.269	24.300	24.330	24.361	24.392	24.423	24.453	24.484	24.515	24.546	830
840	24.546	24.576	24.607	24.638	24.669	24.699	24.730	24.761	24.792	24.822	24.853	840
850	24.853	24.884	24.915	24.946	24.976	25.007	25.038	25.069	25.099	25.130	25.161	850
860	25.161	25.192	25.223	25.254	25.284	25.315	25.346	25.377	25.408	25.438	25.469	860
870	25.469	25.500	25.531	25.562	25.593	25.623	25.654	25.685	25.716	25.747	25.778	870
880	25.778	25.809	25.840	25.870	25.901	25.932	25.963	25.994	26.025	26.056	26.087	880
890	26.087	26.118	26.148	26.179	26.210	26.241	26.272	26.303	26.334	26.365	26.396	890
900	26.396	26.427	26.458	26.489	26.520	26.551	26.582	26.613	26.644	26.675	26.705	900
910	26.705	26.736	26.767	26.798	26.829	26.860	26.891	26.922	26.954	26.985	27.016	910
920	27.016	27.047	27.078	27.109	27.140	27.171	27.202	27.233	27.264	27.295	27.326	920
930	27.326	27.357	27.388	27.419	27.450	27.482	27.513	27.544	27.575	27.606	27.637	930
940	27.637	27.668	27.699	27.731	27.762	27.793	27.824	27.855	27.886	27.917	27.949	940
950	27.949	27.980	28.011	28.042	28.073	28.105	28.136	28.167	28.198	28.230	28.261	950
960	28.261	28.292	28.323	28.355	28.386	28.417	28.448	28.480	28.511	28.542	28.573	960
970	28.573	28.605	28.636	28.667	28.699	28.730	28.761	28.793	28.824	28.855	28.887	970
980	28.887	28.918	28.950	28.981	29.012	29.044	29.075	29.107	29.138	29.169	29.201	980
990	29.201	29.232	29.264	29.295	29.327	29.358	29.390	29.421	29.452	29.484	29.515	990
1,000	29.515	29.547	29.578	29.610	29.642	29.673	29.705	29.736	29.768	29.799	29.831	1,000
1,010	29.831	29.862	29.894	29.926	29.957	29.989	30.020	30.052	30.084	30.115	30.147	1,010
1,020	30.147	30.179	30.210	30.242	30.274	30.305	30.337	30.369	30.400	30.432	30.464	1,020
1,030	30.464	30.496	30.527	30.559	30.591	30.623	30.654	30.686	30.718	30.750	30.782	1,030
1,040	30.782	30.813	30.845	30.877	30.909	30.941	30.973	31.005	31.036	31.068	31.100	1,040
1,050	31.100	31.132	31.164	31.196	31.228	31.260	31.292	31.324	31.356	31.388	31.420	1,050
1,060	31.420	31.452	31.484	31.516	31.548	31.580	31.612	31.644	31.676	31.708	31.740	1,060
1,070	31.740	31.772	31.804	31.836	31.868	31.901	31.933	31.965	31.997	32.029	32.061	1,070
1,080	32.061	32.094	32.126	32.158	32.190	32.222	32.255	32.287	32.319	32.351	32.384	1,080
1,090	32.384	32.416	32.448	32.480	32.513	32.545	32.577	32.610	32.642	32.674	32.707	1,090
1,100	32.707	32.739	32.772	32.804	32.836	32.869	32.901	32.934	32.966	32.999	33.031	1,100
1,110	33.031	33.064	33.096	33.129	33.161	33.194	33.226	33.259	33.291	33.324	33.356	1,110
1,120	33.356	33.389	33.422	33.454	33.487	33.519	33.552	33.585	33.617	33.650	33.683	1,120
1,130	33.683	33.715	33.748	33.781	33.814	33.846	33.879	33.912	33.945	33.977	34.010	1,130
1,140	34.010	34.043	34.076	34.109	34.141	34.174	34.207	34.240	34.273	34.306	34.339	1,140
1,150	34.339	34.372	34.405	34.437	34.470	34.503	34.536	34.569	34.602	34.635	34.668	1,150
1,160	34.668	34.701	34.734	34.767	34.801	34.834	34.867	34.900	34.933	34.966	34.999	1,160
1,170	34.999	35.032	35.065	35.099	35.132	35.165	35.198	35.231	35.265	35.298	35.331	1,170
1,180	35.331	35.364	35.398	35.431	35.464	35.498	35.531	35.564	35.598	35.631	35.664	1,180
1,190	35.664	35.698	35.731	35.764	35.798	35.831	35.865	35.898	35.932	35.965	35.999	1,190
1,200	35.999	36.032	36.066	36.099	36.133	36.166	36.200	36.233	36.267	36.301	36.334	1,200
1,210	36.334	36.368	36.401	36.435	36.469	36.502	36.536	36.570	36.603	36.637	36.671	1,210
1,220	36.671	36.705	36.738	36.772	36.806	36.840	36.873	36.907	36.941	36.975	37.009	1,220
1,230	37.009	37.043	37.076	37.110	37.144	37.178	37.212	37.246	37.280	37.314	37.348	1,230
1,240	37.348	37.382	37.416	37.450	37.484	37.518	37.552	37.586	37.620	37.654	37.688	1,240
1,250	37.688	37.722	37.756	37.790	37.825	37.859	37.893	37.927	37.961	37.995	38.030	1,250
1,260	38.030	38.064	38.098	38.132	38.167	38.201	38.235	38.269	38.304	38.338	38.372	1,260
1,270	38.372	38.407	38.441	38.475	38.510	38.544	38.578	38.613	38.647	38.682	38.716	1,270
1,280	38.716	38.751	38.785	38.819	38.854	38.888	38.923	38.957	38.992	39.027	39.061	1,280
1,290	39.061	39.096	39.130	39.165	39.199	39.234	39.269	39.303	39.338	39.373	39.407	1,290
1,300	39.407	39.442	39.477	39.511	39.546	39.581	39.615	39.650	39.685	39.720	39.754	1,300
1,310	39.754	39.789	39.824	39.859	39.894	39.928	39.963	39.998	40.033	40.068	40.103	1,310
1,320	40.103	40.138	40.172	40.207	40.242	40.277	40.312	40.347	40.382	40.417	40.452	1,320
1,330	40.452	40.487	40.522	40.557	40.592	40.627	40.662	40.697	40.732	40.767	40.802	1,330
1,340	40.802	40.837	40.872	40.908	40.943	40.978	41.013	41.048	41.083	41.118	41.154	1,340
1,350	41.154	41.189	41.224	41.259	41.294	41.329	41.365	41.400	41.435	41.470	41.506	1,350
1,360	41.506	41.541	41.576	41.611	41.647	41.682	41.717	41.753	41.788	41.823	41.859	1,360
1,370	41.859	41.894	41.929	41.965	42.000	42.035	42.071	42.106	42.142	42.177	42.212	1,370
1,380	42.212	42.248	42.283	42.319	42.354	42.390	42.425	42.460	42.496	42.531	42.567	1,380
1,390	42.567	42.602	42.638	42.673	42.709	42.744	42.780	42.815	42.851	42.886	42.922	1,390
DEG F	0	1	2	3	4	5	6	7	8	9	10	DEG F

[a] Converted from degrees Celsius (IPTS 1968).

TABLE 10.6—*Type J thermocouples (continued).*

Temperature in Degrees Fahrenheit[a]

EMF in Absolute Millivolts

Reference Junctions at 32 F

DEG F	0	1	2	3	4	5	6	7	8	9	10	DEG F
					THERMOELECTRIC VOLTAGE IN ABSOLUTE MILLIVOLTS							
1,400	42.922	42.957	42.993	43.029	43.064	43.100	43.135	43.171	43.207	43.242	43.278	1,400
1,410	43.278	43.313	43.349	43.385	43.420	43.456	43.492	43.527	43.563	43.599	43.635	1,410
1,420	43.635	43.670	43.706	43.742	43.777	43.813	43.849	43.885	43.921	43.956	43.992	1,420
1,430	43.992	44.028	44.064	44.099	44.135	44.171	44.207	44.243	44.279	44.314	44.350	1,430
1,440	44.350	44.386	44.422	44.458	44.494	44.529	44.565	44.601	44.637	44.673	44.709	1,440
1,450	44.709	44.745	44.780	44.816	44.852	44.888	44.924	44.960	44.996	45.032	45.067	1,450
1,460	45.067	45.103	45.139	45.175	45.211	45.247	45.283	45.319	45.355	45.391	45.426	1,460
1,470	45.426	45.462	45.498	45.534	45.570	45.606	45.642	45.678	45.714	45.749	45.785	1,470
1,480	45.785	45.821	45.857	45.893	45.929	45.965	46.001	46.037	46.072	46.108	46.144	1,480
1,490	46.144	46.180	46.216	46.252	46.288	46.324	46.359	46.395	46.431	46.467	46.503	1,490
1,500	46.503	46.539	46.575	46.610	46.646	46.682	46.718	46.754	46.790	46.825	46.861	1,500
1,510	46.861	46.897	46.933	46.969	47.005	47.040	47.076	47.112	47.148	47.183	47.219	1,510
1,520	47.219	47.255	47.291	47.327	47.362	47.398	47.434	47.470	47.505	47.541	47.577	1,520
1,530	47.577	47.612	47.648	47.684	47.720	47.755	47.791	47.827	47.862	47.898	47.934	1,530
1,540	47.934	47.969	48.005	48.041	48.076	48.112	48.147	48.183	48.219	48.254	48.290	1,540
1,550	48.290	48.325	48.361	48.397	48.432	48.468	48.503	48.539	48.574	48.610	48.645	1,550
1,560	48.645	48.681	48.716	48.752	48.787	48.823	48.858	48.894	48.929	48.965	49.000	1,560
1,570	49.000	49.036	49.071	49.107	49.142	49.177	49.213	49.248	49.283	49.319	49.354	1,570
1,580	49.354	49.390	49.425	49.460	49.496	49.531	49.566	49.601	49.637	49.672	49.707	1,580
1,590	49.707	49.742	49.778	49.813	49.848	49.883	49.919	49.954	49.989	50.024	50.059	1,590
1,600	50.059	50.095	50.130	50.165	50.200	50.235	50.270	50.305	50.340	50.376	50.411	1,600
1,610	50.411	50.446	50.481	50.516	50.551	50.586	50.621	50.656	50.691	50.726	50.761	1,610
1,620	50.761	50.796	50.831	50.866	50.901	50.936	50.970	51.005	51.040	51.075	51.110	1,620
1,630	51.110	51.145	51.180	51.215	51.249	51.284	51.319	51.354	51.389	51.423	51.458	1,630
1,640	51.458	51.493	51.528	51.562	51.597	51.632	51.667	51.701	51.736	51.771	51.805	1,640
1,650	51.805	51.840	51.875	51.909	51.944	51.978	52.013	52.048	52.082	52.117	52.151	1,650
1,660	52.151	52.186	52.220	52.255	52.289	52.324	52.358	52.393	52.427	52.462	52.496	1,660
1,670	52.496	52.531	52.565	52.600	52.634	52.668	52.703	52.737	52.772	52.806	52.840	1,670
1,680	52.840	52.875	52.909	52.943	52.977	53.012	53.046	53.080	53.115	53.149	53.183	1,680
1,690	53.183	53.217	53.251	53.286	53.320	53.354	53.388	53.422	53.456	53.491	53.525	1,690
1,700	53.525	53.559	53.593	53.627	53.661	53.695	53.729	53.763	53.797	53.831	53.865	1,700
1,710	53.865	53.899	53.933	53.967	54.001	54.035	54.069	54.103	54.137	54.171	54.205	1,710
1,720	54.205	54.239	54.273	54.307	54.341	54.374	54.408	54.442	54.476	54.510	54.544	1,720
1,730	54.544	54.577	54.611	54.645	54.679	54.712	54.746	54.780	54.814	54.847	54.881	1,730
1,740	54.881	54.915	54.948	54.982	55.016	55.049	55.083	55.117	55.150	55.184	55.218	1,740
1,750	55.218	55.251	55.285	55.318	55.352	55.385	55.419	55.453	55.486	55.520	55.553	1,750
1,760	55.553	55.587	55.620	55.654	55.687	55.720	55.754	55.787	55.821	55.854	55.888	1,760
1,770	55.888	55.921	55.954	55.988	56.021	56.055	56.088	56.121	56.155	56.188	56.221	1,770
1,780	56.221	56.255	56.288	56.321	56.354	56.388	56.421	56.454	56.487	56.521	56.554	1,780
1,790	56.554	56.587	56.620	56.654	56.687	56.720	56.753	56.786	56.819	56.853	56.886	1,790
1,800	56.886	56.919	56.952	56.985	57.018	57.051	57.084	57.118	57.151	57.184	57.217	1,800
1,810	57.217	57.250	57.283	57.316	57.349	57.382	57.415	57.448	57.481	57.514	57.547	1,810
1,820	57.547	57.580	57.613	57.646	57.679	57.712	57.745	57.778	57.810	57.843	57.876	1,820
1,830	57.876	57.909	57.942	57.975	58.008	58.041	58.074	58.106	58.139	58.172	58.205	1,830
1,840	58.205	58.238	58.271	58.303	58.336	58.369	58.402	58.435	58.467	58.500	58.533	1,840
1,850	58.533	58.566	58.598	58.631	58.664	58.697	58.729	58.762	58.795	58.827	58.860	1,850
1,860	58.860	58.893	58.926	58.958	58.991	59.024	59.056	59.089	59.121	59.154	59.187	1,860
1,870	59.187	59.219	59.252	59.285	59.317	59.350	59.382	59.415	59.448	59.480	59.513	1,870
1,880	59.513	59.545	59.578	59.610	59.643	59.676	59.708	59.741	59.773	59.806	59.838	1,880
1,890	59.838	59.871	59.903	59.936	59.968	60.001	60.033	60.066	60.098	60.131	60.163	1,890
1,900	60.163	60.196	60.228	60.261	60.293	60.326	60.358	60.390	60.423	60.455	60.488	1,900
1,910	60.488	60.520	60.553	60.585	60.617	60.650	60.682	60.715	60.747	60.779	60.812	1,910
1,920	60.812	60.844	60.876	60.909	60.941	50.974	61.006	61.038	61.071	61.103	61.135	1,920
1,930	61.135	61.168	61.200	61.232	61.265	61.297	61.329	61.362	61.394	61.426	61.459	1,930
1,940	61.459	61.491	61.523	61.555	61.588	61.620	61.652	61.685	61.717	61.749	61.781	1,940
1,950	61.781	61.814	61.846	61.878	61.910	61.943	61.975	62.007	62.039	62.072	62.104	1,950
1,960	62.104	62.136	62.168	62.201	62.233	62.265	62.297	62.330	62.362	62.394	62.426	1,960
1,970	62.426	62.458	62.491	62.523	62.555	62.587	62.619	62.652	62.684	62.716	62.748	1,970
1,980	62.748	62.780	62.813	62.845	62.877	62.909	62.941	62.974	63.006	63.038	63.070	1,980
1,990	63.070	63.102	63.134	63.167	63.199	63.231	63.263	63.295	63.327	63.359	63.392	1,990
DEG F	0	1	2	3	4	5	6	7	8	9	10	DEG F

[a] Converted from degrees Celsius (IPTS 1968).

TABLE 10.6—*Type J thermocouples (continued).*

Temperature in Degrees Fahrenheit[a]

EMF in Absolute Millivolts — Reference Junctions at 32 F

DEG F	0	1	2	3	4	5	6	7	8	9	10	DEG F
					THERMOELECTRIC VOLTAGE IN ABSOLUTE MILLIVOLTS							
2,000	63.392	63.424	63.456	63.488	63.520	63.552	63.584	63.617	63.649	63.681	63.713	2,000
2,010	63.713	63.745	63.777	63.809	63.842	63.874	63.906	63.938	63.970	64.002	64.034	2,010
2,020	64.034	64.066	64.098	64.131	64.163	64.195	64.227	64.259	64.291	64.323	64.355	2,020
2,030	64.355	64.387	64.420	64.452	64.484	64.516	64.548	64.580	64.612	64.644	64.676	2,030
2,040	64.676	64.708	64.740	64.773	64.805	64.837	64.869	64.901	64.933	64.965	64.997	2,040
2,050	64.997	65.029	65.061	65.093	65.125	65.158	65.190	65.222	65.254	65.286	65.318	2,050
2,060	65.318	65.350	65.382	65.414	65.446	65.478	65.510	65.542	65.574	65.606	65.638	2,060
2,070	65.638	65.671	65.703	65.735	65.767	65.799	65.831	65.863	65.895	65.927	65.959	2,070
2,080	65.959	65.991	66.023	66.055	66.087	66.119	66.151	66.183	66.215	66.247	66.279	2,080
2,090	66.279	66.311	66.343	66.375	66.407	66.439	66.472	66.504	66.536	66.568	66.600	2,090
2,100	66.600	66.632	66.664	66.696	66.728	66.760	66.792	66.824	66.856	66.888	66.920	2,100
2,110	66.920	66.952	66.984	67.016	67.048	67.080	67.112	67.144	67.176	67.208	67.240	2,110
2,120	67.240	67.272	67.304	67.336	67.368	67.400	67.432	67.464	67.495	67.527	67.559	2,120
2,130	67.559	67.591	67.623	67.655	67.687	67.719	67.751	67.783	67.815	67.847	67.879	2,130
2,140	67.879	67.911	67.943	67.975	68.007	68.039	68.071	68.103	68.134	68.166	68.198	2,140
2,150	68.198	68.230	68.262	68.294	68.326	68.358	68.390	68.422	68.454	68.486	68.517	2,150
2,160	68.517	68.549	68.581	68.613	68.645	68.677	68.709	68.741	68.772	68.804	68.836	2,160
2,170	68.836	68.868	68.900	68.932	68.964	68.995	69.027	69.059	69.091	69.123	69.155	2,170
2,180	69.155	69.186	69.218	69.250	69.282	69.314	69.345	69.377	69.409	69.441	69.472	2,180
2,190	69.472	69.504	69.536									2,190
DEG F	0	1	2	3	4	5	6	7	8	9	10	DEG F

[a] Converted from degrees Celsius (IPTS 1968).

TABLE 10.7—*Type J thermocouples (deg C-millivolts).*

NOTE—The maximum recommended temperature limit for Type J thermocouples is 1400 F (760 C) as specified in Table 2. Extension of the Type J tables beyond 1400 F gives temperature - electromotive force data to 2192 F (1200 C). This extension is a mathematical extrapolation based on limited calibration data and caution should be exercised in its use. The basis for the extended curve is discussed in *NBS Monograph 125*.

It should be noted that limits of error for Type J thermocouples (Table 1) do not apply above 1400 F (760 C).

Temperature in Degrees Celsius (IPTS 1968)

EMF in Absolute Millivolts — Reference Junctions at 0 C

THERMOELECTRIC VOLTAGE IN ABSOLUTE MILLIVOLTS

DEG C	0	1	2	3	4	5	6	7	8	9	10	DEG C
-210	-8.096											-210
-200	-7.890	-7.912	-7.934	-7.955	-7.976	-7.996	-8.017	-8.037	-8.057	-8.076	-8.096	-200
-190	-7.659	-7.683	-7.707	-7.731	-7.755	-7.778	-7.801	-7.824	-7.846	-7.868	-7.890	-190
-180	-7.402	-7.429	-7.455	-7.482	-7.508	-7.533	-7.559	-7.584	-7.609	-7.634	-7.659	-180
-170	-7.122	-7.151	-7.180	-7.209	-7.237	-7.265	-7.293	-7.321	-7.348	-7.375	-7.402	-170
-160	-6.821	-6.852	-6.883	-6.914	-6.944	-6.974	-7.004	-7.034	-7.064	-7.093	-7.122	-160
-150	-6.499	-6.532	-6.565	-6.598	-6.630	-6.663	-6.695	-6.727	-6.758	-6.790	-6.821	-150
-140	-6.159	-6.194	-6.228	-6.263	-6.297	-6.331	-6.365	-6.399	-6.433	-6.466	-6.499	-140
-130	-5.801	-5.837	-5.874	-5.910	-5.946	-5.982	-6.018	-6.053	-6.089	-6.124	-6.159	-130
-120	-5.426	-5.464	-5.502	-5.540	-5.578	-5.615	-5.653	-5.690	-5.727	-5.764	-5.801	-120
-110	-5.036	-5.076	-5.115	-5.155	-5.194	-5.233	-5.272	-5.311	-5.349	-5.388	-5.426	-110
-100	-4.632	-4.673	-4.714	-4.755	-4.795	-4.836	-4.876	-4.916	-4.956	-4.996	-5.036	-100
-90	-4.215	-4.257	-4.299	-4.341	-4.383	-4.425	-4.467	-4.508	-4.550	-4.591	-4.632	-90
-80	-3.785	-3.829	-3.872	-3.915	-3.958	-4.001	-4.044	-4.087	-4.130	-4.172	-4.215	-80
-70	-3.344	-3.389	-3.433	-3.478	-3.522	-3.566	-3.610	-3.654	-3.698	-3.742	-3.785	-70
-60	-2.892	-2.938	-2.984	-3.029	-3.074	-3.120	-3.165	-3.210	-3.255	-3.299	-3.344	-60
-50	-2.431	-2.478	-2.524	-2.570	-2.617	-2.663	-2.709	-2.755	-2.801	-2.847	-2.892	-50
-40	-1.960	-2.008	-2.055	-2.102	-2.150	-2.197	-2.244	-2.291	-2.338	-2.384	-2.431	-40
-30	-1.481	-1.530	-1.578	-1.626	-1.674	-1.722	-1.770	-1.818	-1.865	-1.913	-1.960	-30
-20	-0.995	-1.044	-1.093	-1.141	-1.190	-1.239	-1.288	-1.336	-1.385	-1.433	-1.481	-20
-10	-0.501	-0.550	-0.600	-0.650	-0.699	-0.748	-0.798	-0.847	-0.896	-0.945	-0.995	-10
- 0	0.000	-0.050	-0.101	-0.151	-0.201	-0.251	-0.301	-0.351	-0.401	-0.451	-0.501	- 0
0	0.000	0.050	0.101	0.151	0.202	0.253	0.303	0.354	0.405	0.456	0.507	0
10	0.507	0.558	0.609	0.660	0.711	0.762	0.813	0.865	0.916	0.967	1.019	10
20	1.019	1.070	1.122	1.174	1.225	1.277	1.329	1.381	1.432	1.484	1.536	20
30	1.536	1.588	1.640	1.693	1.745	1.797	1.849	1.901	1.954	2.006	2.058	30
40	2.058	2.111	2.163	2.216	2.268	2.321	2.374	2.426	2.479	2.532	2.585	40
50	2.585	2.638	2.691	2.743	2.796	2.849	2.902	2.956	3.009	3.062	3.115	50
60	3.115	3.168	3.221	3.275	3.328	3.381	3.435	3.488	3.542	3.595	3.649	60
70	3.649	3.702	3.756	3.809	3.863	3.917	3.971	4.024	4.078	4.132	4.186	70
80	4.186	4.239	4.293	4.347	4.401	4.455	4.509	4.563	4.617	4.671	4.725	80
90	4.725	4.780	4.834	4.888	4.942	4.996	5.050	5.105	5.159	5.213	5.268	90
100	5.268	5.322	5.376	5.431	5.485	5.540	5.594	5.649	5.703	5.758	5.812	100
110	5.812	5.867	5.921	5.976	6.031	6.085	6.140	6.195	6.249	6.304	6.359	110
120	6.359	6.414	6.468	6.523	6.578	6.633	6.688	6.742	6.797	6.852	6.907	120
130	6.907	6.962	7.017	7.072	7.127	7.182	7.237	7.292	7.347	7.402	7.457	130
140	7.457	7.512	7.567	7.622	7.677	7.732	7.787	7.843	7.898	7.953	8.008	140
150	8.008	8.063	8.118	8.174	8.229	8.284	8.339	8.394	8.450	8.505	8.560	150
160	8.560	8.616	8.671	8.726	8.781	8.837	8.892	8.947	9.003	9.058	9.113	160
170	9.113	9.169	9.224	9.279	9.335	9.390	9.446	9.501	9.556	9.612	9.667	170
180	9.667	9.723	9.778	9.834	9.889	9.944	10.000	10.055	10.111	10.166	10.222	180
190	10.222	10.277	10.333	10.388	10.444	10.499	10.555	10.610	10.666	10.721	10.777	190
200	10.777	10.832	10.888	10.943	10.999	11.054	11.110	11.165	11.221	11.276	11.332	200
210	11.332	11.387	11.443	11.498	11.554	11.609	11.665	11.720	11.776	11.831	11.887	210
220	11.887	11.943	11.998	12.054	12.109	12.165	12.220	12.276	12.331	12.387	12.442	220
230	12.442	12.498	12.553	12.609	12.664	12.720	12.776	12.831	12.887	12.942	12.998	230
240	12.998	13.053	13.109	13.164	13.220	13.275	13.331	13.386	13.442	13.497	13.553	240
250	13.553	13.608	13.664	13.719	13.775	13.830	13.886	13.941	13.997	14.052	14.108	250
260	14.108	14.163	14.219	14.274	14.330	14.385	14.441	14.496	14.552	14.607	14.663	260
270	14.663	14.718	14.774	14.829	14.885	14.940	14.995	15.051	15.106	15.162	15.217	270
280	15.217	15.273	15.328	15.383	15.439	15.494	15.550	15.605	15.661	15.716	15.771	280
290	15.771	15.827	15.882	15.938	15.993	16.048	16.104	16.159	16.214	16.270	16.325	290
300	16.325	16.380	16.436	16.491	16.547	16.602	16.657	16.713	16.768	16.823	16.879	300
310	16.879	16.934	16.989	17.044	17.100	17.155	17.210	17.266	17.321	17.376	17.432	310
320	17.432	17.487	17.542	17.597	17.653	17.708	17.763	17.818	17.874	17.929	17.984	320
330	17.984	18.039	18.095	18.150	18.205	18.260	18.316	18.371	18.426	18.481	18.537	330
340	18.537	18.592	18.647	18.702	18.757	18.813	18.868	18.923	18.978	19.033	19.089	340
DEG C	0	1	2	3	4	5	6	7	8	9	10	DEG C

TABLE 10.7—*Type J thermocouples (continued).*

Temperature in Degrees Celsius (IPTS 1968)

EMF in Absolute Millivolts | Reference Junctions at 0 C

DEG C	0	1	2	3	4	5	6	7	8	9	10	DEG C
					THERMOELECTRIC VOLTAGE IN ABSOLUTE MILLIVOLTS							
350	19.089	19.144	19.199	19.254	19.309	19.364	19.420	19.475	19.530	19.585	19.640	350
360	19.640	19.695	19.751	19.806	19.861	19.916	19.971	20.026	20.081	20.137	20.192	360
370	20.192	20.247	20.302	20.357	20.412	20.467	20.523	20.578	20.633	20.688	20.743	370
380	20.743	20.798	20.853	20.909	20.964	21.019	21.074	21.129	21.184	21.239	21.295	380
390	21.295	21.350	21.405	21.460	21.515	21.570	21.625	21.680	21.736	21.791	21.846	390
400	21.846	21.901	21.956	22.011	22.066	22.122	22.177	22.232	22.287	22.342	22.397	400
410	22.397	22.453	22.508	22.563	22.618	22.673	22.728	22.784	22.839	22.894	22.949	410
420	22.949	23.004	23.060	23.115	23.170	23.225	23.280	23.336	23.391	23.446	23.501	420
430	23.501	23.556	23.612	23.667	23.722	23.777	23.833	23.888	23.943	23.999	24.054	430
440	24.054	24.109	24.164	24.220	24.275	24.330	24.386	24.441	24.496	24.552	24.607	440
450	24.607	24.662	24.718	24.773	24.829	24.884	24.939	24.995	25.050	25.106	25.161	450
460	25.161	25.217	25.272	25.327	25.383	25.438	25.494	25.549	25.605	25.661	25.716	460
470	25.716	25.772	25.827	25.883	25.938	25.994	26.050	26.105	26.161	26.216	26.272	470
480	26.272	26.328	26.383	26.439	26.495	26.551	26.606	26.662	26.718	26.774	26.829	480
490	26.829	26.885	26.941	26.997	27.053	27.109	27.165	27.220	27.276	27.332	27.388	490
500	27.388	27.444	27.500	27.556	27.612	27.668	27.724	27.780	27.836	27.893	27.949	500
510	27.949	28.005	28.061	28.117	28.173	28.230	28.286	28.342	28.398	28.455	28.511	510
520	28.511	28.567	28.624	28.680	28.736	28.793	28.849	28.906	28.962	29.019	29.075	520
530	29.075	29.132	29.188	29.245	29.301	29.358	29.415	29.471	29.528	29.585	29.642	530
540	29.642	29.698	29.755	29.812	29.869	29.926	29.983	30.039	30.096	30.153	30.210	540
550	30.210	30.267	30.324	30.381	30.439	30.496	30.553	30.610	30.667	30.724	30.782	550
560	30.782	30.839	30.896	30.954	31.011	31.068	31.126	31.183	31.241	31.298	31.356	560
570	31.356	31.413	31.471	31.528	31.586	31.644	31.702	31.759	31.817	31.875	31.933	570
580	31.933	31.991	32.048	32.106	32.164	32.222	32.280	32.338	32.396	32.455	32.513	580
590	32.513	32.571	32.629	32.687	32.746	32.804	32.862	32.921	32.979	33.038	33.096	590
600	33.096	33.155	33.213	33.272	33.330	33.389	33.448	33.506	33.565	33.624	33.683	600
610	33.683	33.742	33.800	33.859	33.918	33.977	34.036	34.095	34.155	34.214	34.273	610
620	34.273	34.332	34.391	34.451	34.510	34.569	34.629	34.688	34.748	34.807	34.867	620
630	34.867	34.926	34.986	35.046	35.105	35.165	35.225	35.285	35.344	35.404	35.464	630
640	35.464	35.524	35.584	35.644	35.704	35.764	35.825	35.885	35.945	36.005	36.066	640
650	36.066	36.126	36.186	36.247	36.307	36.368	36.428	36.489	36.549	36.610	36.671	650
660	36.671	36.732	36.792	36.853	36.914	36.975	37.036	37.097	37.158	37.219	37.280	660
670	37.280	37.341	37.402	37.463	37.525	37.586	37.647	37.709	37.770	37.831	37.893	670
680	37.893	37.954	38.016	38.078	38.139	38.201	38.262	38.324	38.386	38.448	38.510	680
690	38.510	38.572	38.633	38.695	38.757	38.819	38.882	38.944	39.006	39.068	39.130	690
700	39.130	39.192	39.255	39.317	39.379	39.442	39.504	39.567	39.629	39.692	39.754	700
710	39.754	39.817	39.880	39.942	40.005	40.068	40.131	40.193	40.256	40.319	40.382	710
720	40.382	40.445	40.508	40.571	40.634	40.697	40.760	40.823	40.886	40.950	41.013	720
730	41.013	41.076	41.139	41.203	41.266	41.329	41.393	41.456	41.520	41.583	41.647	730
740	41.647	41.710	41.774	41.837	41.901	41.965	42.028	42.092	42.156	42.219	42.283	740
750	42.283	42.347	42.411	42.475	42.538	42.602	42.666	42.730	42.794	42.858	42.922	750
760	42.922											760
760	42.922	42.986	43.050	43.114	43.178	43.242	43.306	43.370	43.435	43.499	43.563	760
770	43.563	43.627	43.692	43.756	43.820	43.885	43.949	44.014	44.078	44.142	44.207	770
780	44.207	44.271	44.336	44.400	44.465	44.529	44.594	44.658	44.723	44.788	44.852	780
790	44.852	44.917	44.981	45.046	45.111	45.175	45.240	45.304	45.369	45.434	45.498	790
800	45.498	45.563	45.627	45.692	45.757	45.821	45.886	45.950	46.015	46.080	46.144	800
810	46.144	46.209	46.273	46.338	46.403	46.467	46.532	46.596	46.661	46.725	46.790	810
820	46.790	46.854	46.919	46.983	47.047	47.112	47.176	47.241	47.305	47.369	47.434	820
830	47.434	47.498	47.562	47.627	47.691	47.755	47.819	47.884	47.948	48.012	48.076	830
840	48.076	48.140	48.204	48.269	48.333	48.397	48.461	48.525	48.589	48.653	48.716	840
850	48.716	48.780	48.844	48.908	48.972	49.036	49.099	49.163	49.227	49.291	49.354	850
860	49.354	49.418	49.481	49.545	49.608	49.672	49.735	49.799	49.862	49.926	49.989	860
870	49.989	50.052	50.116	50.179	50.242	50.305	50.369	50.432	50.495	50.558	50.621	870
880	50.621	50.684	50.747	50.810	50.873	50.936	50.998	51.061	51.124	51.187	51.249	880
890	51.249	51.312	51.375	51.437	51.500	51.562	51.625	51.687	51.750	51.812	51.875	890
900	51.875	51.937	51.999	52.061	52.124	52.186	52.248	52.310	52.372	52.434	52.496	900
910	52.496	52.558	52.620	52.682	52.744	52.806	52.868	52.929	52.991	53.053	53.115	910
920	53.115	53.176	53.238	53.299	53.361	53.422	53.484	53.545	53.607	53.668	53.729	920
930	53.729	53.791	53.852	53.913	53.974	54.035	54.096	54.157	54.219	54.280	54.341	930
940	54.341	54.401	54.462	54.523	54.584	54.645	54.706	54.766	54.827	54.888	54.948	940
950	54.948	55.009	55.070	55.130	55.191	55.251	55.312	55.372	55.432	55.493	55.553	950
960	55.553	55.613	55.674	55.734	55.794	55.854	55.914	55.974	56.035	56.095	56.155	960
970	56.155	56.215	56.275	56.334	56.394	56.454	56.514	56.574	56.634	56.693	56.753	970
980	56.753	56.813	56.873	56.932	56.992	57.051	57.111	57.170	57.230	57.289	57.349	980
990	57.349	57.408	57.468	57.527	57.586	57.646	57.705	57.764	57.824	57.883	57.942	990
DEG C	0	1	2	3	4	5	6	7	8	9	10	DEG C

TABLE 10.7—*Type J thermocouples (continued).*

Temperature in Degrees Celsius (IPTS 1968)

EMF in Absolute Millivolts — Reference Junctions at 0 C

DEG C	0	1	2	3	4	5	6	7	8	9	10	DEG C
THERMOELECTRIC VOLTAGE IN ABSOLUTE MILLIVOLTS												
1,000	57.942	58.001	58.060	58.120	58.179	58.238	58.297	58.356	58.415	58.474	58.533	1,000
1,010	58.533	58.592	58.651	58.710	58.769	58.827	58.886	58.945	59.004	59.063	59.121	1,010
1,020	59.121	59.180	59.239	59.298	59.356	59.415	59.474	59.532	59.591	59.650	59.708	1,020
1,030	59.708	59.767	59.825	59.884	59.942	60.001	60.059	60.118	60.176	60.235	60.293	1,030
1,040	60.293	60.351	60.410	60.468	60.527	60.585	60.643	60.702	60.760	60.818	60.876	1,040
1,050	60.876	60.935	60.993	61.051	61.109	61.168	61.226	61.284	61.342	61.400	61.459	1,050
1,060	61.459	61.517	61.575	61.633	61.691	61.749	61.807	61.865	61.923	61.981	62.039	1,060
1,070	62.039	62.097	62.156	62.214	62.272	62.330	62.388	62.446	62.504	62.562	62.619	1,070
1,080	62.619	62.677	62.735	62.793	62.851	62.909	62.967	63.025	63.083	63.141	63.199	1,080
1,090	63.199	63.257	63.314	63.372	63.430	63.488	63.546	63.604	63.662	63.719	63.777	1,090
1,100	63.777	63.835	63.893	63.951	64.009	64.066	64.124	64.182	64.240	64.298	64.355	1,100
1,110	64.355	64.413	64.471	64.529	64.586	64.644	64.702	64.760	64.817	64.875	64.933	1,110
1,120	64.933	64.991	65.048	65.106	65.164	65.222	65.279	65.337	65.395	65.453	65.510	1,120
1,130	65.510	65.568	65.626	65.683	65.741	65.799	65.856	65.914	65.972	66.029	66.087	1,130
1,140	66.087	66.145	66.202	66.260	66.318	66.375	66.433	66.491	66.548	66.606	66.664	1,140
1,150	66.664	66.721	66.779	66.836	66.894	66.952	67.009	67.067	67.124	67.182	67.240	1,150
1,160	67.240	67.297	67.355	67.412	67.470	67.527	67.585	67.643	67.700	67.758	67.815	1,160
1,170	67.815	67.873	67.930	67.988	68.045	68.103	68.160	68.217	68.275	68.332	68.390	1,170
1,180	68.390	68.447	68.505	68.562	68.619	68.677	68.734	68.792	68.849	68.906	68.964	1,180
1,190	68.964	69.021	69.078	69.135	69.193	69.250	69.307	69.364	69.422	69.479	69.536	1,190
1,200	69.536											1,200
DEG C	0	1	2	3	4	5	6	7	8	9	10	DEG C

TABLE 10.8—*Type K thermocouples (deg F-millivolts).*

Temperature in Degrees Fahrenheit[a]

EMF in Absolute Millivolts — Reference Junctions at 32 F

DEG F	0	1	2	3	4	5	6	7	8	9	10	DEG F
					THERMOELECTRIC VOLTAGE IN ABSOLUTE MILLIVOLTS							
-450	-6.456	-6.456	-6.457	-6.457	-6.458							-450
-440	-6.447	-6.448	-6.449	-6.450	-6.451	-6.452	-6.453	-6.454	-6.454	-6.455	-6.456	-440
-430	-6.431	-6.433	-6.435	-6.436	-6.438	-6.440	-6.441	-6.443	-6.444	-6.445	-6.447	-430
-420	-6.409	-6.411	-6.414	-6.416	-6.419	-6.421	-6.423	-6.425	-6.427	-6.429	-6.431	-420
-410	-6.380	-6.383	-6.386	-6.389	-6.392	-6.395	-6.398	-6.401	-6.404	-6.406	-6.409	-410
-400	-6.344	-6.348	-6.352	-6.355	-6.359	-6.363	-6.366	-6.370	-6.373	-6.377	-6.380	-400
-390	-6.301	-6.306	-6.310	-6.315	-6.319	-6.323	-6.328	-6.332	-6.336	-6.340	-6.344	-390
-380	-6.251	-6.257	-6.262	-6.267	-6.272	-6.277	-6.282	-6.287	-6.292	-6.296	-6.301	-380
-370	-6.195	-6.201	-6.207	-6.213	-6.219	-6.224	-6.230	-6.235	-6.241	-6.246	-6.251	-370
-360	-6.133	-6.139	-6.146	-6.152	-6.158	-6.165	-6.171	-6.177	-6.183	-6.189	-6.195	-360
-350	-6.064	-6.071	-6.078	-6.085	-6.092	-6.099	-6.106	-6.113	-6.119	-6.126	-6.133	-350
-340	-5.989	-5.997	-6.004	-6.012	-6.020	-6.027	-6.035	-6.042	-6.049	-6.057	-6.064	-340
-330	-5.908	-5.917	-5.925	-5.933	-5.941	-5.949	-5.957	-5.965	-5.973	-5.981	-5.989	-330
-320	-5.822	-5.831	-5.839	-5.848	-5.857	-5.866	-5.874	-5.883	-5.891	-5.900	-5.908	-320
-310	-5.730	-5.739	-5.748	-5.758	-5.767	-5.776	-5.786	-5.795	-5.804	-5.813	-5.822	-310
-300	-5.632	-5.642	-5.652	-5.662	-5.672	-5.682	-5.691	-5.701	-5.711	-5.720	-5.730	-300
-290	-5.529	-5.540	-5.550	-5.561	-5.571	-5.581	-5.592	-5.602	-5.612	-5.622	-5.632	-290
-280	-5.421	-5.432	-5.443	-5.454	-5.465	-5.476	-5.487	-5.497	-5.508	-5.519	-5.529	-280
-270	-5.308	-5.319	-5.331	-5.342	-5.354	-5.365	-5.376	-5.388	-5.399	-5.410	-5.421	-270
-260	-5.190	-5.202	-5.214	-5.226	-5.238	-5.249	-5.261	-5.273	-5.285	-5.296	-5.308	-260
-250	-5.067	-5.079	-5.092	-5.104	-5.116	-5.129	-5.141	-5.153	-5.165	-5.178	-5.190	-250
-240	-4.939	-4.952	-4.965	-4.978	-4.990	-5.003	-5.016	-5.029	-5.041	-5.054	-5.067	-240
-230	-4.806	-4.819	-4.833	-4.846	-4.860	-4.873	-4.886	-4.899	-4.912	-4.926	-4.939	-230
-220	-4.669	-4.683	-4.697	-4.710	-4.724	-4.738	-4.752	-4.765	-4.779	-4.792	-4.806	-220
-210	-4.527	-4.541	-4.556	-4.570	-4.584	-4.598	-4.613	-4.627	-4.641	-4.655	-4.669	-210
-200	-4.381	-4.396	-4.410	-4.425	-4.440	-4.454	-4.469	-4.484	-4.498	-4.512	-4.527	-200
-190	-4.230	-4.245	-4.261	-4.276	-4.291	-4.306	-4.321	-4.336	-4.351	-4.366	-4.381	-190
-180	-4.075	-4.091	-4.107	-4.122	-4.138	-4.153	-4.169	-4.184	-4.200	-4.215	-4.230	-180
-170	-3.917	-3.933	-3.949	-3.965	-3.981	-3.997	-4.012	-4.028	-4.044	-4.060	-4.075	-170
-160	-3.754	-3.770	-3.787	-3.803	-3.819	-3.836	-3.852	-3.868	-3.884	-3.901	-3.917	-160
-150	-3.587	-3.604	-3.621	-3.637	-3.654	-3.671	-3.688	-3.704	-3.721	-3.737	-3.754	-150
-140	-3.417	-3.434	-3.451	-3.468	-3.485	-3.502	-3.519	-3.536	-3.553	-3.570	-3.587	-140
-130	-3.242	-3.260	-3.277	-3.295	-3.312	-3.330	-3.347	-3.365	-3.382	-3.399	-3.417	-130
-120	-3.065	-3.082	-3.100	-3.118	-3.136	-3.154	-3.172	-3.189	-3.207	-3.225	-3.242	-120
-110	-2.883	-2.902	-2.920	-2.938	-2.956	-2.974	-2.992	-3.010	-3.029	-3.047	-3.065	-110
-100	-2.699	-2.717	-2.736	-2.754	-2.773	-2.791	-2.810	-2.828	-2.847	-2.865	-2.883	-100
-90	-2.511	-2.530	-2.549	-2.567	-2.586	-2.605	-2.624	-2.643	-2.661	-2.680	-2.699	-90
-80	-2.320	-2.339	-2.358	-2.377	-2.397	-2.416	-2.435	-2.454	-2.473	-2.492	-2.511	-80
-70	-2.126	-2.145	-2.165	-2.184	-2.204	-2.223	-2.243	-2.262	-2.281	-2.300	-2.320	-70
-60	-1.929	-1.949	-1.968	-1.988	-2.008	-2.028	-2.047	-2.067	-2.087	-2.106	-2.126	-60
-50	-1.729	-1.749	-1.769	-1.789	-1.809	-1.829	-1.849	-1.869	-1.889	-1.909	-1.929	-50
-40	-1.527	-1.547	-1.567	-1.588	-1.608	-1.628	-1.648	-1.669	-1.689	-1.709	-1.729	-40
-30	-1.322	-1.342	-1.363	-1.383	-1.404	-1.424	-1.445	-1.465	-1.486	-1.506	-1.527	-30
-20	-1.114	-1.135	-1.156	-1.177	-1.197	-1.218	-1.239	-1.260	-1.280	-1.301	-1.322	-20
-10	-0.904	-0.925	-0.946	-0.968	-0.989	-1.010	-1.031	-1.051	-1.072	-1.093	-1.114	-10
- 0	-0.692	-0.714	-0.735	-0.756	-0.777	-0.799	-0.820	-0.841	-0.862	-0.883	-0.904	- 0
0	-0.692	-0.671	-0.650	-0.628	-0.607	-0.585	-0.564	-0.543	-0.521	-0.500	-0.478	0
10	-0.478	-0.457	-0.435	-0.413	-0.392	-0.370	-0.349	-0.327	-0.305	-0.284	-0.262	10
20	-0.262	-0.240	-0.218	-0.197	-0.175	-0.153	-0.131	-0.109	-0.088	-0.066	-0.044	20
30	-0.044	-0.022	0.000	0.022	0.044	0.066	0.088	0.110	0.132	0.154	0.176	30
40	0.176	0.198	0.220	0.242	0.264	0.286	0.308	0.331	0.353	0.375	0.397	40
50	0.397	0.419	0.441	0.464	0.486	0.508	0.530	0.553	0.575	0.597	0.619	50
60	0.619	0.642	0.664	0.686	0.709	0.731	0.753	0.776	0.798	0.821	0.843	60
70	0.843	0.865	0.888	0.910	0.933	0.955	0.978	1.000	1.023	1.045	1.068	70
80	1.068	1.090	1.113	1.135	1.158	1.181	1.203	1.226	1.248	1.271	1.294	80
90	1.294	1.316	1.339	1.362	1.384	1.407	1.430	1.452	1.475	1.498	1.520	90
100	1.520	1.543	1.566	1.589	1.611	1.634	1.657	1.680	1.703	1.725	1.748	100
110	1.748	1.771	1.794	1.817	1.839	1.862	1.885	1.908	1.931	1.954	1.977	110
120	1.977	2.000	2.022	2.045	2.068	2.091	2.114	2.137	2.160	2.183	2.206	120
130	2.206	2.229	2.252	2.275	2.298	2.321	2.344	2.367	2.390	2.413	2.436	130
140	2.436	2.459	2.482	2.505	2.528	2.551	2.574	2.597	2.620	2.643	2.666	140
DEG F	0	1	2	3	4	5	6	7	8	9	10	DEG F

[a] Converted from degrees Celsius (IPTS 1968).

TABLE 10.8—*Type K thermocouples (continued).*

Temperature in Degrees Fahrenheit[a]

EMF in Absolute Millivolts

Reference Junctions at 32 F

DEG F	0	1	2	3	4	5	6	7	8	9	10	DEG F
					THERMOELECTRIC VOLTAGE IN ABSOLUTE MILLIVOLTS							
150	2.666	2.689	2.712	2.735	2.758	2.781	2.804	2.827	2.850	2.873	2.896	150
160	2.896	2.920	2.943	2.966	2.989	3.012	3.035	3.058	3.081	3.104	3.127	160
170	3.127	3.150	3.173	3.196	3.220	3.243	3.266	3.289	3.312	3.335	3.358	170
180	3.358	3.381	3.404	3.427	3.450	3.473	3.496	3.519	3.543	3.566	3.589	180
190	3.589	3.612	3.635	3.658	3.681	3.704	3.727	3.750	3.773	3.796	3.819	190
200	3.819	3.842	3.865	3.888	3.911	3.934	3.957	3.980	4.003	4.026	4.049	200
210	4.049	4.072	4.095	4.118	4.141	4.164	4.187	4.210	4.233	4.256	4.279	210
220	4.279	4.302	4.325	4.348	4.371	4.394	4.417	4.439	4.462	4.485	4.508	220
230	4.508	4.531	4.554	4.577	4.600	4.622	4.645	4.668	4.691	4.714	4.737	230
240	4.737	4.759	4.782	4.805	4.828	4.851	4.873	4.896	4.919	4.942	4.964	240
250	4.964	4.987	5.010	5.033	5.055	5.078	5.101	5.124	5.146	5.169	5.192	250
260	5.192	5.214	5.237	5.260	5.282	5.305	5.327	5.350	5.373	5.395	5.418	260
270	5.418	5.440	5.463	5.486	5.508	5.531	5.553	5.576	5.598	5.621	5.643	270
280	5.643	5.666	5.688	5.711	5.733	5.756	5.778	5.801	5.823	5.846	5.868	280
290	5.868	5.891	5.913	5.936	5.958	5.980	6.003	6.025	6.048	6.070	6.092	290
300	6.092	6.115	6.137	6.160	6.182	6.204	6.227	6.249	6.271	6.294	6.316	300
310	6.316	6.338	6.361	6.383	6.405	6.428	6.450	6.472	6.494	6.517	6.539	310
320	6.539	6.561	6.583	6.606	6.628	6.650	6.672	6.695	6.717	6.739	6.761	320
330	6.761	6.784	6.806	6.828	6.850	6.873	6.895	6.917	6.939	6.961	6.984	330
340	6.984	7.006	7.028	7.050	7.072	7.094	7.117	7.139	7.161	7.183	7.205	340
350	7.205	7.228	7.250	7.272	7.294	7.316	7.338	7.361	7.383	7.405	7.427	350
360	7.427	7.449	7.471	7.494	7.516	7.538	7.560	7.582	7.604	7.627	7.649	360
370	7.649	7.671	7.693	7.715	7.737	7.760	7.782	7.804	7.826	7.848	7.870	370
380	7.870	7.893	7.915	7.937	7.959	7.981	8.003	8.026	8.048	8.070	8.092	380
390	8.092	8.114	8.137	8.159	8.181	8.203	8.225	8.248	8.270	8.292	8.314	390
400	8.314	8.336	8.359	8.381	8.403	8.425	8.448	8.470	8.492	8.514	8.537	400
410	8.537	8.559	8.581	8.603	8.626	8.648	8.670	8.692	8.715	8.737	8.759	410
420	8.759	8.782	8.804	8.826	8.849	8.871	8.893	8.916	8.938	8.960	8.983	420
430	8.983	9.005	9.027	9.050	9.072	9.094	9.117	9.139	9.161	9.184	9.206	430
440	9.206	9.229	9.251	9.273	9.296	9.318	9.341	9.363	9.385	9.408	9.430	440
450	9.430	9.453	9.475	9.498	9.520	9.543	9.565	9.588	9.610	9.633	9.655	450
460	9.655	9.678	9.700	9.723	9.745	9.768	9.790	9.813	9.835	9.858	9.880	460
470	9.880	9.903	9.926	9.948	9.971	9.993	10.016	10.038	10.061	10.084	10.106	470
480	10.106	10.129	10.151	10.174	10.197	10.219	10.242	10.265	10.287	10.310	10.333	480
490	10.333	10.355	10.378	10.401	10.423	10.446	10.469	10.491	10.514	10.537	10.560	490
500	10.560	10.582	10.605	10.628	10.650	10.673	10.696	10.719	10.741	10.764	10.787	500
510	10.787	10.810	10.833	10.855	10.878	10.901	10.924	10.947	10.969	10.992	11.015	510
520	11.015	11.038	11.061	11.083	11.106	11.129	11.152	11.175	11.198	11.221	11.243	520
530	11.243	11.266	11.289	11.312	11.335	11.358	11.381	11.404	11.426	11.449	11.472	530
540	11.472	11.495	11.518	11.541	11.564	11.587	11.610	11.633	11.656	11.679	11.702	540
550	11.702	11.725	11.748	11.770	11.793	11.816	11.839	11.862	11.885	11.908	11.931	550
560	11.931	11.954	11.977	12.000	12.023	12.046	12.069	12.092	12.115	12.138	12.161	560
570	12.161	12.184	12.207	12.230	12.254	12.277	12.300	12.323	12.346	12.369	12.392	570
580	12.392	12.415	12.438	12.461	12.484	12.507	12.530	12.553	12.576	12.599	12.623	580
590	12.623	12.646	12.669	12.692	12.715	12.738	12.761	12.784	12.807	12.831	12.854	590
600	12.854	12.877	12.900	12.923	12.946	12.969	12.992	13.016	13.039	13.062	13.085	600
610	13.085	13.108	13.131	13.154	13.178	13.201	13.224	13.247	13.270	13.293	13.317	610
620	13.317	13.340	13.363	13.386	13.409	13.433	13.456	13.479	13.502	13.525	13.549	620
630	13.549	13.572	13.595	13.618	13.641	13.665	13.688	13.711	13.734	13.757	13.781	630
640	13.781	13.804	13.827	13.850	13.874	13.897	13.920	13.943	13.967	13.990	14.013	640
650	14.013	14.036	14.060	14.083	14.106	14.129	14.153	14.176	14.199	14.222	14.246	650
660	14.246	14.269	14.292	14.316	14.339	14.362	14.385	14.409	14.432	14.455	14.479	660
670	14.479	14.502	14.525	14.548	14.572	14.595	14.618	14.642	14.665	14.688	14.712	670
680	14.712	14.735	14.758	14.782	14.805	14.828	14.852	14.875	14.898	14.922	14.945	680
690	14.945	14.968	14.992	15.015	15.038	15.062	15.085	15.108	15.132	15.155	15.178	690
700	15.178	15.202	15.225	15.248	15.272	15.295	15.318	15.342	15.365	15.389	15.412	700
710	15.412	15.435	15.459	15.482	15.505	15.529	15.552	15.576	15.599	15.622	15.646	710
720	15.646	15.669	15.693	15.716	15.739	15.763	15.786	15.810	15.833	15.856	15.880	720
730	15.880	15.903	15.927	15.950	15.974	15.997	16.020	16.044	16.067	16.091	16.114	730
740	16.114	16.138	16.161	16.184	16.208	16.231	16.255	16.278	16.302	16.325	16.349	740
750	16.349	16.372	16.395	16.419	16.442	16.466	16.489	16.513	16.536	16.560	16.583	750
760	16.583	16.607	16.630	16.654	16.677	16.700	16.724	16.747	16.771	16.794	16.818	760
770	16.818	16.841	16.865	16.888	16.912	16.935	16.959	16.982	17.006	17.029	17.053	770
780	17.053	17.076	17.100	17.123	17.147	17.170	17.194	17.217	17.241	17.264	17.288	780
790	17.288	17.311	17.335	17.358	17.382	17.406	17.429	17.453	17.476	17.500	17.523	790
DEG F	0	1	2	3	4	5	6	7	8	9	10	DEG F

[a] Converted from degrees Celsius (IPTS 1968).

TABLE 10.8—*Type K thermocouples (continued).*

Temperature in Degrees Fahrenheit[a]

EMF in Absolute Millivolts | Reference Junctions at 32 F

DEG F	0	1	2	3	4	5	6	7	8	9	10	DEG F
					THERMOELECTRIC VOLTAGE IN ABSOLUTE MILLIVOLTS							
800	17.523	17.547	17.570	17.594	17.617	17.641	17.664	17.688	17.711	17.735	17.759	800
810	17.759	17.782	17.806	17.829	17.853	17.876	17.900	17.923	17.947	17.971	17.994	810
820	17.994	18.018	18.041	18.065	18.088	18.112	18.136	18.159	18.183	18.206	18.230	820
830	18.230	18.253	18.277	18.301	18.324	18.348	18.371	18.395	18.418	18.442	18.466	830
840	18.466	18.489	18.513	18.536	18.560	18.584	18.607	18.631	18.654	18.678	18.702	840
850	18.702	18.725	18.749	18.772	18.796	18.820	18.843	18.867	18.890	18.914	18.938	850
860	18.938	18.961	18.985	19.008	19.032	19.056	19.079	19.103	19.127	19.150	19.174	860
870	19.174	19.197	19.221	19.245	19.268	19.292	19.316	19.339	19.363	19.386	19.410	870
880	19.410	19.434	19.457	19.481	19.505	19.528	19.552	19.576	19.599	19.623	19.646	880
890	19.646	19.670	19.694	19.717	19.741	19.765	19.788	19.812	19.836	19.859	19.883	890
900	19.883	19.907	19.930	19.954	19.978	20.001	20.025	20.049	20.072	20.096	20.120	900
910	20.120	20.143	20.167	20.190	20.214	20.238	20.261	20.285	20.309	20.332	20.356	910
920	20.356	20.380	20.403	20.427	20.451	20.474	20.498	20.522	20.545	20.569	20.593	920
930	20.593	20.616	20.640	20.664	20.688	20.711	20.735	20.759	20.782	20.806	20.830	930
940	20.830	20.853	20.877	20.901	20.924	20.948	20.972	20.995	21.019	21.043	21.066	940
950	21.066	21.090	21.114	21.137	21.161	21.185	21.208	21.232	21.256	21.280	21.303	950
960	21.303	21.327	21.351	21.374	21.398	21.422	21.445	21.469	21.493	21.516	21.540	960
970	21.540	21.564	21.587	21.611	21.635	21.659	21.682	21.706	21.730	21.753	21.777	970
980	21.777	21.801	21.824	21.848	21.872	21.895	21.919	21.943	21.966	21.990	22.014	980
990	22.014	22.038	22.061	22.085	22.109	22.132	22.156	22.180	22.203	22.227	22.251	990
1,000	22.251	22.274	22.298	22.322	22.346	22.369	22.393	22.417	22.440	22.464	22.488	1,000
1,010	22.488	22.511	22.535	22.559	22.582	22.606	22.630	22.654	22.677	22.701	22.725	1,010
1,020	22.725	22.748	22.772	22.796	22.819	22.843	22.867	22.890	22.914	22.938	22.961	1,020
1,030	22.961	22.985	23.009	23.032	23.056	23.080	23.104	23.127	23.151	23.175	23.198	1,030
1,040	23.198	23.222	23.246	23.269	23.293	23.317	23.340	23.364	23.388	23.411	23.435	1,040
1,050	23.435	23.459	23.482	23.506	23.530	23.553	23.577	23.601	23.624	23.648	23.672	1,050
1,060	23.672	23.695	23.719	23.743	23.766	23.790	23.814	23.837	23.861	23.885	23.908	1,060
1,070	23.908	23.932	23.956	23.979	24.003	24.027	24.050	24.074	24.098	24.121	24.145	1,070
1,080	24.145	24.169	24.192	24.216	24.240	24.263	24.287	24.311	24.334	24.358	24.382	1,080
1,090	24.382	24.405	24.429	24.453	24.476	24.500	24.523	24.547	24.571	24.594	24.618	1,090
1,100	24.618	24.642	24.665	24.689	24.713	24.736	24.760	24.783	24.807	24.831	24.854	1,100
1,110	24.854	24.878	24.902	24.925	24.949	24.972	24.996	25.020	25.043	25.067	25.091	1,110
1,120	25.091	25.114	25.138	25.161	25.185	25.209	25.232	25.256	25.279	25.303	25.327	1,120
1,130	25.327	25.350	25.374	25.397	25.421	25.445	25.468	25.492	25.515	25.539	25.563	1,130
1,140	25.563	25.586	25.610	25.633	25.657	25.681	25.704	25.728	25.751	25.775	25.799	1,140
1,150	25.799	25.822	25.846	25.869	25.893	25.916	25.940	25.964	25.987	26.011	26.034	1,150
1,160	26.034	26.058	26.081	26.105	26.128	26.152	26.176	26.199	26.223	26.246	26.270	1,160
1,170	26.270	26.293	26.317	26.340	26.364	26.387	26.411	26.435	26.458	26.482	26.505	1,170
1,180	26.505	26.529	26.552	26.576	26.599	26.623	26.646	26.670	26.693	26.717	26.740	1,180
1,190	26.740	26.764	26.787	26.811	26.834	26.858	26.881	26.905	26.928	26.952	26.975	1,190
1,200	26.975	26.999	27.022	27.046	27.069	27.093	27.116	27.140	27.163	27.187	27.210	1,200
1,210	27.210	27.234	27.257	27.281	27.304	27.328	27.351	27.375	27.398	27.422	27.445	1,210
1,220	27.445	27.468	27.492	27.515	27.539	27.562	27.586	27.609	27.633	27.656	27.679	1,220
1,230	27.679	27.703	27.726	27.750	27.773	27.797	27.820	27.843	27.867	27.890	27.914	1,230
1,240	27.914	27.937	27.961	27.984	28.007	28.031	28.054	28.078	28.101	28.124	28.148	1,240
1,250	28.148	28.171	28.195	28.218	28.241	28.265	28.298	28.311	28.335	28.358	28.382	1,250
1,260	28.382	28.405	28.428	28.452	28.475	28.498	28.522	28.545	28.569	28.592	28.615	1,260
1,270	28.615	28.639	28.662	28.685	28.709	28.732	28.755	28.779	28.802	28.825	28.849	1,270
1,280	28.849	28.872	28.895	28.919	28.942	28.965	28.988	29.012	29.035	29.058	29.082	1,280
1,290	29.082	29.105	29.128	29.152	29.175	29.198	29.221	29.245	29.268	29.291	29.315	1,290
1,300	29.315	29.338	29.361	29.384	29.408	29.431	29.454	29.477	29.501	29.524	29.547	1,300
1,310	29.547	29.570	29.594	29.617	29.640	29.663	29.687	29.710	29.733	29.756	29.780	1,310
1,320	29.780	29.803	29.826	29.849	29.872	29.896	29.919	29.942	29.965	29.989	30.012	1,320
1,330	30.012	30.035	30.058	30.081	30.104	30.128	30.151	30.174	30.197	30.220	30.244	1,330
1,340	30.244	30.267	30.290	30.313	30.336	30.359	30.383	30.406	30.429	30.452	30.475	1,340
1,350	30.475	30.498	30.521	30.545	30.568	30.591	30.614	30.637	30.660	30.683	30.706	1,350
1,360	30.706	30.730	30.753	30.776	30.799	30.822	30.845	30.868	30.891	30.914	30.937	1,360
1,370	30.937	30.961	30.984	31.007	31.030	31.053	31.076	31.099	31.122	31.145	31.168	1,370
1,380	31.168	31.191	31.214	31.237	31.260	31.283	31.306	31.329	31.353	31.376	31.399	1,380
1,390	31.399	31.422	31.445	31.468	31.491	31.514	31.537	31.560	31.583	31.606	31.629	1,390
1,400	31.629	31.652	31.675	31.698	31.721	31.744	31.767	31.790	31.813	31.836	31.859	1,400
1,410	31.859	31.882	31.905	31.927	31.950	31.973	31.996	32.019	32.042	32.065	32.088	1,410
1,420	32.088	32.111	32.134	32.157	32.180	32.203	32.226	32.249	32.272	32.294	32.317	1,420
1,430	32.317	32.340	32.363	32.386	32.409	32.432	32.455	32.478	32.501	32.523	32.546	1,430
1,440	32.546	32.569	32.592	32.615	32.638	32.661	32.683	32.706	32.729	32.752	32.775	1,440
DEG F	0	1	2	3	4	5	6	7	8	9	10	DEG F

[a] Converted from degrees Celsius (IPTS 1968).

TABLE 10.8—*Type K thermocouples (continued).*

Temperature in Degrees Fahrenheit[a]

EMF in Absolute Millivolts

Reference Junctions at 32 F

DEG F	0	1	2	3	4	5	6	7	8	9	10	DEG F
					THERMOELECTRIC VOLTAGE IN ABSOLUTE MILLIVOLTS							
1,450	32.775	32.798	32.821	32.843	32.866	32.889	32.912	32.935	32.958	32.980	33.003	1,450
1,460	33.003	33.026	33.049	33.072	33.094	33.117	33.140	33.163	33.186	33.208	33.231	1,460
1,470	33.231	33.254	33.277	33.300	33.322	33.345	33.368	33.391	33.413	33.436	33.459	1,470
1,480	33.459	33.482	33.504	33.527	33.550	33.573	33.595	33.618	33.641	33.664	33.686	1,480
1,490	33.686	33.709	33.732	33.754	33.777	33.800	33.823	33.845	33.868	33.891	33.913	1,490
1,500	33.913	33.936	33.959	33.981	34.004	34.027	34.049	34.072	34.095	34.117	34.140	1,500
1,510	34.140	34.163	34.185	34.208	34.231	34.253	34.276	34.299	34.321	34.344	34.366	1,510
1,520	34.366	34.389	34.412	34.434	34.457	34.480	34.502	34.525	34.547	34.570	34.593	1,520
1,530	34.593	34.615	34.638	34.660	34.683	34.705	34.728	34.751	34.773	34.796	34.818	1,530
1,540	34.818	34.841	34.863	34.886	34.909	34.931	34.954	34.976	34.999	35.021	35.044	1,540
1,550	35.044	35.066	35.089	35.111	35.134	35.156	35.179	35.201	35.224	35.246	35.269	1,550
1,560	35.269	35.291	35.314	35.336	35.359	35.381	35.404	35.426	35.449	35.471	35.494	1,560
1,570	35.494	35.516	35.539	35.561	35.583	35.606	35.628	35.651	35.673	35.696	35.718	1,570
1,580	35.718	35.741	35.763	35.785	35.808	35.830	35.853	35.875	35.897	35.920	35.942	1,580
1,590	35.942	35.965	35.987	36.009	36.032	36.054	36.077	36.099	36.121	36.144	36.166	1,590
1,600	36.166	36.188	36.211	36.233	36.256	36.278	36.300	36.323	36.345	36.367	36.390	1,600
1,610	36.390	36.412	36.434	36.457	36.479	36.501	36.524	36.546	36.568	36.590	36.613	1,610
1,620	36.613	36.635	36.657	36.680	36.702	36.724	36.746	36.769	36.791	36.813	36.836	1,620
1,630	36.836	36.858	36.880	36.902	36.925	36.947	36.969	36.991	37.014	37.036	37.058	1,630
1,640	37.058	37.080	37.103	37.125	37.147	37.169	37.191	37.214	37.236	37.258	37.280	1,640
1,650	37.280	37.303	37.325	37.347	37.369	37.391	37.413	37.436	37.458	37.480	37.502	1,650
1,660	37.502	37.524	37.547	37.569	37.591	37.613	37.635	37.657	37.679	37.702	37.724	1,660
1,670	37.724	37.746	37.768	37.790	37.812	37.834	37.857	37.879	37.901	37.923	37.945	1,670
1,680	37.945	37.967	37.989	38.011	38.033	38.055	38.078	38.100	38.122	38.144	38.166	1,680
1,690	38.166	38.188	38.210	38.232	38.254	38.276	38.298	38.320	38.342	38.364	38.387	1,690
1,700	38.387	38.409	38.431	38.453	38.475	38.497	38.519	38.541	38.563	38.585	38.607	1,700
1,710	38.607	38.629	38.651	38.673	38.695	38.717	38.739	38.761	38.783	38.805	38.827	1,710
1,720	38.827	38.849	38.871	38.893	38.915	38.937	38.959	38.981	39.003	39.024	39.046	1,720
1,730	39.046	39.068	39.090	39.112	39.134	39.156	39.178	39.200	39.222	39.244	39.266	1,730
1,740	39.266	39.288	39.310	39.331	39.353	39.375	39.397	39.419	39.441	39.463	39.485	1,740
1,750	39.485	39.507	39.529	39.550	39.572	39.594	39.616	39.638	39.660	39.682	39.703	1,750
1,760	39.703	39.725	39.747	39.769	39.791	39.813	39.835	39.856	39.878	39.900	39.922	1,760
1,770	39.922	39.944	39.965	39.987	40.009	40.031	40.053	40.075	40.096	40.118	40.140	1,770
1,780	40.140	40.162	40.183	40.205	40.227	40.249	40.271	40.292	40.314	40.336	40.358	1,780
1,790	40.358	40.379	40.401	40.423	40.445	40.466	40.488	40.510	40.532	40.553	40.575	1,790
1,800	40.575	40.597	40.619	40.640	40.662	40.684	40.705	40.727	40.749	40.770	40.792	1,800
1,810	40.792	40.814	40.836	40.857	40.879	40.901	40.922	40.944	40.966	40.987	41.009	1,810
1,820	41.009	41.031	41.052	41.074	41.096	41.117	41.139	41.161	41.182	41.204	41.225	1,820
1,830	41.225	41.247	41.269	41.290	41.312	41.334	41.355	41.377	41.398	41.420	41.442	1,830
1,840	41.442	41.463	41.485	41.506	41.528	41.550	41.571	41.593	41.614	41.636	41.657	1,840
1,850	41.657	41.679	41.701	41.722	41.744	41.765	41.787	41.808	41.830	41.851	41.873	1,850
1,860	41.873	41.895	41.916	41.938	41.959	41.981	42.002	42.024	42.045	42.067	42.088	1,860
1,870	42.088	42.110	42.131	42.153	42.174	42.196	42.217	42.239	42.260	42.282	42.303	1,870
1,880	42.303	42.325	42.346	42.367	42.389	42.410	42.432	42.453	42.475	42.496	42.518	1,880
1,890	42.518	42.539	42.560	42.582	42.603	42.625	42.646	42.668	42.689	42.710	42.732	1,890
1,900	42.732	42.753	42.775	42.796	42.817	42.839	42.860	42.882	42.903	42.924	42.946	1,900
1,910	42.946	42.967	42.989	43.010	43.031	43.053	43.074	43.095	43.117	43.138	43.159	1,910
1,920	43.159	43.181	43.202	43.223	43.245	43.266	43.287	43.309	43.330	43.351	43.373	1,920
1,930	43.373	43.394	43.415	43.436	43.458	43.479	43.500	43.522	43.543	43.564	43.585	1,930
1,940	43.585	43.607	43.628	43.649	43.671	43.692	43.713	43.734	43.756	43.777	43.798	1,940
1,950	43.798	43.819	43.841	43.862	43.883	43.904	43.925	43.947	43.968	43.989	44.010	1,950
1,960	44.010	44.031	44.053	44.074	44.095	44.116	44.137	44.159	44.180	44.201	44.222	1,960
1,970	44.222	44.243	44.265	44.286	44.307	44.328	44.349	44.370	44.391	44.413	44.434	1,970
1,980	44.434	44.455	44.476	44.497	44.518	44.539	44.560	44.582	44.603	44.624	44.645	1,980
1,990	44.645	44.666	44.687	44.708	44.729	44.750	44.771	44.793	44.814	44.835	44.856	1,990
2,000	44.856	44.877	44.898	44.919	44.940	44.961	44.982	45.003	45.024	45.045	45.066	2,000
2,010	45.066	45.087	45.108	45.129	45.150	45.171	45.192	45.213	45.234	45.255	45.276	2,010
2,020	45.276	45.297	45.318	45.339	45.360	45.381	45.402	45.423	45.444	45.465	45.486	2,020
2,030	45.486	45.507	45.528	45.549	45.570	45.591	45.612	45.633	45.654	45.675	45.695	2,030
2,040	45.695	45.716	45.737	45.758	45.779	45.800	45.821	45.842	45.863	45.884	45.904	2,040
2,050	45.904	45.925	45.946	45.967	45.988	46.009	46.030	46.051	46.071	46.092	46.113	2,050
2,060	46.113	46.134	46.155	46.176	46.196	46.217	46.238	46.259	46.280	46.300	46.321	2,060
2,070	46.321	46.342	46.363	46.384	46.404	46.425	46.446	46.467	46.488	46.508	46.529	2,070
2,080	46.529	46.550	46.571	46.591	46.612	46.633	46.654	46.674	46.695	46.716	46.737	2,080
2,090	46.737	46.757	46.778	46.799	46.819	46.840	46.861	46.881	46.902	46.923	46.944	2,090
DEG C	0	1	2	3	4	5	6	7	8	9	10	DEG C

[a] Converted from degrees Celsius (IPTS 1968).

TABLE 10.8—*Type K thermocouples (continued).*

Temperature in Degrees Fahrenheit[a]

EMF in Absolute Millivolts — Reference Junctions at 32 F

DEG F	0	1	2	3	4	5	6	7	8	9	10	DEG F
					THERMOELECTRIC VOLTAGE IN ABSOLUTE MILLIVOLTS							
2,100	46.944	46.964	46.985	47.006	47.026	47.047	47.068	47.088	47.109	47.130	47.150	2,100
2,110	47.150	47.171	47.191	47.212	47.233	47.253	47.274	47.295	47.315	47.336	47.356	2,110
2,120	47.356	47.377	47.398	47.418	47.439	47.459	47.480	47.500	47.521	47.542	47.562	2,120
2,130	47.562	47.583	47.603	47.624	47.644	47.665	47.685	47.706	47.726	47.747	47.767	2,130
2,140	47.767	47.788	47.808	47.829	47.849	47.870	47.890	47.911	47.931	47.952	47.972	2,140
2,150	47.972	47.993	48.013	48.034	48.054	48.075	48.095	48.116	48.136	48.156	48.177	2,150
2,160	48.177	48.197	48.218	48.238	48.258	48.279	48.299	48.320	48.340	48.360	48.381	2,160
2,170	48.381	48.401	48.422	48.442	48.462	48.483	48.503	48.523	48.544	48.564	48.584	2,170
2,180	48.584	48.605	48.625	48.645	48.666	48.686	48.706	48.727	48.747	48.767	48.787	2,180
2,190	48.787	48.808	48.828	48.848	48.869	48.889	48.909	48.929	48.950	48.970	48.990	2,190
2,200	48.990	49.010	49.031	49.051	49.071	49.091	49.111	49.132	49.152	49.172	49.192	2,200
2,210	49.192	49.212	49.233	49.253	49.273	49.293	49.313	49.333	49.354	49.374	49.394	2,210
2,220	49.394	49.414	49.434	49.454	49.474	49.495	49.515	49.535	49.555	49.575	49.595	2,220
2,230	49.595	49.615	49.635	49.655	49.675	49.696	49.716	49.736	49.756	49.776	49.796	2,230
2,240	49.796	49.816	49.836	49.856	49.876	49.896	49.916	49.936	49.956	49.976	49.996	2,240
2,250	49.996	50.016	50.036	50.056	50.076	50.096	50.116	50.136	50.156	50.176	50.196	2,250
2,260	50.196	50.216	50.236	50.256	50.276	50.296	50.315	50.335	50.355	50.375	50.395	2,260
2,270	50.395	50.415	50.435	50.455	50.475	50.494	50.514	50.534	50.554	50.574	50.594	2,270
2,280	50.594	50.614	50.633	50.653	50.673	50.693	50.713	50.733	50.752	50.772	50.792	2,280
2,290	50.792	50.812	50.832	50.851	50.871	50.891	50.911	50.930	50.950	50.970	50.990	2,290
2,300	50.990	51.009	51.029	51.049	51.069	51.088	51.108	51.128	51.148	51.167	51.187	2,300
2,310	51.187	51.207	51.226	51.246	51.266	51.285	51.305	51.325	51.344	51.364	51.384	2,310
2,320	51.384	51.403	51.423	51.443	51.462	51.482	51.501	51.521	51.541	51.560	51.580	2,320
2,330	51.580	51.599	51.619	51.639	51.658	51.678	51.697	51.717	51.736	51.756	51.776	2,330
2,340	51.776	51.795	51.815	51.834	51.854	51.873	51.893	51.912	51.932	51.951	51.971	2,340
2,350	51.971	51.990	52.010	52.029	52.049	52.068	52.088	52.107	52.127	52.146	52.165	2,350
2,360	52.165	52.185	52.204	52.224	52.243	52.263	52.282	52.301	52.321	52.340	52.360	2,360
2,370	52.360	52.379	52.398	52.418	52.437	52.457	52.476	52.495	52.515	52.534	52.553	2,370
2,380	52.553	52.573	52.592	52.611	52.631	52.650	52.669	52.689	52.708	52.727	52.747	2,380
2,390	52.747	52.766	52.785	52.805	52.824	52.843	52.862	52.882	52.901	52.920	52.939	2,390
2,400	52.939	52.959	52.978	52.997	53.016	53.036	53.055	53.074	53.093	53.113	53.132	2,400
2,410	53.132	53.151	53.170	53.189	53.209	53.228	53.247	53.266	53.285	53.304	53.324	2,410
2,420	53.324	53.343	53.362	53.381	53.400	53.419	53.439	53.458	53.477	53.496	53.515	2,420
2,430	53.515	53.534	53.553	53.572	53.592	53.611	53.630	53.649	53.668	53.687	53.706	2,430
2,440	53.706	53.725	53.744	53.763	53.782	53.801	53.821	53.840	53.859	53.878	53.897	2,440
2,450	53.897	53.916	53.935	53.954	53.973	53.992	54.011	54.030	54.049	54.068	54.087	2,450
2,460	54.087	54.106	54.125	54.144	54.163	54.182	54.201	54.220	54.239	54.258	54.277	2,460
2,470	54.277	54.296	54.315	54.334	54.353	54.372	54.391	54.410	54.429	54.447	54.466	2,470
2,480	54.466	54.485	54.504	54.523	54.542	54.561	54.580	54.599	54.618	54.637	54.656	2,480
2,490	54.656	54.675	54.694	54.712	54.731	54.750	54.769	54.788	54.807	54.826	54.845	2,490
2,500	54.845											2,500
DEG F	0	1	2	3	4	5	6	7	8	9	10	DEG F

[a] Converted from degrees Celsius (IPTS 1968).

TABLE 10.9—*Type K thermocouples (deg C-millivolts).*

Temperature in Degrees Celsius (IPTS 1968)

EMF in Absolute Millivolts — Reference Junctions at 0 C

DEG C	0	1	2	3	4	5	6	7	8	9	10	DEG C
	THERMOELECTRIC VOLTAGE IN ABSOLUTE MILLIVOLTS											
-270	-6.458											-270
-260	-6.441	-6.444	-6.446	-6.448	-6.450	-6.452	-6.453	-6.455	-6.456	-6.457	-6.458	-260
-250	-6.404	-6.408	-6.413	-6.417	-6.421	-6.425	-6.429	-6.432	-6.435	-6.438	-6.441	-250
-240	-6.344	-6.351	-6.358	-6.364	-6.371	-6.377	-6.382	-6.388	-6.394	-6.399	-6.404	-240
-230	-6.262	-6.271	-6.280	-6.289	-6.297	-6.306	-6.314	-6.322	-6.329	-6.337	-6.344	-230
-220	-6.158	-6.170	-6.181	-6.192	-6.202	-6.213	-6.223	-6.233	-6.243	-6.253	-6.262	-220
-210	-6.035	-6.048	-6.061	-6.074	-6.087	-6.099	-6.111	-6.123	-6.135	-6.147	-6.158	-210
-200	-5.891	-5.907	-5.922	-5.936	-5.951	-5.965	-5.980	-5.994	-6.007	-6.021	-6.035	-200
-190	-5.730	-5.747	-5.763	-5.780	-5.796	-5.813	-5.829	-5.845	-5.860	-5.876	-5.891	-190
-180	-5.550	-5.569	-5.587	-5.606	-5.624	-5.642	-5.660	-5.678	-5.695	-5.712	-5.730	-180
-170	-5.354	-5.374	-5.394	-5.414	-5.434	-5.454	-5.474	-5.493	-5.512	-5.531	-5.550	-170
-160	-5.141	-5.163	-5.185	-5.207	-5.228	-5.249	-5.271	-5.292	-5.313	-5.333	-5.354	-160
-150	-4.912	-4.936	-4.959	-4.983	-5.006	-5.029	-5.051	-5.074	-5.097	-5.119	-5.141	-150
-140	-4.669	-4.694	-4.719	-4.743	-4.768	-4.792	-4.817	-4.841	-4.865	-4.889	-4.912	-140
-130	-4.410	-4.437	-4.463	-4.489	-4.515	-4.541	-4.567	-4.593	-4.618	-4.644	-4.669	-130
-120	-4.138	-4.166	-4.193	-4.221	-4.248	-4.276	-4.303	-4.330	-4.357	-4.384	-4.410	-120
-110	-3.852	-3.881	-3.910	-3.939	-3.968	-3.997	-4.025	-4.053	-4.082	-4.110	-4.138	-110
-100	-3.553	-3.584	-3.614	-3.644	-3.674	-3.704	-3.734	-3.764	-3.793	-3.823	-3.852	-100
-90	-3.242	-3.274	-3.305	-3.337	-3.368	-3.399	-3.430	-3.461	-3.492	-3.523	-3.553	-90
-80	-2.920	-2.953	-2.985	-3.018	-3.050	-3.082	-3.115	-3.147	-3.179	-3.211	-3.242	-80
-70	-2.586	-2.620	-2.654	-2.687	-2.721	-2.754	-2.788	-2.821	-2.854	-2.887	-2.920	-70
-60	-2.243	-2.277	-2.312	-2.347	-2.381	-2.416	-2.450	-2.484	-2.518	-2.552	-2.586	-60
-50	-1.889	-1.925	-1.961	-1.996	-2.032	-2.067	-2.102	-2.137	-2.173	-2.208	-2.243	-50
-40	-1.527	-1.563	-1.600	-1.636	-1.673	-1.709	-1.745	-1.781	-1.817	-1.853	-1.889	-40
-30	-1.156	-1.193	-1.231	-1.268	-1.305	-1.342	-1.379	-1.416	-1.453	-1.490	-1.527	-30
-20	-0.777	-0.816	-0.854	-0.892	-0.930	-0.968	-1.005	-1.043	-1.081	-1.118	-1.156	-20
-10	-0.392	-0.431	-0.469	-0.508	-0.547	-0.585	-0.624	-0.662	-0.701	-0.739	-0.777	-10
- 0	0.000	-0.039	-0.079	-0.118	-0.157	-0.197	-0.236	-0.275	-0.314	-0.353	-0.392	- 0
0	0.000	0.039	0.079	0.119	0.158	0.198	0.238	0.277	0.317	0.357	0.397	0
10	0.397	0.437	0.477	0.517	0.557	0.597	0.637	0.677	0.718	0.758	0.798	10
20	0.798	0.838	0.879	0.919	0.960	1.000	1.041	1.081	1.122	1.162	1.203	20
30	1.203	1.244	1.285	1.325	1.366	1.407	1.448	1.489	1.529	1.570	1.611	30
40	1.611	1.652	1.693	1.734	1.776	1.817	1.858	1.899	1.940	1.981	2.022	40
50	2.022	2.064	2.105	2.146	2.188	2.229	2.270	2.312	2.353	2.394	2.436	50
60	2.436	2.477	2.519	2.560	2.601	2.643	2.684	2.726	2.767	2.809	2.850	60
70	2.850	2.892	2.933	2.975	3.016	3.058	3.100	3.141	3.183	3.224	3.266	70
80	3.266	3.307	3.349	3.390	3.432	3.473	3.515	3.556	3.598	3.639	3.681	80
90	3.681	3.722	3.764	3.805	3.847	3.888	3.930	3.971	4.012	4.054	4.095	90
100	4.095	4.137	4.178	4.219	4.261	4.302	4.343	4.384	4.426	4.467	4.508	100
110	4.508	4.549	4.590	4.632	4.673	4.714	4.755	4.796	4.837	4.878	4.919	110
120	4.919	4.960	5.001	5.042	5.083	5.124	5.164	5.205	5.246	5.287	5.327	120
130	5.327	5.368	5.409	5.450	5.490	5.531	5.571	5.612	5.652	5.693	5.733	130
140	5.733	5.774	5.814	5.855	5.895	5.936	5.976	6.016	6.057	6.097	6.137	140
150	6.137	6.177	6.218	6.258	6.298	6.338	6.378	6.419	6.459	6.499	6.539	150
160	6.539	6.579	6.619	6.659	6.699	6.739	6.779	6.819	6.859	6.899	6.939	160
170	6.939	6.979	7.019	7.059	7.099	7.139	7.179	7.219	7.259	7.299	7.338	170
180	7.338	7.378	7.418	7.458	7.498	7.538	7.578	7.618	7.658	7.697	7.737	180
190	7.737	7.777	7.817	7.857	7.897	7.937	7.977	8.017	8.057	8.097	8.137	190
200	8.137	8.177	8.216	8.256	8.296	8.336	8.376	8.416	8.456	8.497	8.537	200
210	8.537	8.577	8.617	8.657	8.697	8.737	8.777	8.817	8.857	8.898	8.938	210
220	8.938	8.978	9.018	9.058	9.099	9.139	9.179	9.220	9.260	9.300	9.341	220
230	9.341	9.381	9.421	9.462	9.502	9.543	9.583	9.624	9.664	9.705	9.745	230
240	9.745	9.786	9.826	9.867	9.907	9.948	9.989	10.029	10.070	10.111	10.151	240
250	10.151	10.192	10.233	10.274	10.315	10.355	10.396	10.437	10.478	10.519	10.560	250
260	10.560	10.600	10.641	10.682	10.723	10.764	10.805	10.846	10.887	10.928	10.969	260
270	10.969	11.010	11.051	11.093	11.134	11.175	11.216	11.257	11.298	11.339	11.381	270
280	11.381	11.422	11.463	11.504	11.546	11.587	11.628	11.669	11.711	11.752	11.793	280
290	11.793	11.835	11.876	11.918	11.959	12.000	12.042	12.083	12.125	12.166	12.207	290
300	12.207	12.249	12.290	12.332	12.373	12.415	12.456	12.498	12.539	12.581	12.623	300
310	12.623	12.664	12.706	12.747	12.789	12.831	12.872	12.914	12.955	12.997	13.039	310
320	13.039	13.080	13.122	13.164	13.205	13.247	13.289	13.331	13.372	13.414	13.456	320
330	13.456	13.497	13.539	13.581	13.623	13.665	13.706	13.748	13.790	13.832	13.874	330
340	13.874	13.915	13.957	13.999	14.041	14.083	14.125	14.167	14.208	14.250	14.292	340
DEG C	0	1	2	3	4	5	6	7	8	9	10	DEG C

TABLE 10.9—*Type K thermocouples (continued).*

Temperature in Degrees Celsius (IPTS 1968)

EMF in Absolute Millivolts

Reference Junctions at 0 C

DEG C	0	1	2	3	4	5	6	7	8	9	10	DEG C
					THERMOELECTRIC VOLTAGE IN ABSOLUTE MILLIVOLTS							
350	14.292	14.334	14.376	14.418	14.460	14.502	14.544	14.586	14.628	14.670	14.712	350
360	14.712	14.754	14.796	14.838	14.880	14.922	14.964	15.006	15.048	15.090	15.132	360
370	15.132	15.174	15.216	15.258	15.300	15.342	15.384	15.426	15.468	15.510	15.552	370
380	15.552	15.594	15.636	15.679	15.721	15.763	15.805	15.847	15.889	15.931	15.974	380
390	15.974	16.016	16.058	16.100	16.142	16.184	16.227	16.269	16.311	16.353	16.395	390
400	16.395	16.438	16.480	16.522	16.564	16.607	16.649	16.691	16.733	16.776	16.818	400
410	16.818	16.860	16.902	16.945	16.987	17.029	17.072	17.114	17.156	17.199	17.241	410
420	17.241	17.283	17.326	17.368	17.410	17.453	17.495	17.537	17.580	17.622	17.664	420
430	17.664	17.707	17.749	17.792	17.834	17.876	17.919	17.961	18.004	18.046	18.088	430
440	18.088	18.131	18.173	18.216	18.258	18.301	18.343	18.385	18.428	18.470	18.513	440
450	18.513	18.555	18.598	18.640	18.683	18.725	18.768	18.810	18.853	18.895	18.938	450
460	18.938	18.980	19.023	19.065	19.108	19.150	19.193	19.235	19.278	19.320	19.363	460
470	19.363	19.405	19.448	19.490	19.533	19.576	19.618	19.661	19.703	19.746	19.788	470
480	19.788	19.831	19.873	19.916	19.959	20.001	20.044	20.086	20.129	20.172	20.214	480
490	20.214	20.257	20.299	20.342	20.385	20.427	20.470	20.512	20.555	20.598	20.640	490
500	20.640	20.683	20.725	20.768	20.811	20.853	20.896	20.938	20.981	21.024	21.066	500
510	21.066	21.109	21.152	21.194	21.237	21.280	21.322	21.365	21.407	21.450	21.493	510
520	21.493	21.535	21.578	21.621	21.663	21.706	21.749	21.791	21.834	21.876	21.919	520
530	21.919	21.962	22.004	22.047	22.090	22.132	22.175	22.218	22.260	22.303	22.346	530
540	22.346	22.388	22.431	22.473	22.516	22.559	22.601	22.644	22.687	22.729	22.772	540
550	22.772	22.815	22.857	22.900	22.942	22.985	23.028	23.070	23.113	23.156	23.198	550
560	23.198	23.241	23.284	23.326	23.369	23.411	23.454	23.497	23.539	23.582	23.624	560
570	23.624	23.667	23.710	23.752	23.795	23.837	23.880	23.923	23.965	24.008	24.050	570
580	24.050	24.093	24.136	24.178	24.221	24.263	24.306	24.348	24.391	24.434	24.476	580
590	24.476	24.519	24.561	24.604	24.646	24.689	24.731	24.774	24.817	24.859	24.902	590
600	24.902	24.944	24.987	25.029	25.072	25.114	25.157	25.199	25.242	25.284	25.327	600
610	25.327	25.369	25.412	25.454	25.497	25.539	25.582	25.624	25.666	25.709	25.751	610
620	25.751	25.794	25.836	25.879	25.921	25.964	26.006	26.048	26.091	26.133	26.176	620
630	26.176	26.218	26.260	26.303	26.345	26.387	26.430	26.472	26.515	26.557	26.599	630
640	26.599	26.642	26.684	26.726	26.769	26.811	26.853	26.896	26.938	26.980	27.022	640
650	27.022	27.065	27.107	27.149	27.192	27.234	27.276	27.318	27.361	27.403	27.445	650
660	27.445	27.487	27.529	27.572	27.614	27.656	27.698	27.740	27.783	27.825	27.867	660
670	27.867	27.909	27.951	27.993	28.035	28.078	28.120	28.162	28.204	28.246	28.288	670
680	28.288	28.330	28.372	28.414	28.456	28.498	28.540	28.583	28.625	28.667	28.709	680
690	28.709	28.751	28.793	28.835	28.877	28.919	28.961	29.002	29.044	29.086	29.128	690
700	29.128	29.170	29.212	29.254	29.296	29.338	29.380	29.422	29.464	29.505	29.547	700
710	29.547	29.589	29.631	29.673	29.715	29.756	29.798	29.840	29.882	29.924	29.965	710
720	29.965	30.007	30.049	30.091	30.132	30.174	30.216	30.257	30.299	30.341	30.383	720
730	30.383	30.424	30.466	30.508	30.549	30.591	30.632	30.674	30.716	30.757	30.799	730
740	30.799	30.840	30.882	30.924	30.965	31.007	31.048	31.090	31.131	31.173	31.214	740
750	31.214	31.256	31.297	31.339	31.380	31.422	31.463	31.504	31.546	31.587	31.629	750
760	31.629	31.670	31.712	31.753	31.794	31.836	31.877	31.918	31.960	32.001	32.042	760
770	32.042	32.084	32.125	32.166	32.207	32.249	32.290	32.331	32.372	32.414	32.455	770
780	32.455	32.496	32.537	32.578	32.619	32.661	32.702	32.743	32.784	32.825	32.866	780
790	32.866	32.907	32.948	32.990	33.031	33.072	33.113	33.154	33.195	33.236	33.277	790
800	33.277	33.318	33.359	33.400	33.441	33.482	33.523	33.564	33.604	33.645	33.686	800
810	33.686	33.727	33.768	33.809	33.850	33.891	33.931	33.972	34.013	34.054	34.095	810
820	34.095	34.136	34.176	34.217	34.258	34.299	34.339	34.380	34.421	34.461	34.502	820
830	34.502	34.543	34.583	34.624	34.665	34.705	34.746	34.787	34.827	34.868	34.909	830
840	34.909	34.949	34.990	35.030	35.071	35.111	35.152	35.192	35.233	35.273	35.314	840
850	35.314	35.354	35.395	35.435	35.476	35.516	35.557	35.597	35.637	35.678	35.718	850
860	35.718	35.758	35.799	35.839	35.880	35.920	35.960	36.000	36.041	36.081	36.121	860
870	36.121	36.162	36.202	36.242	36.282	36.323	36.363	36.403	36.443	36.483	36.524	870
880	36.524	36.564	36.604	36.644	36.684	36.724	36.764	36.804	36.844	36.885	36.925	880
890	36.925	36.965	37.005	37.045	37.085	37.125	37.165	37.205	37.245	37.285	37.325	890
900	37.325	37.365	37.405	37.445	37.484	37.524	37.564	37.604	37.644	37.684	37.724	900
910	37.724	37.764	37.803	37.843	37.883	37.923	37.963	38.002	38.042	38.082	38.122	910
920	38.122	38.162	38.201	38.241	38.281	38.320	38.360	38.400	38.439	38.479	38.519	920
930	38.519	38.558	38.598	38.638	38.677	38.717	38.756	38.796	38.836	38.875	38.915	930
940	38.915	38.954	38.994	39.033	39.073	39.112	39.152	39.191	39.231	39.270	39.310	940
950	39.310	39.349	39.388	39.428	39.467	39.507	39.546	39.585	39.625	39.664	39.703	950
960	39.703	39.743	39.782	39.821	39.861	39.900	39.939	39.979	40.018	40.057	40.096	960
970	40.096	40.136	40.175	40.214	40.253	40.292	40.332	40.371	40.410	40.449	40.488	970
980	40.488	40.527	40.566	40.605	40.645	40.684	40.723	40.762	40.801	40.840	40.879	980
990	40.879	40.918	40.957	40.996	41.035	41.074	41.113	41.152	41.191	41.230	41.269	990
DEG C	0	1	2	3	4	5	6	7	8	9	10	DEG C

TABLE 10.9—*Type K thermocouples (continued).*

Temperature in Degrees Celsius (IPTS 1968)

EMF in Absolute Millivolts — Reference Junctions at 0 C

DEG C	0	1	2	3	4	5	6	7	8	9	10	DEG C
THERMOELECTRIC VOLTAGE IN ABSOLUTE MILLIVOLTS												
1,000	41.269	41.308	41.347	41.385	41.424	41.463	41.502	41.541	41.580	41.619	41.657	1,000
1,010	41.657	41.696	41.735	41.774	41.813	41.851	41.890	41.929	41.968	42.006	42.045	1,010
1,020	42.045	42.084	42.123	42.161	42.200	42.239	42.277	42.316	42.355	42.393	42.432	1,020
1,030	42.432	42.470	42.509	42.548	42.586	42.625	42.663	42.702	42.740	42.779	42.817	1,030
1,040	42.817	42.856	42.894	42.933	42.971	43.010	43.048	43.087	43.125	43.164	43.202	1,040
1,050	43.202	43.240	43.279	43.317	43.356	43.394	43.432	43.471	43.509	43.547	43.585	1,050
1,060	43.585	43.624	43.662	43.700	43.739	43.777	43.815	43.853	43.891	43.930	43.968	1,060
1,070	43.968	44.006	44.044	44.082	44.121	44.159	44.197	44.235	44.273	44.311	44.349	1,070
1,080	44.349	44.387	44.425	44.463	44.501	44.539	44.577	44.615	44.653	44.691	44.729	1,080
1,090	44.729	44.767	44.805	44.843	44.881	44.919	44.957	44.995	45.033	45.070	45.108	1,090
1,100	45.108	45.146	45.184	45.222	45.260	45.297	45.335	45.373	45.411	45.448	45.486	1,100
1,110	45.486	45.524	45.561	45.599	45.637	45.675	45.712	45.750	45.787	45.825	45.863	1,110
1,120	45.863	45.900	45.938	45.975	46.013	46.051	46.088	46.126	46.163	46.201	46.238	1,120
1,130	46.238	46.275	46.313	46.350	46.388	46.425	46.463	46.500	46.537	46.575	46.612	1,130
1,140	46.612	46.649	46.687	46.724	46.761	46.799	46.836	46.873	46.910	46.948	46.985	1,140
1,150	46.985	47.022	47.059	47.096	47.134	47.171	47.208	47.245	47.282	47.319	47.356	1,150
1,160	47.356	47.393	47.430	47.468	47.505	47.542	47.579	47.616	47.653	47.689	47.726	1,160
1,170	47.726	47.763	47.800	47.837	47.874	47.911	47.948	47.985	48.021	48.058	48.095	1,170
1,180	48.095	48.132	48.169	48.205	48.242	48.279	48.316	48.352	48.389	48.426	48.462	1,180
1,190	48.462	48.499	48.536	48.572	48.609	48.645	48.682	48.718	48.755	48.792	48.828	1,190
1,200	48.828	48.865	48.901	48.937	48.974	49.010	49.047	49.083	49.120	49.156	49.192	1,200
1,210	49.192	49.229	49.265	49.301	49.338	49.374	49.410	49.446	49.483	49.519	49.555	1,210
1,220	49.555	49.591	49.627	49.663	49.700	49.736	49.772	49.808	49.844	49.880	49.916	1,220
1,230	49.916	49.952	49.988	50.024	50.060	50.096	50.132	50.168	50.204	50.240	50.276	1,230
1,240	50.276	50.311	50.347	50.383	50.419	50.455	50.491	50.526	50.562	50.598	50.633	1,240
1,250	50.633	50.669	50.705	50.741	50.776	50.812	50.847	50.883	50.919	50.954	50.990	1,250
1,260	50.990	51.025	51.061	51.096	51.132	51.167	51.203	51.238	51.274	51.309	51.344	1,260
1,270	51.344	51.380	51.415	51.450	51.486	51.521	51.556	51.592	51.627	51.662	51.697	1,270
1,280	51.697	51.733	51.768	51.803	51.838	51.873	51.908	51.943	51.979	52.014	52.049	1,280
1,290	52.049	52.084	52.119	52.154	52.189	52.224	52.259	52.294	52.329	52.364	52.398	1,290
1,300	52.398	52.433	52.468	52.503	52.538	52.573	52.608	52.642	52.677	52.712	52.747	1,300
1,310	52.747	52.781	52.816	52.851	52.886	52.920	52.955	52.989	53.024	53.059	53.093	1,310
1,320	53.093	53.128	53.162	53.197	53.232	53.266	53.301	53.335	53.370	53.404	53.439	1,320
1,330	53.439	53.473	53.507	53.542	53.576	53.611	53.645	53.679	53.714	53.748	53.782	1,330
1,340	53.782	53.817	53.851	53.885	53.920	53.954	53.988	54.022	54.057	54.091	54.125	1,340
1,350	54.125	54.159	54.193	54.228	54.262	54.296	54.330	54.364	54.398	54.432	54.466	1,350
1,360	54.466	54.501	54.535	54.569	54.603	54.637	54.671	54.705	54.739	54.773	54.807	1,360
1,370	54.807	54.841	54.875									1,370
DEG C	0	1	2	3	4	5	6	7	8	9	10	DEG C

TABLE 10.10—*Type R thermocouples (deg F-millivolts).*

Temperature in Degrees Fahrenheit[a]

EMF in Absolute Millivolts | Reference Junctions at 32 F

DEG F	0	1	2	3	4	5	6	7	8	9	10	DEG F
					THERMOELECTRIC VOLTAGE IN ABSOLUTE MILLIVOLTS							
-50	-0.210	-0.212	-0.214	-0.216	-0.218	-0.220	-0.222	-0.224	-0.226			-50
-40	-0.188	-0.190	-0.192	-0.194	-0.197	-0.199	-0.201	-0.203	-0.205	-0.207	-0.210	-40
-30	-0.165	-0.167	-0.169	-0.172	-0.174	-0.176	-0.179	-0.181	-0.183	-0.185	-0.188	-30
-20	-0.141	-0.143	-0.145	-0.148	-0.150	-0.153	-0.155	-0.158	-0.160	-0.162	-0.165	-20
-10	-0.116	-0.118	-0.121	-0.123	-0.126	-0.128	-0.131	-0.133	-0.136	-0.138	-0.141	-10
- 0	-0.089	-0.092	-0.095	-0.097	-0.100	-0.103	-0.105	-0.108	-0.110	-0.113	-0.116	- 0
0	-0.089	-0.087	-0.084	-0.082	-0.079	-0.076	-0.073	-0.071	-0.068	-0.065	-0.063	0
10	-0.063	-0.060	-0.057	-0.054	-0.051	-0.049	-0.046	-0.043	-0.040	-0.037	-0.035	10
20	-0.035	-0.032	-0.029	-0.026	-0.023	-0.020	-0.017	-0.015	-0.012	-0.009	-0.006	20
30	-0.006	-0.003	0.000	0.003	0.006	0.009	0.012	0.015	0.018	0.021	0.024	30
40	0.024	0.027	0.030	0.033	0.036	0.039	0.042	0.045	0.048	0.051	0.054	40
50	0.054	0.057	0.060	0.064	0.067	0.070	0.073	0.076	0.079	0.082	0.086	50
60	0.086	0.089	0.092	0.095	0.098	0.101	0.105	0.108	0.111	0.114	0.118	60
70	0.118	0.121	0.124	0.127	0.131	0.134	0.137	0.141	0.144	0.147	0.150	70
80	0.150	0.154	0.157	0.161	0.164	0.167	0.171	0.174	0.177	0.181	0.184	80
90	0.184	0.188	0.191	0.194	0.198	0.201	0.205	0.208	0.212	0.215	0.218	90
100	0.218	0.222	0.225	0.229	0.232	0.236	0.239	0.243	0.246	0.250	0.253	100
110	0.253	0.257	0.261	0.264	0.268	0.271	0.275	0.278	0.282	0.286	0.289	110
120	0.289	0.293	0.296	0.300	0.304	0.307	0.311	0.315	0.318	0.322	0.326	120
130	0.326	0.329	0.333	0.337	0.340	0.344	0.348	0.351	0.355	0.359	0.363	130
140	0.363	0.366	0.370	0.374	0.378	0.381	0.385	0.389	0.393	0.397	0.400	140
150	0.400	0.404	0.408	0.412	0.416	0.419	0.423	0.427	0.431	0.435	0.439	150
160	0.439	0.443	0.446	0.450	0.454	0.458	0.462	0.466	0.470	0.474	0.478	160
170	0.478	0.482	0.485	0.489	0.493	0.497	0.501	0.505	0.509	0.513	0.517	170
180	0.517	0.521	0.525	0.529	0.533	0.537	0.541	0.545	0.549	0.553	0.557	180
190	0.557	0.561	0.565	0.569	0.573	0.577	0.581	0.586	0.590	0.594	0.598	190
200	0.598	0.602	0.606	0.610	0.614	0.618	0.622	0.627	0.631	0.635	0.639	200
210	0.639	0.643	0.647	0.651	0.656	0.660	0.664	0.668	0.672	0.676	0.681	210
220	0.681	0.685	0.689	0.693	0.697	0.702	0.706	0.710	0.714	0.719	0.723	220
230	0.723	0.727	0.731	0.736	0.740	0.744	0.748	0.753	0.757	0.761	0.766	230
240	0.766	0.770	0.774	0.778	0.783	0.787	0.791	0.796	0.800	0.804	0.809	240
250	0.809	0.813	0.817	0.822	0.826	0.830	0.835	0.839	0.844	0.848	0.852	250
260	0.852	0.857	0.861	0.866	0.870	0.874	0.879	0.883	0.888	0.892	0.897	260
270	0.897	0.901	0.905	0.910	0.914	0.919	0.923	0.928	0.932	0.937	0.941	270
280	0.941	0.946	0.950	0.955	0.959	0.964	0.968	0.973	0.977	0.982	0.986	280
290	0.986	0.991	0.995	1.000	1.004	1.009	1.013	1.018	1.022	1.027	1.032	290
300	1.032	1.036	1.041	1.045	1.050	1.054	1.059	1.064	1.068	1.073	1.077	300
310	1.077	1.082	1.087	1.091	1.096	1.101	1.105	1.110	1.114	1.119	1.124	310
320	1.124	1.128	1.133	1.138	1.142	1.147	1.152	1.156	1.161	1.166	1.170	320
330	1.170	1.175	1.180	1.184	1.189	1.194	1.199	1.203	1.208	1.213	1.217	330
340	1.217	1.222	1.227	1.232	1.236	1.241	1.246	1.251	1.255	1.260	1.265	340
350	1.265	1.270	1.274	1.279	1.284	1.289	1.294	1.298	1.303	1.308	1.313	350
360	1.313	1.318	1.322	1.327	1.332	1.337	1.342	1.346	1.351	1.356	1.361	360
370	1.361	1.366	1.371	1.375	1.380	1.385	1.390	1.395	1.400	1.405	1.409	370
380	1.409	1.414	1.419	1.424	1.429	1.434	1.439	1.444	1.449	1.453	1.458	380
390	1.458	1.463	1.468	1.473	1.478	1.483	1.488	1.493	1.498	1.503	1.508	390
400	1.508	1.512	1.517	1.522	1.527	1.532	1.537	1.542	1.547	1.552	1.557	400
410	1.557	1.562	1.567	1.572	1.577	1.582	1.587	1.592	1.597	1.602	1.607	410
420	1.607	1.612	1.617	1.622	1.627	1.632	1.637	1.642	1.647	1.652	1.657	420
430	1.657	1.662	1.667	1.672	1.677	1.682	1.687	1.692	1.698	1.703	1.708	430
440	1.708	1.713	1.718	1.723	1.728	1.733	1.738	1.743	1.748	1.753	1.758	440
450	1.758	1.764	1.769	1.774	1.779	1.784	1.789	1.794	1.799	1.804	1.810	450
460	1.810	1.815	1.820	1.825	1.830	1.835	1.840	1.845	1.851	1.856	1.861	460
470	1.861	1.866	1.871	1.876	1.882	1.887	1.892	1.897	1.902	1.907	1.913	470
480	1.913	1.918	1.923	1.928	1.933	1.938	1.944	1.949	1.954	1.959	1.964	480
490	1.964	1.970	1.975	1.980	1.985	1.991	1.996	2.001	2.006	2.011	2.017	490
500	2.017	2.022	2.027	2.032	2.038	2.043	2.048	2.053	2.059	2.064	2.069	500
510	2.069	2.074	2.080	2.085	2.090	2.095	2.101	2.106	2.111	2.117	2.122	510
520	2.122	2.127	2.132	2.138	2.143	2.148	2.154	2.159	2.164	2.170	2.175	520
530	2.175	2.180	2.186	2.191	2.196	2.201	2.207	2.212	2.217	2.223	2.228	530
540	2.228	2.233	2.239	2.244	2.249	2.255	2.260	2.266	2.271	2.276	2.282	540
DEG F	0	1	2	3	4	5	6	7	8	9	10	DEG F

[a] Converted from degrees Celsius (IPTS 1968).

TABLE 10.10—*Type R thermocouples (continued).*

Temperature in Degrees Fahrenheit[a]

EMF in Absolute Millivolts

Reference Junctions at 32 F

DEG F	0	1	2	3	4	5	6	7	8	9	10	DEG F
					THERMOELECTRIC VOLTAGE IN ABSOLUTE MILLIVOLTS							
550	2.282	2.287	2.292	2.298	2.303	2.308	2.314	2.319	2.325	2.330	2.335	550
560	2.335	2.341	2.346	2.351	2.357	2.362	2.368	2.373	2.378	2.384	2.389	560
570	2.389	2.395	2.400	2.405	2.411	2.416	2.422	2.427	2.433	2.438	2.443	570
580	2.443	2.449	2.454	2.460	2.465	2.471	2.476	2.481	2.487	2.492	2.498	580
590	2.498	2.503	2.509	2.514	2.520	2.525	2.531	2.536	2.541	2.547	2.552	590
600	2.552	2.558	2.563	2.569	2.574	2.580	2.585	2.591	2.596	2.602	2.607	600
610	2.607	2.613	2.618	2.624	2.629	2.635	2.640	2.646	2.651	2.657	2.662	610
620	2.662	2.668	2.673	2.679	2.684	2.690	2.695	2.701	2.706	2.712	2.718	620
630	2.718	2.723	2.729	2.734	2.740	2.745	2.751	2.756	2.762	2.767	2.773	630
640	2.773	2.779	2.784	2.790	2.795	2.801	2.806	2.812	2.818	2.823	2.829	640
650	2.829	2.834	2.840	2.845	2.851	2.857	2.862	2.868	2.873	2.879	2.885	650
660	2.885	2.890	2.896	2.901	2.907	2.913	2.918	2.924	2.929	2.935	2.941	660
670	2.941	2.946	2.952	2.957	2.963	2.969	2.974	2.980	2.986	2.991	2.997	670
680	2.997	3.002	3.008	3.014	3.019	3.025	3.031	3.036	3.042	3.048	3.053	680
690	3.053	3.059	3.065	3.070	3.076	3.082	3.087	3.093	3.099	3.104	3.110	690
700	3.110	3.116	3.121	3.127	3.133	3.138	3.144	3.150	3.155	3.161	3.167	700
710	3.167	3.172	3.178	3.184	3.189	3.195	3.201	3.207	3.212	3.218	3.224	710
720	3.224	3.229	3.235	3.241	3.247	3.252	3.258	3.264	3.269	3.275	3.281	720
730	3.281	3.287	3.292	3.298	3.304	3.309	3.315	3.321	3.327	3.332	3.338	730
740	3.338	3.344	3.350	3.355	3.361	3.367	3.373	3.378	3.384	3.390	3.396	740
750	3.396	3.401	3.407	3.413	3.419	3.424	3.430	3.436	3.442	3.448	3.453	750
760	3.453	3.459	3.465	3.471	3.476	3.482	3.488	3.494	3.500	3.505	3.511	760
770	3.511	3.517	3.523	3.529	3.534	3.540	3.546	3.552	3.558	3.563	3.569	770
780	3.569	3.575	3.581	3.587	3.592	3.598	3.604	3.610	3.616	3.622	3.627	780
790	3.627	3.633	3.639	3.645	3.651	3.657	3.662	3.668	3.674	3.680	3.686	790
800	3.686	3.692	3.697	3.703	3.709	3.715	3.721	3.727	3.733	3.738	3.744	800
810	3.744	3.750	3.756	3.762	3.768	3.774	3.779	3.785	3.791	3.797	3.803	810
820	3.803	3.809	3.815	3.821	3.826	3.832	3.838	3.844	3.850	3.856	3.862	820
830	3.862	3.868	3.874	3.879	3.885	3.891	3.897	3.903	3.909	3.915	3.921	830
840	3.921	3.927	3.933	3.938	3.944	3.950	3.956	3.962	3.968	3.974	3.980	840
850	3.980	3.986	3.992	3.998	4.004	4.009	4.015	4.021	4.027	4.033	4.039	850
860	4.039	4.045	4.051	4.057	4.063	4.069	4.075	4.081	4.087	4.093	4.099	860
870	4.099	4.105	4.110	4.116	4.122	4.128	4.134	4.140	4.146	4.152	4.158	870
880	4.158	4.164	4.170	4.176	4.182	4.188	4.194	4.200	4.206	4.212	4.218	880
890	4.218	4.224	4.230	4.236	4.242	4.248	4.254	4.260	4.266	4.272	4.278	890
900	4.278	4.284	4.290	4.296	4.302	4.308	4.314	4.320	4.326	4.332	4.338	900
910	4.338	4.344	4.350	4.356	4.362	4.368	4.374	4.380	4.386	4.392	4.398	910
920	4.398	4.404	4.410	4.416	4.422	4.428	4.434	4.440	4.446	4.452	4.458	920
930	4.458	4.465	4.471	4.477	4.483	4.489	4.495	4.501	4.507	4.513	4.519	930
940	4.519	4.525	4.531	4.537	4.543	4.549	4.555	4.561	4.567	4.574	4.580	940
950	4.580	4.586	4.592	4.598	4.604	4.610	4.616	4.622	4.628	4.634	4.640	950
960	4.640	4.647	4.653	4.659	4.665	4.671	4.677	4.683	4.689	4.695	4.701	960
970	4.701	4.707	4.714	4.720	4.726	4.732	4.738	4.744	4.750	4.756	4.762	970
980	4.762	4.769	4.775	4.781	4.787	4.793	4.799	4.805	4.811	4.818	4.824	980
990	4.824	4.830	4.836	4.842	4.848	4.854	4.860	4.867	4.873	4.879	4.885	990
1,000	4.885	4.891	4.897	4.904	4.910	4.916	4.922	4.928	4.934	4.940	4.947	1,000
1,010	4.947	4.953	4.959	4.965	4.971	4.977	4.984	4.990	4.996	5.002	5.008	1,010
1,020	5.008	5.014	5.021	5.027	5.033	5.039	5.045	5.052	5.058	5.064	5.070	1,020
1,030	5.070	5.076	5.082	5.089	5.095	5.101	5.107	5.113	5.120	5.126	5.132	1,030
1,040	5.132	5.138	5.144	5.151	5.157	5.163	5.169	5.175	5.182	5.188	5.194	1,040
1,050	5.194	5.200	5.207	5.213	5.219	5.225	5.231	5.238	5.244	5.250	5.256	1,050
1,060	5.256	5.263	5.269	5.275	5.281	5.288	5.294	5.300	5.306	5.313	5.319	1,060
1,070	5.319	5.325	5.331	5.337	5.344	5.350	5.356	5.362	5.369	5.375	5.381	1,070
1,080	5.381	5.388	5.394	5.400	5.406	5.413	5.419	5.425	5.431	5.438	5.444	1,080
1,090	5.444	5.450	5.456	5.463	5.469	5.475	5.482	5.488	5.494	5.500	5.507	1,090
1,100	5.507	5.513	5.519	5.526	5.532	5.538	5.544	5.551	5.557	5.563	5.570	1,100
1,110	5.570	5.576	5.582	5.589	5.595	5.601	5.607	5.614	5.620	5.626	5.633	1,110
1,120	5.633	5.639	5.645	5.652	5.658	5.664	5.671	5.677	5.683	5.690	5.696	1,120
1,130	5.696	5.702	5.709	5.715	5.721	5.728	5.734	5.740	5.747	5.753	5.759	1,130
1,140	5.759	5.766	5.772	5.778	5.785	5.791	5.797	5.804	5.810	5.816	5.823	1,140
1,150	5.823	5.829	5.835	5.842	5.848	5.855	5.861	5.867	5.874	5.880	5.886	1,150
1,160	5.886	5.893	5.899	5.905	5.912	5.918	5.925	5.931	5.937	5.944	5.950	1,160
1,170	5.950	5.957	5.963	5.969	5.976	5.982	5.988	5.995	6.001	6.008	6.014	1,170
1,180	6.014	6.021	6.027	6.033	6.040	6.046	6.053	6.059	6.065	6.072	6.078	1,180
1,190	6.078	6.085	6.091	6.098	6.104	6.110	6.117	6.123	6.130	6.136	6.143	1,190
DEG F	0	1	2	3	4	5	6	7	8	9	10	DEG F

[a] Converted from degrees Celsius (IPTS 1968).

TABLE 10.10—*Type R thermocouples (continued).*

Temperature in Degrees Fahrenheit[a]

EMF in Absolute Millivolts — Reference Junctions at 32 F

DEG F	0	1	2	3	4	5	6	7	8	9	10	DEG F
	THERMOELECTRIC VOLTAGE IN ABSOLUTE MILLIVOLTS											
1,200	6.143	6.149	6.155	6.162	6.168	6.175	6.181	6.188	6.194	6.201	6.207	1,200
1,210	6.207	6.213	6.220	6.226	6.233	6.239	6.246	6.252	6.259	6.265	6.272	1,210
1,220	6.272	6.278	6.285	6.291	6.297	6.304	6.310	6.317	6.323	6.330	6.336	1,220
1,230	6.336	6.343	6.349	6.356	6.362	6.369	6.375	6.382	6.388	6.395	6.401	1,230
1,240	6.401	6.408	6.414	6.421	6.427	6.434	6.440	6.447	6.453	6.460	6.466	1,240
1,250	6.466	6.473	6.479	6.486	6.492	6.499	6.505	6.512	6.518	6.525	6.532	1,250
1,260	6.532	6.538	6.545	6.551	6.558	6.564	6.571	6.577	6.584	6.590	6.597	1,260
1,270	6.597	6.603	6.610	6.616	6.623	6.630	6.636	6.643	6.649	6.656	6.662	1,270
1,280	6.662	6.669	6.675	6.682	6.689	6.695	6.702	6.708	6.715	6.721	6.728	1,280
1,290	6.728	6.735	6.741	6.748	6.754	6.761	6.767	6.774	6.781	6.787	6.794	1,290
1,300	6.794	6.800	6.807	6.814	6.820	6.827	6.833	6.840	6.847	6.853	6.860	1,300
1,310	6.860	6.866	6.873	6.880	6.886	6.893	6.899	6.906	6.913	6.919	6.926	1,310
1,320	6.926	6.932	6.939	6.946	6.952	6.959	6.966	6.972	6.979	6.985	6.992	1,320
1,330	6.992	6.999	7.005	7.012	7.019	7.025	7.032	7.039	7.045	7.052	7.059	1,330
1,340	7.059	7.065	7.072	7.078	7.085	7.092	7.098	7.105	7.112	7.118	7.125	1,340
1,350	7.125	7.132	7.138	7.145	7.152	7.158	7.165	7.172	7.178	7.185	7.192	1,350
1,360	7.192	7.198	7.205	7.212	7.218	7.225	7.232	7.239	7.245	7.252	7.259	1,360
1,370	7.259	7.265	7.272	7.279	7.285	7.292	7.299	7.305	7.312	7.319	7.326	1,370
1,380	7.326	7.332	7.339	7.346	7.352	7.359	7.366	7.373	7.379	7.386	7.393	1,380
1,390	7.393	7.399	7.406	7.413	7.420	7.426	7.433	7.440	7.447	7.453	7.460	1,390
1,400	7.460	7.467	7.474	7.480	7.487	7.494	7.500	7.507	7.514	7.521	7.527	1,400
1,410	7.527	7.534	7.541	7.548	7.554	7.561	7.568	7.575	7.582	7.588	7.595	1,410
1,420	7.595	7.602	7.609	7.615	7.622	7.629	7.636	7.642	7.649	7.656	7.663	1,420
1,430	7.663	7.670	7.676	7.683	7.690	7.697	7.703	7.710	7.717	7.724	7.731	1,430
1,440	7.731	7.737	7.744	7.751	7.758	7.765	7.771	7.778	7.785	7.792	7.799	1,440
1,450	7.799	7.805	7.812	7.819	7.826	7.833	7.840	7.846	7.853	7.860	7.867	1,450
1,460	7.867	7.874	7.880	7.887	7.894	7.901	7.908	7.915	7.921	7.928	7.935	1,460
1,470	7.935	7.942	7.949	7.956	7.963	7.969	7.976	7.983	7.990	7.997	8.004	1,470
1,480	8.004	8.010	8.017	8.024	8.031	8.038	8.045	8.052	8.058	8.065	8.072	1,480
1,490	8.072	8.079	8.086	8.093	8.100	8.107	8.113	8.120	8.127	8.134	8.141	1,490
1,500	8.141	8.148	8.155	8.162	8.168	8.175	8.182	8.189	8.196	8.203	8.210	1,500
1,510	8.210	8.217	8.224	8.231	8.237	8.244	8.251	8.258	8.265	8.272	8.279	1,510
1,520	8.279	8.286	8.293	8.300	8.306	8.313	8.320	8.327	8.334	8.341	8.348	1,520
1,530	8.348	8.355	8.362	8.369	8.376	8.383	8.390	8.397	8.403	8.410	8.417	1,530
1,540	8.417	8.424	8.431	8.438	8.445	8.452	8.459	8.466	8.473	8.480	8.487	1,540
1,550	8.487	8.494	8.501	8.508	8.515	8.522	8.529	8.535	8.542	8.549	8.556	1,550
1,560	8.556	8.563	8.570	8.577	8.584	8.591	8.598	8.605	8.612	8.619	8.626	1,560
1,570	8.626	8.633	8.640	8.647	8.654	8.661	8.668	8.675	8.682	8.689	8.696	1,570
1,580	8.696	8.703	8.710	8.717	8.724	8.731	8.738	8.745	8.752	8.759	8.766	1,580
1,590	8.766	8.773	8.780	8.787	8.794	8.801	8.808	8.815	8.822	8.829	8.836	1,590
1,600	8.836	8.843	8.850	8.857	8.864	8.871	8.878	8.885	8.892	8.899	8.907	1,600
1,610	8.907	8.914	8.921	8.928	8.935	8.942	8.949	8.956	8.963	8.970	8.977	1,610
1,620	8.977	8.984	8.991	8.998	9.005	9.012	9.019	9.026	9.033	9.040	9.048	1,620
1,630	9.048	9.055	9.062	9.069	9.076	9.083	9.090	9.097	9.104	9.111	9.118	1,630
1,640	9.118	9.125	9.132	9.140	9.147	9.154	9.161	9.168	9.175	9.182	9.189	1,640
1,650	9.189	9.196	9.203	9.210	9.218	9.225	9.232	9.239	9.246	9.253	9.260	1,650
1,660	9.260	9.267	9.274	9.282	9.289	9.296	9.303	9.310	9.317	9.324	9.331	1,660
1,670	9.331	9.338	9.346	9.353	9.360	9.367	9.374	9.381	9.388	9.395	9.403	1,670
1,680	9.403	9.410	9.417	9.424	9.431	9.438	9.445	9.453	9.460	9.467	9.474	1,680
1,690	9.474	9.481	9.488	9.495	9.503	9.510	9.517	9.524	9.531	9.538	9.546	1,690
1,700	9.546	9.553	9.560	9.567	9.574	9.581	9.589	9.596	9.603	9.610	9.617	1,700
1,710	9.617	9.624	9.632	9.639	9.646	9.653	9.660	9.668	9.675	9.682	9.689	1,710
1,720	9.689	9.696	9.704	9.711	9.718	9.725	9.732	9.740	9.747	9.754	9.761	1,720
1,730	9.761	9.768	9.776	9.783	9.790	9.797	9.804	9.812	9.819	9.826	9.833	1,730
1,740	9.833	9.840	9.848	9.855	9.862	9.869	9.877	9.884	9.891	9.898	9.906	1,740
1,750	9.906	9.913	9.920	9.927	9.934	9.942	9.949	9.956	9.963	9.971	9.978	1,750
1,760	9.978	9.985	9.992	10.000	10.007	10.014	10.021	10.029	10.036	10.043	10.050	1,760
1,770	10.050	10.058	10.065	10.072	10.079	10.087	10.094	10.101	10.109	10.116	10.123	1,770
1,780	10.123	10.130	10.138	10.145	10.152	10.159	10.167	10.174	10.181	10.189	10.196	1,780
1,790	10.196	10.203	10.210	10.218	10.225	10.232	10.240	10.247	10.254	10.262	10.269	1,790
1,800	10.269	10.276	10.283	10.291	10.298	10.305	10.313	10.320	10.327	10.335	10.342	1,800
1,810	10.342	10.349	10.357	10.364	10.371	10.379	10.386	10.393	10.400	10.408	10.415	1,810
1,820	10.415	10.422	10.430	10.437	10.444	10.452	10.459	10.466	10.474	10.481	10.488	1,820
1,830	10.488	10.496	10.503	10.511	10.518	10.525	10.533	10.540	10.547	10.555	10.562	1,830
1,840	10.562	10.569	10.577	10.584	10.591	10.599	10.606	10.613	10.621	10.628	10.636	1,840
DEG F.	0	1	2	3	4	5	6	7	8	9	10	DEG F

[a] Converted from degrees Celsius (IPTS 1968).

TABLE 10.10—*Type R thermocouples (continued).*

Temperature in Degrees Fahrenheit[a]

EMF in Absolute Millivolts | **Reference Junctions at 32 F**

THERMOELECTRIC VOLTAGE IN ABSOLUTE MILLIVOLTS

DEG F	0	1	2	3	4	5	6	7	8	9	10	DEG F
1,850	10.636	10.643	10.650	10.658	10.665	10.672	10.680	10.687	10.695	10.702	10.709	1,850
1,860	10.709	10.717	10.724	10.731	10.739	10.746	10.754	10.761	10.768	10.776	10.783	1,860
1,870	10.783	10.791	10.798	10.805	10.813	10.820	10.828	10.835	10.842	10.850	10.857	1,870
1,880	10.857	10.865	10.872	10.879	10.887	10.894	10.902	10.909	10.917	10.924	10.931	1,880
1,890	10.931	10.939	10.946	10.954	10.961	10.968	10.976	10.983	10.991	10.998	11.006	1,890
1,900	11.006	11.013	11.021	11.028	11.035	11.043	11.050	11.058	11.065	11.073	11.080	1,900
1,910	11.080	11.088	11.095	11.102	11.110	11.117	11.125	11.132	11.140	11.147	11.155	1,910
1,920	11.155	11.162	11.170	11.177	11.184	11.192	11.199	11.207	11.214	11.222	11.229	1,920
1,930	11.229	11.237	11.244	11.252	11.259	11.267	11.274	11.282	11.289	11.297	11.304	1,930
1,940	11.304	11.312	11.319	11.327	11.334	11.342	11.349	11.357	11.364	11.372	11.379	1,940
1,950	11.379	11.387	11.394	11.402	11.409	11.417	11.424	11.432	11.439	11.447	11.454	1,950
1,960	11.454	11.462	11.469	11.477	11.484	11.492	11.499	11.507	11.514	11.522	11.529	1,960
1,970	11.529	11.537	11.544	11.552	11.559	11.567	11.574	11.582	11.590	11.597	11.605	1,970
1,980	11.605	11.612	11.620	11.627	11.635	11.642	11.650	11.657	11.665	11.672	11.680	1,980
1,990	11.680	11.688	11.695	11.703	11.710	11.718	11.725	11.733	11.740	11.748	11.756	1,990
2,000	11.756	11.763	11.771	11.778	11.786	11.793	11.801	11.808	11.816	11.824	11.831	2,000
2,010	11.831	11.839	11.846	11.854	11.861	11.869	11.877	11.884	11.892	11.899	11.907	2,010
2,020	11.907	11.914	11.922	11.930	11.937	11.945	11.952	11.960	11.968	11.975	11.983	2,020
2,030	11.983	11.990	11.998	12.005	12.013	12.021	12.028	12.036	12.043	12.051	12.059	2,030
2,040	12.059	12.066	12.074	12.081	12.089	12.097	12.104	12.112	12.119	12.127	12.135	2,040
2,050	12.135	12.142	12.150	12.157	12.165	12.173	12.180	12.188	12.196	12.203	12.211	2,050
2,060	12.211	12.218	12.226	12.234	12.241	12.249	12.257	12.264	12.272	12.279	12.287	2,060
2,070	12.287	12.295	12.302	12.310	12.318	12.325	12.333	12.340	12.348	12.356	12.363	2,070
2,080	12.363	12.371	12.379	12.386	12.394	12.402	12.409	12.417	12.424	12.432	12.440	2,080
2,090	12.440	12.447	12.455	12.463	12.470	12.478	12.486	12.493	12.501	12.509	12.516	2,090
2,100	12.516	12.524	12.532	12.539	12.547	12.555	12.562	12.570	12.577	12.585	12.593	2,100
2,110	12.593	12.600	12.608	12.616	12.623	12.631	12.639	12.646	12.654	12.662	12.669	2,110
2,120	12.669	12.677	12.685	12.693	12.700	12.708	12.716	12.723	12.731	12.739	12.746	2,120
2,130	12.746	12.754	12.762	12.769	12.777	12.785	12.792	12.800	12.808	12.815	12.823	2,130
2,140	12.823	12.831	12.838	12.846	12.854	12.862	12.869	12.877	12.885	12.892	12.900	2,140
2,150	12.900	12.908	12.915	12.923	12.931	12.938	12.946	12.954	12.962	12.969	12.977	2,150
2,160	12.977	12.985	12.992	13.000	13.008	13.016	13.023	13.031	13.039	13.046	13.054	2,160
2,170	13.054	13.062	13.069	13.077	13.085	13.093	13.100	13.108	13.116	13.123	13.131	2,170
2,180	13.131	13.139	13.147	13.154	13.162	13.170	13.178	13.185	13.193	13.201	13.208	2,180
2,190	13.208	13.216	13.224	13.232	13.239	13.247	13.255	13.263	13.270	13.278	13.286	2,190
2,200	13.286	13.293	13.301	13.309	13.317	13.324	13.332	13.340	13.348	13.355	13.363	2,200
2,210	13.363	13.371	13.379	13.386	13.394	13.402	13.409	13.417	13.425	13.433	13.440	2,210
2,220	13.440	13.448	13.456	13.464	13.471	13.479	13.487	13.495	13.502	13.510	13.518	2,220
2,230	13.518	13.526	13.533	13.541	13.549	13.557	13.564	13.572	13.580	13.588	13.595	2,230
2,240	13.595	13.603	13.611	13.619	13.627	13.634	13.642	13.650	13.658	13.665	13.673	2,240
2,250	13.673	13.681	13.689	13.696	13.704	13.712	13.720	13.727	13.735	13.743	13.751	2,250
2,260	13.751	13.759	13.766	13.774	13.782	13.790	13.797	13.805	13.813	13.821	13.828	2,260
2,270	13.828	13.836	13.844	13.852	13.860	13.867	13.875	13.883	13.891	13.898	13.906	2,270
2,280	13.906	13.914	13.922	13.930	13.937	13.945	13.953	13.961	13.968	13.976	13.984	2,280
2,290	13.984	13.992	14.000	14.007	14.015	14.023	14.031	14.039	14.046	14.054	14.062	2,290
2,300	14.062	14.070	14.078	14.085	14.093	14.101	14.109	14.116	14.124	14.132	14.140	2,300
2,310	14.140	14.148	14.155	14.163	14.171	14.179	14.187	14.194	14.202	14.210	14.218	2,310
2,320	14.218	14.226	14.233	14.241	14.249	14.257	14.265	14.272	14.280	14.288	14.296	2,320
2,330	14.296	14.304	14.311	14.319	14.327	14.335	14.343	14.350	14.358	14.366	14.374	2,330
2,340	14.374	14.382	14.389	14.397	14.405	14.413	14.421	14.429	14.436	14.444	14.452	2,340
2,350	14.452	14.460	14.468	14.475	14.483	14.491	14.499	14.507	14.514	14.522	14.530	2,350
2,360	14.530	14.538	14.546	14.554	14.561	14.569	14.577	14.585	14.593	14.600	14.608	2,360
2,370	14.608	14.616	14.624	14.632	14.640	14.647	14.655	14.663	14.671	14.679	14.686	2,370
2,380	14.686	14.694	14.702	14.710	14.718	14.726	14.733	14.741	14.749	14.757	14.765	2,380
2,390	14.765	14.772	14.780	14.788	14.796	14.804	14.812	14.819	14.827	14.835	14.843	2,390
2,400	14.843	14.851	14.859	14.866	14.874	14.882	14.890	14.898	14.906	14.913	14.921	2,400
2,410	14.921	14.929	14.937	14.945	14.953	14.960	14.968	14.976	14.984	14.992	15.000	2,410
2,420	15.000	15.007	15.015	15.023	15.031	15.039	15.047	15.054	15.062	15.070	15.078	2,420
2,430	15.078	15.086	15.094	15.101	15.109	15.117	15.125	15.133	15.141	15.148	15.156	2,430
2,440	15.156	15.164	15.172	15.180	15.188	15.195	15.203	15.211	15.219	15.227	15.235	2,440
2,450	15.235	15.242	15.250	15.258	15.266	15.274	15.282	15.289	15.297	15.305	15.313	2,450
2,460	15.313	15.321	15.329	15.337	15.344	15.352	15.360	15.368	15.376	15.384	15.391	2,460
2,470	15.391	15.399	15.407	15.415	15.423	15.431	15.438	15.446	15.454	15.462	15.470	2,470
2,480	15.470	15.478	15.486	15.493	15.501	15.509	15.517	15.525	15.533	15.540	15.548	2,480
2,490	15.548	15.556	15.564	15.572	15.580	15.587	15.595	15.603	15.611	15.619	15.627	2,490
DEG F	0	1	2	3	4	5	6	7	8	9	10	DEG F

[a] Converted from degrees Celsius (IPTS 1968).

TABLE 10.10—*Type R thermocouples (continued).*

EMF in Absolute Millivolts — **Temperature in Degrees Fahrenheit**[a] — **Reference Junctions at 32 F**

DEG F	0	1	2	3	4	5	6	7	8	9	10	DEG F
					THERMOELECTRIC VOLTAGE IN ABSOLUTE MILLIVOLTS							
2,500	15.627	15.635	15.642	15.650	15.658	15.666	15.674	15.682	15.689	15.697	15.705	2,500
2,510	15.705	15.713	15.721	15.729	15.737	15.744	15.752	15.760	15.768	15.776	15.784	2,510
2,520	15.784	15.791	15.799	15.807	15.815	15.823	15.831	15.839	15.846	15.854	15.862	2,520
2,530	15.862	15.870	15.878	15.886	15.893	15.901	15.909	15.917	15.925	15.933	15.941	2,530
2,540	15.941	15.948	15.956	15.964	15.972	15.980	15.988	15.995	16.003	16.011	16.019	2,540
2,550	16.019	16.027	16.035	16.043	16.050	16.058	16.066	16.074	16.082	16.090	16.097	2,550
2,560	16.097	16.105	16.113	16.121	16.129	16.137	16.145	16.152	16.160	16.168	16.176	2,560
2,570	16.176	16.184	16.192	16.199	16.207	16.215	16.223	16.231	16.239	16.247	16.254	2,570
2,580	16.254	16.262	16.270	16.278	16.286	16.294	16.301	16.309	16.317	16.325	16.333	2,580
2,590	16.333	16.341	16.349	16.356	16.364	16.372	16.380	16.388	16.396	16.403	16.411	2,590
2,600	16.411	16.419	16.427	16.435	16.443	16.450	16.458	16.466	16.474	16.482	16.490	2,600
2,610	16.490	16.498	16.505	16.513	16.521	16.529	16.537	16.545	16.552	16.560	16.568	2,610
2,620	16.568	16.576	16.584	16.592	16.599	16.607	16.615	16.623	16.631	16.639	16.646	2,620
2,630	16.646	16.654	16.662	16.670	16.678	16.686	16.694	16.701	16.709	16.717	16.725	2,630
2,640	16.725	16.733	16.741	16.748	16.756	16.764	16.772	16.780	16.788	16.795	16.803	2,640
2,650	16.803	16.811	16.819	16.827	16.835	16.842	16.850	16.858	16.866	16.874	16.882	2,650
2,660	16.882	16.889	16.897	16.905	16.913	16.921	16.929	16.936	16.944	16.952	16.960	2,660
2,670	16.960	16.968	16.976	16.983	16.991	16.999	17.007	17.015	17.022	17.030	17.038	2,670
2,680	17.038	17.046	17.054	17.062	17.069	17.077	17.085	17.093	17.101	17.109	17.116	2,680
2,690	17.116	17.124	17.132	17.140	17.148	17.156	17.163	17.171	17.179	17.187	17.195	2,690
2,700	17.195	17.202	17.210	17.218	17.226	17.234	17.242	17.249	17.257	17.265	17.273	2,700
2,710	17.273	17.281	17.288	17.296	17.304	17.312	17.320	17.328	17.335	17.343	17.351	2,710
2,720	17.351	17.359	17.367	17.374	17.382	17.390	17.398	17.406	17.413	17.421	17.429	2,720
2,730	17.429	17.437	17.445	17.453	17.460	17.468	17.476	17.484	17.492	17.499	17.507	2,730
2,740	17.507	17.515	17.523	17.531	17.538	17.546	17.554	17.562	17.570	17.577	17.585	2,740
2,750	17.585	17.593	17.601	17.609	17.616	17.624	17.632	17.640	17.648	17.655	17.663	2,750
2,760	17.663	17.671	17.679	17.687	17.694	17.702	17.710	17.718	17.726	17.733	17.741	2,760
2,770	17.741	17.749	17.757	17.765	17.772	17.780	17.788	17.796	17.804	17.811	17.819	2,770
2,780	17.819	17.827	17.835	17.842	17.850	17.858	17.866	17.874	17.881	17.889	17.897	2,780
2,790	17.897	17.905	17.913	17.920	17.928	17.936	17.944	17.951	17.959	17.967	17.975	2,790
2,800	17.975	17.983	17.990	17.998	18.006	18.014	18.021	18.029	18.037	18.045	18.053	2,800
2,810	18.053	18.060	18.068	18.076	18.084	18.091	18.099	18.107	18.115	18.123	18.130	2,810
2,820	18.130	18.138	18.146	18.154	18.161	18.169	18.177	18.185	18.192	18.200	18.208	2,820
2,830	18.208	18.216	18.223	18.231	18.239	18.247	18.255	18.262	18.270	18.278	18.286	2,830
2,840	18.286	18.293	18.301	18.309	18.317	18.324	18.332	18.340	18.348	18.355	18.363	2,840
2,850	18.363	18.371	18.379	18.386	18.394	18.402	18.410	18.417	18.425	18.433	18.441	2,850
2,860	18.441	18.448	18.456	18.464	18.472	18.479	18.487	18.495	18.502	18.510	18.518	2,860
2,870	18.518	18.526	18.533	18.541	18.549	18.557	18.564	18.572	18.580	18.588	18.595	2,870
2,880	18.595	18.603	18.611	18.619	18.626	18.634	18.642	18.649	18.657	18.665	18.673	2,880
2,890	18.673	18.680	18.688	18.696	18.703	18.711	18.719	18.727	18.734	18.742	18.750	2,890
2,900	18.750	18.758	18.765	18.773	18.781	18.788	18.796	18.804	18.812	18.819	18.827	2,900
2,910	18.827	18.835	18.842	18.850	18.858	18.865	18.873	18.881	18.889	18.896	18.904	2,910
2,920	18.904	18.912	18.919	18.927	18.935	18.943	18.950	18.958	18.966	18.973	18.981	2,920
2,930	18.981	18.989	18.996	19.004	19.012	19.019	19.027	19.035	19.043	19.050	19.058	2,930
2,940	19.058	19.066	19.073	19.081	19.089	19.096	19.104	19.112	19.119	19.127	19.135	2,940
2,950	19.135	19.142	19.150	19.158	19.165	19.173	19.181	19.188	19.196	19.204	19.211	2,950
2,960	19.211	19.219	19.227	19.234	19.242	19.250	19.257	19.265	19.273	19.280	19.288	2,960
2,970	19.288	19.296	19.303	19.311	19.319	19.326	19.334	19.342	19.349	19.357	19.365	2,970
2,980	19.365	19.372	19.380	19.388	19.395	19.403	19.411	19.418	19.426	19.434	19.441	2,980
2,990	19.441	19.449	19.457	19.464	19.472	19.479	19.487	19.495	19.502	19.510	19.518	2,990
3,000	19.518	19.525	19.533	19.541	19.548	19.556	19.563	19.571	19.579	19.586	19.594	3,000
3,010	19.594	19.602	19.609	19.617	19.624	19.632	19.640	19.647	19.655	19.663	19.670	3,010
3,020	19.670	19.678	19.685	19.693	19.701	19.708	19.716	19.723	19.731	19.739	19.746	3,020
3,030	19.746	19.754	19.761	19.769	19.777	19.784	19.792	19.800	19.807	19.815	19.822	3,030
3,040	19.822	19.830	19.837	19.845	19.853	19.860	19.868	19.875	19.883	19.891	19.898	3,040
3,050	19.898	19.906	19.913	19.921	19.929	19.936	19.944	19.951	19.959	19.966	19.974	3,050
3,060	19.974	19.982	19.989	19.997	20.004	20.012	20.019	20.027	20.034	20.042	20.050	3,060
3,070	20.050	20.057	20.065	20.072	20.080	20.087	20.095	20.102	20.110	20.117	20.125	3,070
3,080	20.125	20.132	20.140	20.148	20.155	20.163	20.170	20.178	20.185	20.193	20.200	3,080
3,090	20.200	20.208	20.215	20.223	20.230	20.238	20.245	20.253	20.260	20.268	20.275	3,090
3,100	20.275	20.283	20.290	20.297	20.305	20.312	20.320	20.327	20.335	20.342	20.350	3,100
3,110	20.350	20.357	20.365	20.372	20.380	20.387	20.394	20.402	20.409	20.417	20.424	3,110
3,120	20.424	20.432	20.439	20.446	20.454	20.461	20.469	20.476	20.483	20.491	20.498	3,120
3,130	20.498	20.506	20.513	20.520	20.528	20.535	20.543	20.550	20.557	20.565	20.572	3,130
3,140	20.572	20.579	20.587	20.594	20.601	20.609	20.616	20.623	20.631	20.638	20.645	3,140
DEG F	0	1	2	3	4	5	6	7	8	9	10	DEG F

[a] Converted from degrees Celsius (IPTS 1968).

TABLE 10.10—*Type R thermocouples (continued).*

Temperature in Degrees Fahrenheit[a]

EMF in Absolute Millivolts — Reference Junctions at 32 F

DEG F	0	1	2	3	4	5	6	7	8	9	10	DEG F
					THERMOELECTRIC VOLTAGE IN ABSOLUTE MILLIVOLTS							
3,150	20.645	20.653	20.660	20.667	20.675	20.682	20.689	20.697	20.704	20.711	20.718	3,150
3,160	20.718	20.726	20.733	20.740	20.748	20.755	20.762	20.769	20.777	20.784	20.791	3,160
3,170	20.791	20.798	20.806	20.813	20.820	20.827	20.834	20.842	20.849	20.856	20.863	3,170
3,180	20.863	20.870	20.878	20.885	20.892	20.899	20.906	20.914	20.921	20.928	20.935	3,180
3,190	20.935	20.942	20.949	20.956	20.964	20.971	20.978	20.985	20.992	20.999	21.006	3,190
3,200	21.006	21.013	21.021	21.028	21.035	21.042	21.049	21.056	21.063	21.070	21.077	3,200
3,210	21.077	21.084	21.091	21.098	21.105							3,210
DEG F	0	1	2	3	4	5	6	7	8	9	10	DEG F

[a] Converted from degrees Celsius (IPTS 1968).

TABLE 10.11—*Type R thermocouples (deg C-millivolts).*

Temperature in Degrees Celsius (IPTS 1968)

EMF in Absolute Millivolts | Reference Junctions at 0 C

DEG C	0	1	2	3	4	5	6	7	8	9	10	DEG C
	THERMOELECTRIC VOLTAGE IN ABSOLUTE MILLIVOLTS											
-50	-0.226											-50
-40	-0.188	-0.192	-0.196	-0.200	-0.204	-0.207	-0.211	-0.215	-0.219	-0.223	-0.226	-40
-30	-0.145	-0.150	-0.154	-0.158	-0.163	-0.167	-0.171	-0.175	-0.180	-0.184	-0.188	-30
-20	-0.100	-0.105	-0.109	-0.114	-0.119	-0.123	-0.128	-0.132	-0.137	-0.141	-0.145	-20
-10	-0.051	-0.056	-0.061	-0.066	-0.071	-0.076	-0.081	-0.086	-0.091	-0.095	-0.100	-10
-0	0.000	-0.005	-0.011	-0.016	-0.021	-0.026	-0.031	-0.036	-0.041	-0.046	-0.051	-0
0	0.000	0.005	0.011	0.016	0.021	0.027	0.032	0.038	0.043	0.049	0.054	0
10	0.054	0.060	0.065	0.071	0.077	0.082	0.088	0.094	0.100	0.105	0.111	10
20	0.111	0.117	0.123	0.129	0.135	0.141	0.147	0.152	0.158	0.165	0.171	20
30	0.171	0.177	0.183	0.189	0.195	0.201	0.207	0.214	0.220	0.226	0.232	30
40	0.232	0.239	0.245	0.251	0.258	0.264	0.271	0.277	0.283	0.290	0.296	40
50	0.296	0.303	0.310	0.316	0.323	0.329	0.336	0.343	0.349	0.356	0.363	50
60	0.363	0.369	0.376	0.383	0.390	0.397	0.403	0.410	0.417	0.424	0.431	60
70	0.431	0.438	0.445	0.452	0.459	0.466	0.473	0.480	0.487	0.494	0.501	70
80	0.501	0.508	0.515	0.523	0.530	0.537	0.544	0.552	0.559	0.566	0.573	80
90	0.573	0.581	0.588	0.595	0.603	0.610	0.617	0.625	0.632	0.640	0.647	90
100	0.647	0.655	0.662	0.670	0.677	0.685	0.692	0.700	0.708	0.715	0.723	100
110	0.723	0.730	0.738	0.746	0.754	0.761	0.769	0.777	0.784	0.792	0.800	110
120	0.800	0.808	0.816	0.824	0.831	0.839	0.847	0.855	0.863	0.871	0.879	120
130	0.879	0.887	0.895	0.903	0.911	0.919	0.927	0.935	0.943	0.951	0.959	130
140	0.959	0.967	0.975	0.983	0.992	1.000	1.008	1.016	1.024	1.032	1.041	140
150	1.041	1.049	1.057	1.065	1.074	1.082	1.090	1.099	1.107	1.115	1.124	150
160	1.124	1.132	1.140	1.149	1.157	1.166	1.174	1.183	1.191	1.200	1.208	160
170	1.208	1.217	1.225	1.234	1.242	1.251	1.259	1.268	1.276	1.285	1.294	170
180	1.294	1.302	1.311	1.319	1.328	1.337	1.345	1.354	1.363	1.372	1.380	180
190	1.380	1.389	1.398	1.407	1.415	1.424	1.433	1.442	1.450	1.459	1.468	190
200	1.468	1.477	1.486	1.495	1.504	1.512	1.521	1.530	1.539	1.548	1.557	200
210	1.557	1.566	1.575	1.584	1.593	1.602	1.611	1.620	1.629	1.638	1.647	210
220	1.647	1.656	1.665	1.674	1.683	1.692	1.702	1.711	1.720	1.729	1.738	220
230	1.738	1.747	1.756	1.766	1.775	1.784	1.793	1.802	1.812	1.821	1.830	230
240	1.830	1.839	1.849	1.858	1.867	1.876	1.886	1.895	1.904	1.914	1.923	240
250	1.923	1.932	1.942	1.951	1.960	1.970	1.979	1.988	1.998	2.007	2.017	250
260	2.017	2.026	2.036	2.045	2.054	2.064	2.073	2.083	2.092	2.102	2.111	260
270	2.111	2.121	2.130	2.140	2.149	2.159	2.169	2.178	2.188	2.197	2.207	270
280	2.207	2.216	2.226	2.236	2.245	2.255	2.264	2.274	2.284	2.293	2.303	280
290	2.303	2.313	2.322	2.332	2.342	2.351	2.361	2.371	2.381	2.390	2.400	290
300	2.400	2.410	2.420	2.429	2.439	2.449	2.459	2.468	2.478	2.488	2.498	300
310	2.498	2.508	2.517	2.527	2.537	2.547	2.557	2.567	2.577	2.586	2.596	310
320	2.596	2.606	2.616	2.626	2.636	2.646	2.656	2.666	2.676	2.685	2.695	320
330	2.695	2.705	2.715	2.725	2.735	2.745	2.755	2.765	2.775	2.785	2.795	330
340	2.795	2.805	2.815	2.825	2.835	2.845	2.855	2.866	2.876	2.886	2.896	340
350	2.896	2.906	2.916	2.926	2.936	2.946	2.956	2.966	2.977	2.987	2.997	350
360	2.997	3.007	3.017	3.027	3.037	3.048	3.058	3.068	3.078	3.088	3.099	360
370	3.099	3.109	3.119	3.129	3.139	3.150	3.160	3.170	3.180	3.191	3.201	370
380	3.201	3.211	3.221	3.232	3.242	3.252	3.263	3.273	3.283	3.293	3.304	380
390	3.304	3.314	3.324	3.335	3.345	3.355	3.366	3.376	3.386	3.397	3.407	390
400	3.407	3.418	3.428	3.438	3.449	3.459	3.470	3.480	3.490	3.501	3.511	400
410	3.511	3.522	3.532	3.543	3.553	3.563	3.574	3.584	3.595	3.605	3.616	410
420	3.616	3.626	3.637	3.647	3.658	3.668	3.679	3.689	3.700	3.710	3.721	420
430	3.721	3.731	3.742	3.752	3.763	3.774	3.784	3.795	3.805	3.816	3.826	430
440	3.826	3.837	3.848	3.858	3.869	3.879	3.890	3.901	3.911	3.922	3.933	440
450	3.933	3.943	3.954	3.964	3.975	3.986	3.996	4.007	4.018	4.028	4.039	450
460	4.039	4.050	4.061	4.071	4.082	4.093	4.103	4.114	4.125	4.136	4.146	460
470	4.146	4.157	4.168	4.178	4.189	4.200	4.211	4.222	4.232	4.243	4.254	470
480	4.254	4.265	4.275	4.286	4.297	4.308	4.319	4.329	4.340	4.351	4.362	480
490	4.362	4.373	4.384	4.394	4.405	4.416	4.427	4.438	4.449	4.460	4.471	490
500	4.471	4.481	4.492	4.503	4.514	4.525	4.536	4.547	4.558	4.569	4.580	500
510	4.580	4.591	4.601	4.612	4.623	4.634	4.645	4.656	4.667	4.678	4.689	510
520	4.689	4.700	4.711	4.722	4.733	4.744	4.755	4.766	4.777	4.788	4.799	520
530	4.799	4.810	4.821	4.832	4.843	4.854	4.865	4.876	4.888	4.899	4.910	530
540	4.910	4.921	4.932	4.943	4.954	4.965	4.976	4.987	4.998	5.009	5.021	540
550	5.021	5.032	5.043	5.054	5.065	5.076	5.087	5.099	5.110	5.121	5.132	550
560	5.132	5.143	5.154	5.166	5.177	5.188	5.199	5.210	5.221	5.233	5.244	560
570	5.244	5.255	5.266	5.278	5.289	5.300	5.311	5.322	5.334	5.345	5.356	570
580	5.356	5.368	5.379	5.390	5.401	5.413	5.424	5.435	5.446	5.458	5.469	580
590	5.469	5.480	5.492	5.503	5.514	5.526	5.537	5.548	5.560	5.571	5.582	590
DEG C	0	1	2	3	4	5	6	7	8	9	10	DEG C

TABLE 10.11—*Type R thermocouples (continued).*

Temperature in Degrees Celsius (IPTS 1968)

EMF in Absolute Millivolts — Reference Junctions at 0 C

THERMOELECTRIC VOLTAGE IN ABSOLUTE MILLIVOLTS

DEG C	0	1	2	3	4	5	6	7	8	9	10	DEG C
600	5.582	5.594	5.605	5.616	5.628	5.639	5.650	5.662	5.673	5.685	5.696	600
610	5.696	5.707	5.719	5.730	5.742	5.753	5.764	5.776	5.787	5.799	5.810	610
620	5.810	5.821	5.833	5.844	5.856	5.867	5.879	5.890	5.902	5.913	5.925	620
630	5.925	5.936	5.948	5.959	5.971	5.982	5.994	6.005	6.017	6.028	6.040	630
640	6.040	6.051	6.063	6.074	6.086	6.098	6.109	6.121	6.132	6.144	6.155	640
650	6.155	6.167	6.179	6.190	6.202	6.213	6.225	6.237	6.248	6.260	6.272	650
660	6.272	6.283	6.295	6.307	6.318	6.330	6.342	6.353	6.365	6.377	6.388	660
670	6.388	6.400	6.412	6.423	6.435	6.447	6.458	6.470	6.482	6.494	6.505	670
680	6.505	6.517	6.529	6.541	6.552	6.564	6.576	6.588	6.599	6.611	6.623	680
690	6.623	6.635	6.647	6.658	6.670	6.682	6.694	6.706	6.718	6.729	6.741	690
700	6.741	6.753	6.765	6.777	6.789	6.800	6.812	6.824	6.836	6.848	6.860	700
710	6.860	6.872	6.884	6.895	6.907	6.919	6.931	6.943	6.955	6.967	6.979	710
720	6.979	6.991	7.003	7.015	7.027	7.039	7.051	7.063	7.074	7.086	7.098	720
730	7.098	7.110	7.122	7.134	7.146	7.158	7.170	7.182	7.194	7.206	7.218	730
740	7.218	7.231	7.243	7.255	7.267	7.279	7.291	7.303	7.315	7.327	7.339	740
750	7.339	7.351	7.363	7.375	7.387	7.399	7.412	7.424	7.436	7.448	7.460	750
760	7.460	7.472	7.484	7.496	7.509	7.521	7.533	7.545	7.557	7.569	7.582	760
770	7.582	7.594	7.606	7.618	7.630	7.642	7.655	7.667	7.679	7.691	7.703	770
780	7.703	7.716	7.728	7.740	7.752	7.765	7.777	7.789	7.801	7.814	7.826	780
790	7.826	7.838	7.850	7.863	7.875	7.887	7.900	7.912	7.924	7.937	7.949	790
800	7.949	7.961	7.973	7.986	7.998	8.010	8.023	8.035	8.047	8.060	8.072	800
810	8.072	8.085	8.097	8.109	8.122	8.134	8.146	8.159	8.171	8.184	8.196	810
820	8.196	8.208	8.221	8.233	8.246	8.258	8.271	8.283	8.295	8.308	8.320	820
830	8.320	8.333	8.345	8.358	8.370	8.383	8.395	8.408	8.420	8.433	8.445	830
840	8.445	8.458	8.470	8.483	8.495	8.508	8.520	8.533	8.545	8.558	8.570	840
850	8.570	8.583	8.595	8.608	8.621	8.633	8.646	8.658	8.671	8.683	8.696	850
860	8.696	8.709	8.721	8.734	8.746	8.759	8.772	8.784	8.797	8.810	8.822	860
870	8.822	8.835	8.847	8.860	8.873	8.885	8.898	8.911	8.923	8.936	8.949	870
880	8.949	8.961	8.974	8.987	9.000	9.012	9.025	9.038	9.050	9.063	9.076	880
890	9.076	9.089	9.101	9.114	9.127	9.140	9.152	9.165	9.178	9.191	9.203	890
900	9.203	9.216	9.229	9.242	9.254	9.267	9.280	9.293	9.306	9.319	9.331	900
910	9.331	9.344	9.357	9.370	9.383	9.395	9.408	9.421	9.434	9.447	9.460	910
920	9.460	9.473	9.485	9.498	9.511	9.524	9.537	9.550	9.563	9.576	9.589	920
930	9.589	9.602	9.614	9.627	9.640	9.653	9.666	9.679	9.692	9.705	9.718	930
940	9.718	9.731	9.744	9.757	9.770	9.783	9.796	9.809	9.822	9.835	9.848	940
950	9.848	9.861	9.874	9.887	9.900	9.913	9.926	9.939	9.952	9.965	9.978	950
960	9.978	9.991	10.004	10.017	10.030	10.043	10.056	10.069	10.082	10.095	10.109	960
970	10.109	10.122	10.135	10.148	10.161	10.174	10.187	10.200	10.213	10.227	10.240	970
980	10.240	10.253	10.266	10.279	10.292	10.305	10.319	10.332	10.345	10.358	10.371	980
990	10.371	10.384	10.398	10.411	10.424	10.437	10.450	10.464	10.477	10.490	10.503	990
1,000	10.503	10.516	10.530	10.543	10.556	10.569	10.583	10.596	10.609	10.622	10.636	1,000
1,010	10.636	10.649	10.662	10.675	10.689	10.702	10.715	10.729	10.742	10.755	10.768	1,010
1,020	10.768	10.782	10.795	10.808	10.822	10.835	10.848	10.862	10.875	10.888	10.902	1,020
1,030	10.902	10.915	10.928	10.942	10.955	10.968	10.982	10.995	11.009	11.022	11.035	1,030
1,040	11.035	11.049	11.062	11.076	11.089	11.102	11.116	11.129	11.143	11.156	11.170	1,040
1,050	11.170	11.183	11.196	11.210	11.223	11.237	11.250	11.264	11.277	11.291	11.304	1,050
1,060	11.304	11.318	11.331	11.345	11.358	11.372	11.385	11.399	11.412	11.426	11.439	1,060
1,070	11.439	11.453	11.466	11.480	11.493	11.507	11.520	11.534	11.547	11.561	11.574	1,070
1,080	11.574	11.588	11.602	11.615	11.629	11.642	11.656	11.669	11.683	11.697	11.710	1,080
1,090	11.710	11.724	11.737	11.751	11.765	11.778	11.792	11.805	11.819	11.833	11.846	1,090
1,100	11.846	11.860	11.874	11.887	11.901	11.914	11.928	11.942	11.955	11.969	11.983	1,100
1,110	11.983	11.996	12.010	12.024	12.037	12.051	12.065	12.078	12.092	12.106	12.119	1,110
1,120	12.119	12.133	12.147	12.161	12.174	12.188	12.202	12.215	12.229	12.243	12.257	1,120
1,130	12.257	12.270	12.284	12.298	12.311	12.325	12.339	12.353	12.366	12.380	12.394	1,130
1,140	12.394	12.408	12.421	12.435	12.449	12.463	12.476	12.490	12.504	12.518	12.532	1,140
1,150	12.532	12.545	12.559	12.573	12.587	12.600	12.614	12.628	12.642	12.656	12.669	1,150
1,160	12.669	12.683	12.697	12.711	12.725	12.739	12.752	12.766	12.780	12.794	12.808	1,160
1,170	12.808	12.822	12.835	12.849	12.863	12.877	12.891	12.905	12.918	12.932	12.946	1,170
1,180	12.946	12.960	12.974	12.988	13.002	13.016	13.029	13.043	13.057	13.071	13.085	1,180
1,190	13.085	13.099	13.113	13.127	13.140	13.154	13.168	13.182	13.196	13.210	13.224	1,190
1,200	13.224	13.238	13.252	13.266	13.280	13.293	13.307	13.321	13.335	13.349	13.363	1,200
1,210	13.363	13.377	13.391	13.405	13.419	13.433	13.447	13.461	13.475	13.489	13.502	1,210
1,220	13.502	13.516	13.530	13.544	13.558	13.572	13.586	13.600	13.614	13.628	13.642	1,220
1,230	13.642	13.656	13.670	13.684	13.698	13.712	13.726	13.740	13.754	13.768	13.782	1,230
1,240	13.782	13.796	13.810	13.824	13.838	13.852	13.866	13.880	13.894	13.908	13.922	1,240
DEG C	0	1	2	3	4	5	6	7	8	9	10	DEG C

TABLE 10.11—*Type R thermocouples (continued).*

Temperature in Degrees Celsius (IPTS 1968)

EMF in Absolute Millivolts | **Reference Junctions at 0 C**

THERMOELECTRIC VOLTAGE IN ABSOLUTE MILLIVOLTS

DEG C	0	1	2	3	4	5	6	7	8	9	10	DEG C
1,250	13.922	13.936	13.950	13.964	13.978	13.992	14.006	14.020	14.034	14.048	14.062	1,250
1,260	14.062	14.076	14.090	14.104	14.118	14.132	14.146	14.160	14.174	14.188	14.202	1,260
1,270	14.202	14.216	14.230	14.244	14.258	14.272	14.286	14.301	14.315	14.329	14.343	1,270
1,280	14.343	14.357	14.371	14.385	14.399	14.413	14.427	14.441	14.455	14.469	14.483	1,280
1,290	14.483	14.497	14.511	14.525	14.539	14.554	14.568	14.582	14.596	14.610	14.624	1,290
1,300	14.624	14.638	14.652	14.666	14.680	14.694	14.708	14.722	14.737	14.751	14.765	1,300
1,310	14.765	14.779	14.793	14.807	14.821	14.835	14.849	14.863	14.877	14.891	14.906	1,310
1,320	14.906	14.920	14.934	14.948	14.962	14.976	14.990	15.004	15.018	15.032	15.047	1,320
1,330	15.047	15.061	15.075	15.089	15.103	15.117	15.131	15.145	15.159	15.173	15.188	1,330
1,340	15.188	15.202	15.216	15.230	15.244	15.258	15.272	15.286	15.300	15.315	15.329	1,340
1,350	15.329	15.343	15.357	15.371	15.385	15.399	15.413	15.427	15.442	15.456	15.470	1,350
1,360	15.470	15.484	15.498	15.512	15.526	15.540	15.555	15.569	15.583	15.597	15.611	1,360
1,370	15.611	15.625	15.639	15.653	15.667	15.682	15.696	15.710	15.724	15.738	15.752	1,370
1,380	15.752	15.766	15.780	15.795	15.809	15.823	15.837	15.851	15.865	15.879	15.893	1,380
1,390	15.893	15.908	15.922	15.936	15.950	15.964	15.978	15.992	16.006	16.021	16.035	1,390
1,400	16.035	16.049	16.063	16.077	16.091	16.105	16.119	16.134	16.148	16.162	16.176	1,400
1,410	16.176	16.190	16.204	16.218	16.232	16.247	16.261	16.275	16.289	16.303	16.317	1,410
1,420	16.317	16.331	16.345	16.360	16.374	16.388	16.402	16.416	16.430	16.444	16.458	1,420
1,430	16.458	16.472	15.487	16.501	16.515	16.529	16.543	16.557	16.571	16.585	16.599	1,430
1,440	16.599	16.614	16.628	16.642	16.656	16.670	16.684	16.698	16.712	16.726	16.741	1,440
1,450	16.741	16.755	16.769	16.783	16.797	16.811	16.825	16.839	16.853	16.867	16.882	1,450
1,460	16.882	16.896	16.910	16.924	16.938	16.952	16.966	16.980	16.994	17.008	17.022	1,460
1,470	17.022	17.037	17.051	17.065	17.079	17.093	17.107	17.121	17.135	17.149	17.163	1,470
1,480	17.163	17.177	17.192	17.206	17.220	17.234	17.248	17.262	17.276	17.290	17.304	1,480
1,490	17.304	17.318	17.332	17.346	17.360	17.374	17.388	17.403	17.417	17.431	17.445	1,490
1,500	17.445	17.459	17.473	17.487	17.501	17.515	17.529	17.543	17.557	17.571	17.585	1,500
1,510	17.585	17.599	17.613	17.627	17.641	17.655	17.669	17.684	17.698	17.712	17.726	1,510
1,520	17.726	17.740	17.754	17.768	17.782	17.796	17.810	17.824	17.838	17.852	17.866	1,520
1,530	17.866	17.880	17.894	17.908	17.922	17.936	17.950	17.964	17.978	17.992	18.006	1,530
1,540	18.006	18.020	18.034	18.048	18.062	18.076	18.090	18.104	18.118	18.132	18.146	1,540
1,550	18.146	18.160	18.174	18.188	18.202	18.216	18.230	18.244	18.258	18.272	18.286	1,550
1,560	18.286	18.299	18.313	18.327	18.341	18.355	18.369	18.383	18.397	18.411	18.425	1,560
1,570	18.425	18.439	18.453	18.467	18.481	18.495	18.509	18.523	18.537	18.550	18.564	1,570
1,580	18.564	18.578	18.592	18.606	18.620	18.634	18.648	18.662	18.676	18.690	18.703	1,580
1,590	18.703	18.717	18.731	18.745	18.759	18.773	18.787	18.801	18.815	18.828	18.842	1,590
1,600	18.842	18.856	18.870	18.884	18.898	18.912	18.926	18.939	18.953	18.967	18.981	1,600
1,610	18.981	18.995	19.009	19.023	19.036	19.050	19.064	19.078	19.092	19.106	19.119	1,610
1,620	19.119	19.133	19.147	19.161	19.175	19.188	19.202	19.216	19.230	19.244	19.257	1,620
1,630	19.257	19.271	19.285	19.299	19.313	19.326	19.340	19.354	19.368	19.382	19.395	1,630
1,640	19.395	19.409	19.423	19.437	19.450	19.464	19.478	19.492	19.505	19.519	19.533	1,640
1,650	19.533	19.547	19.560	19.574	19.588	19.602	19.615	19.629	19.643	19.656	19.670	1,650
1,660	19.670	19.684	19.698	19.711	19.725	19.739	19.752	19.766	19.780	19.793	19.807	1,660
1,670	19.807	19.821	19.834	19.848	19.862	19.875	19.889	19.903	19.916	19.930	19.944	1,670
1,680	19.944	19.957	19.971	19.985	19.998	20.012	20.025	20.039	20.053	20.066	20.080	1,680
1,690	20.080	20.093	20.107	20.120	20.134	20.148	20.161	20.175	20.188	20.202	20.215	1,690
1,700	20.215	20.229	20.242	20.256	20.269	20.283	20.296	20.309	20.323	20.336	20.350	1,700
1,710	20.350	20.363	20.377	20.390	20.403	20.417	20.430	20.443	20.457	20.470	20.483	1,710
1,720	20.483	20.497	20.510	20.523	20.537	20.550	20.563	20.576	20.590	20.603	20.616	1,720
1,730	20.616	20.629	20.642	20.656	20.669	20.682	20.695	20.708	20.721	20.734	20.748	1,730
1,740	20.748	20.761	20.774	20.787	20.800	20.813	20.826	20.839	20.852	20.865	20.878	1,740
1,750	20.878	20.891	20.904	20.916	20.929	20.942	20.955	20.968	20.981	20.994	21.006	1,750
1,760	21.006	21.019	21.032	21.045	21.057	21.070	21.083	21.096	21.108			1,760
DEG C	0	1	2	3	4	5	6	7	8	9	10	DEG C

TABLE 10.12—*Type S thermocouples (deg F-millivolts).*

Temperature in Degrees Fahrenheit[a]

EMF in Absolute Millivolts — Reference Junctions at 32 F

DEG F	0	1	2	3	4	5	6	7	8	9	10	DEG F
					THERMOELECTRIC VOLTAGE IN ABSOLUTE MILLIVOLTS							
-50	-0.218	-0.220	-0.222	-0.225	-0.227	-0.229	-0.231	-0.233	-0.236			-50
-40	-0.194	-0.197	-0.199	-0.202	-0.204	-0.206	-0.209	-0.211	-0.213	-0.215	-0.218	-40
-30	-0.170	-0.173	-0.175	-0.178	-0.180	-0.182	-0.185	-0.187	-0.190	-0.192	-0.194	-30
-20	-0.145	-0.148	-0.150	-0.153	-0.155	-0.158	-0.160	-0.163	-0.165	-0.168	-0.170	-20
-10	-0.119	-0.122	-0.124	-0.127	-0.129	-0.132	-0.135	-0.137	-0.140	-0.142	-0.145	-10
- 0	-0.092	-0.095	-0.097	-0.100	-0.103	-0.106	-0.108	-0.111	-0.114	-0.116	-0.119	- 0
0	-0.092	-0.089	-0.086	-0.084	-0.081	-0.078	-0.075	-0.073	-0.070	-0.067	-0.064	0
10	-0.064	-0.061	-0.058	-0.056	-0.053	-0.050	-0.047	-0.044	-0.041	-0.038	-0.035	10
20	-0.035	-0.033	-0.030	-0.027	-0.024	-0.021	-0.018	-0.015	-0.012	-0.009	-0.006	20
30	-0.006	-0.003	0.000	0.003	0.006	0.009	0.012	0.015	0.018	0.021	0.024	30
40	0.024	0.027	0.030	0.033	0.037	0.040	0.043	0.046	0.049	0.052	0.055	40
50	0.055	0.058	0.062	0.065	0.068	0.071	0.074	0.077	0.081	0.084	0.087	50
60	0.087	0.090	0.093	0.097	0.100	0.103	0.106	0.110	0.113	0.116	0.119	60
70	0.119	0.123	0.126	0.129	0.133	0.136	0.139	0.142	0.146	0.149	0.152	70
80	0.152	0.156	0.159	0.163	0.166	0.169	0.173	0.176	0.179	0.183	0.186	80
90	0.186	0.190	0.193	0.197	0.200	0.203	0.207	0.210	0.214	0.217	0.221	90
100	0.221	0.224	0.228	0.231	0.235	0.238	0.242	0.245	0.249	0.252	0.256	100
110	0.256	0.259	0.263	0.266	0.270	0.274	0.277	0.281	0.284	0.288	0.291	110
120	0.291	0.295	0.299	0.302	0.306	0.309	0.313	0.317	0.320	0.324	0.328	120
130	0.328	0.331	0.335	0.339	0.342	0.346	0.350	0.353	0.357	0.361	0.365	130
140	0.365	0.368	0.372	0.376	0.379	0.383	0.387	0.391	0.394	0.398	0.402	140
150	0.402	0.406	0.409	0.413	0.417	0.421	0.425	0.428	0.432	0.436	0.440	150
160	0.440	0.444	0.448	0.451	0.455	0.459	0.463	0.467	0.471	0.474	0.478	160
170	0.478	0.482	0.486	0.490	0.494	0.498	0.502	0.506	0.510	0.513	0.517	170
180	0.517	0.521	0.525	0.529	0.533	0.537	0.541	0.545	0.549	0.553	0.557	180
190	0.557	0.561	0.565	0.569	0.573	0.577	0.581	0.585	0.589	0.593	0.597	190
200	0.597	0.601	0.605	0.609	0.613	0.617	0.621	0.625	0.629	0.633	0.637	200
210	0.637	0.641	0.645	0.649	0.653	0.658	0.662	0.666	0.670	0.674	0.678	210
220	0.678	0.682	0.686	0.690	0.695	0.699	0.703	0.707	0.711	0.715	0.719	220
230	0.719	0.724	0.728	0.732	0.736	0.740	0.744	0.749	0.753	0.757	0.761	230
240	0.761	0.765	0.770	0.774	0.778	0.782	0.786	0.791	0.795	0.799	0.803	240
250	0.803	0.808	0.812	0.816	0.820	0.824	0.829	0.833	0.837	0.842	0.846	250
260	0.846	0.850	0.854	0.859	0.863	0.867	0.872	0.876	0.880	0.884	0.889	260
270	0.889	0.893	0.897	0.902	0.906	0.910	0.915	0.919	0.923	0.928	0.932	270
280	0.932	0.936	0.941	0.945	0.950	0.954	0.958	0.963	0.967	0.971	0.976	280
290	0.976	0.980	0.985	0.989	0.993	0.998	1.002	1.007	1.011	1.015	1.020	290
300	1.020	1.024	1.029	1.033	1.038	1.042	1.046	1.051	1.055	1.060	1.064	300
310	1.064	1.069	1.073	1.078	1.082	1.087	1.091	1.095	1.100	1.104	1.109	310
320	1.109	1.113	1.118	1.122	1.127	1.131	1.136	1.140	1.145	1.149	1.154	320
330	1.154	1.158	1.163	1.168	1.172	1.177	1.181	1.186	1.190	1.195	1.199	330
340	1.199	1.204	1.208	1.213	1.218	1.222	1.227	1.231	1.236	1.240	1.245	340
350	1.245	1.250	1.254	1.259	1.263	1.268	1.273	1.277	1.282	1.286	1.291	350
360	1.291	1.296	1.300	1.305	1.309	1.314	1.319	1.323	1.328	1.333	1.337	360
370	1.337	1.342	1.347	1.351	1.356	1.360	1.365	1.370	1.374	1.379	1.384	370
380	1.384	1.388	1.393	1.398	1.402	1.407	1.412	1.417	1.421	1.426	1.431	380
390	1.431	1.435	1.440	1.445	1.449	1.454	1.459	1.464	1.468	1.473	1.478	390
400	1.478	1.482	1.487	1.492	1.497	1.501	1.506	1.511	1.516	1.520	1.525	400
410	1.525	1.530	1.535	1.539	1.544	1.549	1.554	1.558	1.563	1.568	1.573	410
420	1.573	1.577	1.582	1.587	1.592	1.597	1.601	1.606	1.611	1.616	1.620	420
430	1.620	1.625	1.630	1.635	1.640	1.644	1.649	1.654	1.659	1.664	1.669	430
440	1.669	1.673	1.678	1.683	1.688	1.693	1.698	1.702	1.707	1.712	1.717	440
450	1.717	1.722	1.727	1.731	1.736	1.741	1.746	1.751	1.756	1.761	1.765	450
460	1.765	1.770	1.775	1.780	1.785	1.790	1.795	1.799	1.804	1.809	1.814	460
470	1.814	1.819	1.824	1.829	1.834	1.839	1.843	1.848	1.853	1.858	1.863	470
480	1.863	1.868	1.873	1.878	1.883	1.888	1.893	1.898	1.902	1.907	1.912	480
490	1.912	1.917	1.922	1.927	1.932	1.937	1.942	1.947	1.952	1.957	1.962	490
500	1.962	1.967	1.972	1.977	1.981	1.986	1.991	1.996	2.001	2.006	2.011	500
510	2.011	2.016	2.021	2.026	2.031	2.036	2.041	2.046	2.051	2.056	2.061	510
520	2.061	2.066	2.071	2.076	2.081	2.086	2.091	2.096	2.101	2.106	2.111	520
530	2.111	2.116	2.121	2.126	2.131	2.136	2.141	2.146	2.151	2.156	2.161	530
540	2.161	2.166	2.171	2.176	2.181	2.186	2.191	2.196	2.201	2.206	2.211	540
DEG F	0	1	2	3	4	5	6	7	8	9	10	DEG F

[a] Converted from degrees Celsius (IPTS 1968).

TABLE 10.12—*Type S thermocouples (continued).*

Temperature in Degrees Fahrenheit[a]

EMF in Absolute Millivolts **Reference Junctions at 32 F**

DEG F	0	1	2	3	4	5	6	7	8	9	10	DEG F
					THERMOELECTRIC VOLTAGE IN ABSOLUTE MILLIVOLTS							
550	2.211	2.216	2.221	2.227	2.232	2.237	2.242	2.247	2.252	2.257	2.262	550
560	2.262	2.267	2.272	2.277	2.282	2.287	2.292	2.297	2.302	2.307	2.313	560
570	2.313	2.318	2.323	2.328	2.333	2.338	2.343	2.348	2.353	2.358	2.363	570
580	2.363	2.368	2.374	2.379	2.384	2.389	2.394	2.399	2.404	2.409	2.414	580
590	2.414	2.419	2.425	2.430	2.435	2.440	2.445	2.450	2.455	2.460	2.465	590
600	2.465	2.471	2.476	2.481	2.486	2.491	2.496	2.501	2.506	2.512	2.517	600
610	2.517	2.522	2.527	2.532	2.537	2.542	2.548	2.553	2.558	2.563	2.568	610
620	2.568	2.573	2.578	2.584	2.589	2.594	2.599	2.604	2.609	2.615	2.620	620
630	2.620	2.625	2.630	2.635	2.640	2.646	2.651	2.656	2.661	2.666	2.672	630
640	2.672	2.677	2.682	2.687	2.692	2.697	2.703	2.708	2.713	2.718	2.723	640
650	2.723	2.729	2.734	2.739	2.744	2.749	2.755	2.760	2.765	2.770	2.775	650
660	2.775	2.781	2.786	2.791	2.796	2.801	2.807	2.812	2.817	2.822	2.828	660
670	2.828	2.833	2.838	2.843	2.848	2.854	2.859	2.864	2.869	2.875	2.880	670
680	2.880	2.885	2.890	2.895	2.901	2.906	2.911	2.916	2.922	2.927	2.932	680
690	2.932	2.937	2.943	2.948	2.953	2.958	2.964	2.969	2.974	2.979	2.985	690
700	2.985	2.990	2.995	3.000	3.006	3.011	3.016	3.022	3.027	3.032	3.037	700
710	3.037	3.043	3.048	3.053	3.058	3.064	3.069	3.074	3.080	3.085	3.090	710
720	3.090	3.095	3.101	3.106	3.111	3.117	3.122	3.127	3.132	3.138	3.143	720
730	3.143	3.148	3.154	3.159	3.164	3.169	3.175	3.180	3.185	3.191	3.196	730
740	3.196	3.201	3.207	3.212	3.217	3.223	3.228	3.233	3.238	3.244	3.249	740
750	3.249	3.254	3.260	3.265	3.270	3.276	3.281	3.286	3.292	3.297	3.302	750
760	3.302	3.308	3.313	3.318	3.324	3.329	3.334	3.340	3.345	3.350	3.356	760
770	3.356	3.361	3.366	3.372	3.377	3.382	3.388	3.393	3.398	3.404	3.409	770
780	3.409	3.414	3.420	3.425	3.430	3.436	3.441	3.447	3.452	3.457	3.463	780
790	3.463	3.468	3.473	3.479	3.484	3.489	3.495	3.500	3.506	3.511	3.516	790
800	3.516	3.522	3.527	3.532	3.538	3.543	3.549	3.554	3.559	3.565	3.570	800
810	3.570	3.575	3.581	3.586	3.592	3.597	3.602	3.608	3.613	3.619	3.624	810
820	3.624	3.629	3.635	3.640	3.645	3.651	3.656	3.662	3.667	3.672	3.678	820
830	3.678	3.683	3.689	3.694	3.699	3.705	3.710	3.716	3.721	3.726	3.732	830
840	3.732	3.737	3.743	3.748	3.754	3.759	3.764	3.770	3.775	3.781	3.786	840
850	3.786	3.791	3.797	3.802	3.808	3.813	3.819	3.824	3.829	3.835	3.840	850
860	3.840	3.846	3.851	3.857	3.862	3.867	3.873	3.878	3.884	3.889	3.895	860
870	3.895	3.900	3.906	3.911	3.916	3.922	3.927	3.933	3.938	3.944	3.949	870
880	3.949	3.955	3.960	3.965	3.971	3.976	3.982	3.987	3.993	3.998	4.004	880
890	4.004	4.009	4.015	4.020	4.025	4.031	4.036	4.042	4.047	4.053	4.058	890
900	4.058	4.064	4.069	4.075	4.080	4.086	4.091	4.096	4.102	4.107	4.113	900
910	4.113	4.118	4.124	4.129	4.135	4.140	4.146	4.151	4.157	4.162	4.168	910
920	4.168	4.173	4.179	4.184	4.190	4.195	4.201	4.206	4.212	4.217	4.223	920
930	4.223	4.228	4.234	4.239	4.245	4.250	4.256	4.261	4.267	4.272	4.278	930
940	4.278	4.283	4.289	4.294	4.300	4.305	4.311	4.316	4.322	4.327	4.333	940
950	4.333	4.338	4.344	4.349	4.355	4.360	4.366	4.371	4.377	4.382	4.388	950
960	4.388	4.393	4.399	4.404	4.410	4.415	4.421	4.426	4.432	4.438	4.443	960
970	4.443	4.449	4.454	4.460	4.465	4.471	4.476	4.482	4.487	4.493	4.498	970
980	4.498	4.504	4.509	4.515	4.521	4.526	4.532	4.537	4.543	4.548	4.554	980
990	4.554	4.559	4.565	4.570	4.576	4.582	4.587	4.593	4.598	4.604	4.609	990
1,000	4.609	4.615	4.620	4.626	4.632	4.637	4.643	4.648	4.654	4.659	4.665	1,000
1,010	4.665	4.670	4.676	4.682	4.687	4.693	4.698	4.704	4.709	4.715	4.721	1,010
1,020	4.721	4.726	4.732	4.737	4.743	4.748	4.754	4.760	4.765	4.771	4.776	1,020
1,030	4.776	4.782	4.788	4.793	4.799	4.804	4.810	4.815	4.821	4.827	4.832	1,030
1,040	4.832	4.838	4.843	4.849	4.855	4.860	4.866	4.871	4.877	4.883	4.888	1,040
1,050	4.888	4.894	4.899	4.905	4.911	4.916	4.922	4.927	4.933	4.939	4.944	1,050
1,060	4.944	4.950	4.956	4.961	4.967	4.972	4.978	4.984	4.989	4.995	5.000	1,060
1,070	5.000	5.006	5.012	5.017	5.023	5.029	5.034	5.040	5.045	5.051	5.057	1,070
1,080	5.057	5.062	5.068	5.074	5.079	5.085	5.090	5.096	5.102	5.107	5.113	1,080
1,090	5.113	5.119	5.124	5.130	5.136	5.141	5.147	5.153	5.158	5.164	5.169	1,090
1,100	5.169	5.175	5.181	5.186	5.192	5.198	5.203	5.209	5.215	5.220	5.226	1,100
1,110	5.226	5.232	5.237	5.243	5.249	5.254	5.260	5.266	5.271	5.277	5.283	1,110
1,120	5.283	5.288	5.294	5.300	5.305	5.311	5.317	5.322	5.328	5.334	5.339	1,120
1,130	5.339	5.345	5.351	5.356	5.362	5.368	5.373	5.379	5.385	5.391	5.396	1,130
1,140	5.396	5.402	5.408	5.413	5.419	5.425	5.430	5.436	5.442	5.447	5.453	1,140
1,150	5.453	5.459	5.465	5.470	5.476	5.482	5.487	5.493	5.499	5.504	5.510	1,150
1,160	5.510	5.516	5.522	5.527	5.533	5.539	5.544	5.550	5.556	5.562	5.567	1,160
1,170	5.567	5.573	5.579	5.585	5.590	5.596	5.602	5.608	5.613	5.619	5.625	1,170
1,180	5.625	5.631	5.636	5.642	5.648	5.653	5.659	5.665	5.671	5.676	5.682	1,180
1,190	5.682	5.688	5.694	5.700	5.705	5.711	5.717	5.723	5.728	5.734	5.740	1,190
DEG F	0	1	2	3	4	5	6	7	8	9	10	DEG F

[a] Converted from degrees Celsius (IPTS 1968).

TABLE 10.12—*Type S thermocouples (continued).*

Temperature in Degrees Fahrenheit[a]

EMF in Absolute Millivolts — **Reference Junctions at 32 F**

DEG F	0	1	2	3	4	5	6	7	8	9	10	DEG F
					THERMOELECTRIC VOLTAGE IN ABSOLUTE MILLIVOLTS							
1,200	5.740	5.746	5.751	5.757	5.763	5.769	5.774	5.780	5.786	5.792	5.797	1,200
1,210	5.797	5.803	5.809	5.815	5.821	5.826	5.832	5.838	5.844	5.849	5.855	1,210
1,220	5.855	5.861	5.867	5.873	5.878	5.884	5.890	5.896	5.902	5.907	5.913	1,220
1,230	5.913	5.919	5.925	5.931	5.936	5.942	5.948	5.954	5.960	5.965	5.971	1,230
1,240	5.971	5.977	5.983	5.989	5.994	6.000	6.006	6.012	6.018	6.023	6.029	1,240
1,250	6.029	6.035	6.041	6.047	6.052	6.058	6.064	6.070	6.076	6.082	6.087	1,250
1,260	6.087	6.093	6.099	6.105	6.111	6.117	6.122	6.128	6.134	6.140	6.146	1,260
1,270	6.146	6.152	6.157	6.163	6.169	6.175	6.181	6.187	6.192	6.198	6.204	1,270
1,280	6.204	6.210	6.216	6.222	6.227	6.233	6.239	6.245	6.251	6.257	6.263	1,280
1,290	6.263	6.268	6.274	6.280	6.286	6.292	6.298	6.304	6.309	6.315	6.321	1,290
1,300	6.321	6.327	6.333	6.339	6.345	6.350	6.356	6.362	6.368	6.374	6.380	1,300
1,310	6.380	6.386	6.392	6.397	6.403	6.409	6.415	6.421	6.427	6.433	6.439	1,310
1,320	6.439	6.445	6.450	6.456	6.462	6.468	6.474	6.480	6.486	6.492	6.498	1,320
1,330	6.498	6.503	6.509	6.515	6.521	6.527	6.533	6.539	6.545	6.551	6.557	1,330
1,340	6.557	6.562	6.568	6.574	6.580	6.586	6.592	6.598	6.604	6.610	6.616	1,340
1,350	6.616	6.622	6.627	6.633	6.639	6.645	6.651	6.657	6.663	6.669	6.675	1,350
1,360	6.675	6.681	6.687	6.693	6.699	6.704	6.710	6.716	6.722	6.728	6.734	1,360
1,370	6.734	6.740	6.746	6.752	6.758	6.764	6.770	6.776	6.782	6.788	6.794	1,370
1,380	6.794	6.800	6.805	6.811	6.817	6.823	6.829	6.835	6.841	6.847	6.853	1,380
1,390	6.853	6.859	6.865	6.871	6.877	6.883	6.889	6.895	6.901	6.907	6.913	1,390
1,400	6.913	6.919	6.925	6.931	6.937	6.943	6.948	6.954	6.960	6.966	6.972	1,400
1,410	6.972	6.978	6.984	6.990	6.996	7.002	7.008	7.014	7.020	7.026	7.032	1,410
1,420	7.032	7.038	7.044	7.050	7.056	7.062	7.068	7.074	7.080	7.086	7.092	1,420
1,430	7.092	7.098	7.104	7.110	7.116	7.122	7.128	7.134	7.140	7.146	7.152	1,430
1,440	7.152	7.158	7.164	7.170	7.176	7.182	7.188	7.194	7.200	7.206	7.212	1,440
1,450	7.212	7.218	7.224	7.230	7.236	7.242	7.248	7.254	7.260	7.266	7.272	1,450
1,460	7.272	7.278	7.285	7.291	7.297	7.303	7.309	7.315	7.321	7.327	7.333	1,460
1,470	7.333	7.339	7.345	7.351	7.357	7.363	7.369	7.375	7.381	7.387	7.393	1,470
1,480	7.393	7.399	7.405	7.411	7.417	7.423	7.429	7.436	7.442	7.448	7.454	1,480
1,490	7.454	7.460	7.466	7.472	7.478	7.484	7.490	7.496	7.502	7.508	7.514	1,490
1,500	7.514	7.520	7.526	7.533	7.539	7.545	7.551	7.557	7.563	7.569	7.575	1,500
1,510	7.575	7.581	7.587	7.593	7.599	7.605	7.612	7.618	7.624	7.630	7.636	1,510
1,520	7.636	7.642	7.648	7.654	7.660	7.666	7.672	7.679	7.685	7.691	7.697	1,520
1,530	7.697	7.703	7.709	7.715	7.721	7.727	7.733	7.740	7.746	7.752	7.758	1,530
1,540	7.758	7.764	7.770	7.776	7.782	7.788	7.795	7.801	7.807	7.813	7.819	1,540
1,550	7.819	7.825	7.831	7.837	7.843	7.850	7.856	7.862	7.868	7.874	7.880	1,550
1,560	7.880	7.886	7.892	7.899	7.905	7.911	7.917	7.923	7.929	7.935	7.942	1,560
1,570	7.942	7.948	7.954	7.960	7.966	7.972	7.978	7.985	7.991	7.997	8.003	1,570
1,580	8.003	8.009	8.015	8.021	8.028	8.034	8.040	8.046	8.052	8.058	8.065	1,580
1,590	8.065	8.071	8.077	8.083	8.089	8.095	8.101	8.108	8.114	8.120	8.126	1,590
1,600	8.126	8.132	8.138	8.145	8.151	8.157	8.163	8.169	8.176	8.182	8.188	1,600
1,610	8.188	8.194	8.200	8.206	8.213	8.219	8.225	8.231	8.237	8.244	8.250	1,610
1,620	8.250	8.256	8.262	8.268	8.275	8.281	8.287	8.293	8.299	8.305	8.312	1,620
1,630	8.312	8.318	8.324	8.330	8.336	8.343	8.349	8.355	8.361	8.368	8.374	1,630
1,640	8.374	8.380	8.386	8.392	8.399	8.405	8.411	8.417	8.423	8.430	8.436	1,640
1,650	8.436	8.442	8.448	8.455	8.461	8.467	8.473	8.479	8.486	8.492	8.498	1,650
1,660	8.498	8.504	8.511	8.517	8.523	8.529	8.536	8.542	8.548	8.554	8.560	1,660
1,670	8.560	8.567	8.573	8.579	8.585	8.592	8.598	8.604	8.610	8.617	8.623	1,670
1,680	8.623	8.629	8.635	8.642	8.648	8.654	8.660	8.667	8.673	8.679	8.685	1,680
1,690	8.685	8.692	8.698	8.704	8.711	8.717	8.723	8.729	8.736	8.742	8.748	1,690
1,700	8.748	8.754	8.761	8.767	8.773	8.780	8.786	8.792	8.798	8.805	8.811	1,700
1,710	8.811	8.817	8.823	8.830	8.836	8.842	8.849	8.855	8.861	8.867	8.874	1,710
1,720	8.874	8.880	8.886	8.893	8.899	8.905	8.912	8.918	8.924	8.930	8.937	1,720
1,730	8.937	8.943	8.949	8.956	8.962	8.968	8.975	8.981	8.987	8.993	9.000	1,730
1,740	9.000	9.006	9.012	9.019	9.025	9.031	9.038	9.044	9.050	9.057	9.063	1,740
1,750	9.063	9.069	9.076	9.082	9.088	9.095	9.101	9.107	9.114	9.120	9.126	1,750
1,760	9.126	9.133	9.139	9.145	9.152	9.158	9.164	9.171	9.177	9.183	9.190	1,760
1,770	9.190	9.196	9.202	9.209	9.215	9.221	9.228	9.234	9.240	9.247	9.253	1,770
1,780	9.253	9.259	9.266	9.272	9.278	9.285	9.291	9.298	9.304	9.310	9.317	1,780
1,790	9.317	9.323	9.329	9.336	9.342	9.348	9.355	9.361	9.368	9.374	9.380	1,790
1,800	9.380	9.387	9.393	9.399	9.406	9.412	9.419	9.425	9.431	9.438	9.444	1,800
1,810	9.444	9.450	9.457	9.463	9.470	9.476	9.482	9.489	9.495	9.502	9.508	1,810
1,820	9.508	9.514	9.521	9.527	9.533	9.540	9.546	9.553	9.559	9.565	9.572	1,820
1,830	9.572	9.578	9.585	9.591	9.598	9.604	9.610	9.617	9.623	9.630	9.636	1,830
1,840	9.636	9.642	9.649	9.655	9.662	9.668	9.674	9.681	9.687	9.694	9.700	1,840
DEG F	0	1	2	3	4	5	6	7	8	9	10	DEG F

[a] Converted from degrees Celsius (IPTS 1968).

TABLE 10.12—*Type S thermocouples (continued).*

Temperature in Degrees Fahrenheit[a]

EMF in Absolute Millivolts | **Reference Junctions at 32 F**

DEG F	0	1	2	3	4	5	6	7	8	9	10	DEG F
						THERMOELECTRIC VOLTAGE IN ABSOLUTE MILLIVOLTS						
1,850	9.700	9.707	9.713	9.719	9.726	9.732	9.739	9.745	9.752	9.758	9.764	1,850
1,860	9.764	9.771	9.777	9.784	9.790	9.797	9.803	9.809	9.816	9.822	9.829	1,860
1,870	9.829	9.835	9.842	9.848	9.855	9.861	9.867	9.874	9.880	9.887	9.893	1,870
1,880	9.893	9.900	9.906	9.913	9.919	9.926	9.932	9.938	9.945	9.951	9.958	1,880
1,890	9.958	9.964	9.971	9.977	9.984	9.990	9.997	10.003	10.010	10.016	10.023	1,890
1,900	10.023	10.029	10.036	10.042	10.048	10.055	10.061	10.068	10.074	10.081	10.087	1,900
1,910	10.087	10.094	10.100	10.107	10.113	10.120	10.126	10.133	10.139	10.146	10.152	1,910
1,920	10.152	10.159	10.165	10.172	10.178	10.185	10.191	10.198	10.204	10.211	10.217	1,920
1,930	10.217	10.224	10.230	10.237	10.243	10.250	10.256	10.263	10.269	10.276	10.282	1,930
1,940	10.282	10.289	10.295	10.302	10.308	10.315	10.321	10.328	10.334	10.341	10.348	1,940
1,950	10.348	10.354	10.361	10.367	10.374	10.380	10.387	10.393	10.400	10.406	10.413	1,950
1,960	10.413	10.419	10.426	10.432	10.439	10.445	10.452	10.459	10.465	10.472	10.478	1,960
1,970	10.478	10.485	10.491	10.498	10.504	10.511	10.517	10.524	10.531	10.537	10.544	1,970
1,980	10.544	10.550	10.557	10.563	10.570	10.576	10.583	10.589	10.596	10.603	10.609	1,980
1,990	10.609	10.616	10.622	10.629	10.635	10.642	10.648	10.655	10.662	10.668	10.675	1,990
2,000	10.675	10.681	10.688	10.694	10.701	10.708	10.714	10.721	10.727	10.734	10.740	2,000
2,010	10.740	10.747	10.754	10.760	10.767	10.773	10.780	10.786	10.793	10.800	10.806	2,010
2,020	10.806	10.813	10.819	10.826	10.832	10.839	10.846	10.852	10.859	10.865	10.872	2,020
2,030	10.872	10.879	10.885	10.892	10.898	10.905	10.912	10.918	10.925	10.931	10.938	2,030
2,040	10.938	10.944	10.951	10.958	10.964	10.971	10.977	10.984	10.991	10.997	11.004	2,040
2,050	11.004	11.010	11.017	11.024	11.030	11.037	11.043	11.050	11.057	11.063	11.070	2,050
2,060	11.070	11.076	11.083	11.090	11.096	11.103	11.110	11.116	11.123	11.129	11.136	2,060
2,070	11.136	11.143	11.149	11.156	11.162	11.169	11.176	11.182	11.189	11.196	11.202	2,070
2,080	11.202	11.209	11.215	11.222	11.229	11.235	11.242	11.248	11.255	11.262	11.268	2,080
2,090	11.268	11.275	11.282	11.288	11.295	11.301	11.308	11.315	11.321	11.328	11.335	2,090
2,100	11.335	11.341	11.348	11.355	11.361	11.368	11.374	11.381	11.388	11.394	11.401	2,100
2,110	11.401	11.408	11.414	11.421	11.428	11.434	11.441	11.447	11.454	11.461	11.467	2,110
2,120	11.467	11.474	11.481	11.487	11.494	11.501	11.507	11.514	11.521	11.527	11.534	2,120
2,130	11.534	11.541	11.547	11.554	11.560	11.567	11.574	11.580	11.587	11.594	11.600	2,130
2,140	11.600	11.607	11.614	11.620	11.627	11.634	11.640	11.647	11.654	11.660	11.667	2,140
2,150	11.667	11.674	11.680	11.687	11.694	11.700	11.707	11.714	11.720	11.727	11.734	2,150
2,160	11.734	11.740	11.747	11.754	11.760	11.767	11.774	11.780	11.787	11.794	11.800	2,160
2,170	11.800	11.807	11.814	11.820	11.827	11.834	11.840	11.847	11.854	11.860	11.867	2,170
2,180	11.867	11.874	11.880	11.887	11.894	11.900	11.907	11.914	11.920	11.927	11.934	2,180
2,190	11.934	11.940	11.947	11.954	11.960	11.967	11.974	11.980	11.987	11.994	12.001	2,190
2,200	12.001	12.007	12.014	12.021	12.027	12.034	12.041	12.047	12.054	12.061	12.067	2,200
2,210	12.067	12.074	12.081	12.087	12.094	12.101	12.107	12.114	12.121	12.128	12.134	2,210
2,220	12.134	12.141	12.148	12.154	12.161	12.168	12.174	12.181	12.188	12.194	12.201	2,220
2,230	12.201	12.208	12.215	12.221	12.228	12.235	12.241	12.248	12.255	12.261	12.268	2,230
2,240	12.268	12.275	12.282	12.288	12.295	12.302	12.308	12.315	12.322	12.328	12.335	2,240
2,250	12.335	12.342	12.349	12.355	12.362	12.369	12.375	12.382	12.389	12.395	12.402	2,250
2,260	12.402	12.409	12.416	12.422	12.429	12.436	12.442	12.449	12.456	12.463	12.469	2,260
2,270	12.469	12.476	12.483	12.489	12.496	12.503	12.510	12.516	12.523	12.530	12.536	2,270
2,280	12.536	12.543	12.550	12.557	12.563	12.570	12.577	12.583	12.590	12.597	12.604	2,280
2,290	12.604	12.610	12.617	12.624	12.630	12.637	12.644	12.651	12.657	12.664	12.671	2,290
2,300	12.671	12.677	12.684	12.691	12.698	12.704	12.711	12.718	12.724	12.731	12.738	2,300
2,310	12.738	12.745	12.751	12.758	12.765	12.771	12.778	12.785	12.792	12.798	12.805	2,310
2,320	12.805	12.812	12.819	12.825	12.832	12.839	12.845	12.852	12.859	12.866	12.872	2,320
2,330	12.872	12.879	12.886	12.893	12.899	12.906	12.913	12.919	12.926	12.933	12.940	2,330
2,340	12.940	12.946	12.953	12.960	12.967	12.973	12.980	12.987	12.993	13.000	13.007	2,340
2,350	13.007	13.014	13.020	13.027	13.034	13.041	13.047	13.054	13.061	13.067	13.074	2,350
2,360	13.074	13.081	13.088	13.094	13.101	13.108	13.115	13.121	13.128	13.135	13.142	2,360
2,370	13.142	13.148	13.155	13.162	13.168	13.175	13.182	13.189	13.195	13.202	13.209	2,370
2,380	13.209	13.216	13.222	13.229	13.236	13.243	13.249	13.256	13.263	13.269	13.276	2,380
2,390	13.276	13.283	13.290	13.296	13.303	13.310	13.317	13.323	13.330	13.337	13.344	2,390
2,400	13.344	13.350	13.357	13.364	13.371	13.377	13.384	13.391	13.397	13.404	13.411	2,400
2,410	13.411	13.418	13.424	13.431	13.438	13.445	13.451	13.458	13.465	13.472	13.478	2,410
2,420	13.478	13.485	13.492	13.499	13.505	13.512	13.519	13.526	13.532	13.539	13.546	2,420
2,430	13.546	13.552	13.559	13.566	13.573	13.579	13.586	13.593	13.600	13.606	13.613	2,430
2,440	13.613	13.620	13.627	13.633	13.640	13.647	13.654	13.660	13.667	13.674	13.681	2,440
2,450	13.681	13.687	13.694	13.701	13.708	13.714	13.721	13.728	13.734	13.741	13.748	2,450
2,460	13.748	13.755	13.761	13.768	13.775	13.782	13.788	13.795	13.802	13.809	13.815	2,460
2,470	13.815	13.822	13.829	13.836	13.842	13.849	13.856	13.863	13.869	13.876	13.883	2,470
2,480	13.883	13.890	13.896	13.903	13.910	13.916	13.923	13.930	13.937	13.943	13.950	2,480
2,490	13.950	13.957	13.964	13.970	13.977	13.984	13.991	13.997	14.004	14.011	14.018	2,490
DEG F	0	1	2	3	4	5	6	7	8	9	10	DEG F

[a] Converted from degrees Celsius (IPTS 1968).

TABLE 10.12—*Type S thermocouples (continued).*

Temperature in Degrees Fahrenheit[a]

EMF in Absolute Millivolts — Reference Junctions at 32 F

DEG F	0	1	2	3	4	5	6	7	8	9	10	DEG F
					THERMOELECTRIC VOLTAGE IN ABSOLUTE MILLIVOLTS							
2,500	14.018	14.024	14.031	14.038	14.045	14.051	14.058	14.065	14.072	14.078	14.085	2,500
2,510	14.085	14.092	14.098	14.105	14.112	14.119	14.125	14.132	14.139	14.146	14.152	2,510
2,520	14.152	14.159	14.166	14.173	14.179	14.186	14.193	14.200	14.206	14.213	14.220	2,520
2,530	14.220	14.226	14.233	14.240	14.247	14.253	14.260	14.267	14.274	14.280	14.287	2,530
2,540	14.287	14.294	14.301	14.307	14.314	14.321	14.328	14.334	14.341	14.348	14.354	2,540
2,550	14.354	14.361	14.368	14.375	14.381	14.388	14.395	14.402	14.408	14.415	14.422	2,550
2,560	14.422	14.429	14.435	14.442	14.449	14.455	14.462	14.469	14.476	14.482	14.489	2,560
2,570	14.489	14.496	14.503	14.509	14.516	14.523	14.530	14.536	14.543	14.550	14.556	2,570
2,580	14.556	14.563	14.570	14.577	14.583	14.590	14.597	14.604	14.610	14.617	14.624	2,580
2,590	14.624	14.631	14.637	14.644	14.651	14.657	14.664	14.671	14.678	14.684	14.691	2,590
2,600	14.691	14.698	14.705	14.711	14.718	14.725	14.731	14.738	14.745	14.752	14.758	2,600
2,610	14.758	14.765	14.772	14.778	14.785	14.792	14.799	14.805	14.812	14.819	14.826	2,610
2,620	14.826	14.832	14.839	14.846	14.852	14.859	14.866	14.873	14.879	14.886	14.893	2,620
2,630	14.893	14.899	14.906	14.913	14.920	14.926	14.933	14.940	14.946	14.953	14.960	2,630
2,640	14.960	14.967	14.973	14.980	14.987	14.994	15.000	15.007	15.014	15.020	15.027	2,640
2,650	15.027	15.034	15.041	15.047	15.054	15.061	15.067	15.074	15.081	15.088	15.094	2,650
2,660	15.094	15.101	15.108	15.114	15.121	15.128	15.134	15.141	15.148	15.155	15.161	2,660
2,670	15.161	15.168	15.175	15.181	15.188	15.195	15.202	15.208	15.215	15.222	15.228	2,670
2,680	15.228	15.235	15.242	15.248	15.255	15.262	15.269	15.275	15.282	15.289	15.295	2,680
2,690	15.295	15.302	15.309	15.315	15.322	15.329	15.336	15.342	15.349	15.356	15.362	2,690
2,700	15.362	15.369	15.376	15.382	15.389	15.396	15.403	15.409	15.416	15.423	15.429	2,700
2,710	15.429	15.436	15.443	15.449	15.456	15.463	15.469	15.476	15.483	15.490	15.496	2,710
2,720	15.496	15.503	15.510	15.516	15.523	15.530	15.536	15.543	15.550	15.556	15.563	2,720
2,730	15.563	15.570	15.576	15.583	15.590	15.597	15.603	15.610	15.617	15.623	15.630	2,730
2,740	15.630	15.637	15.643	15.650	15.657	15.663	15.670	15.677	15.683	15.690	15.697	2,740
2,750	15.697	15.703	15.710	15.717	15.723	15.730	15.737	15.743	15.750	15.757	15.763	2,750
2,760	15.763	15.770	15.777	15.783	15.790	15.797	15.804	15.810	15.817	15.824	15.830	2,760
2,770	15.830	15.837	15.844	15.850	15.857	15.864	15.870	15.877	15.883	15.890	15.897	2,770
2,780	15.897	15.903	15.910	15.917	15.923	15.930	15.937	15.943	15.950	15.957	15.963	2,780
2,790	15.963	15.970	15.977	15.983	15.990	15.997	16.003	16.010	16.017	16.023	16.030	2,790
2,800	16.030	16.037	16.043	16.050	16.057	16.063	16.070	16.077	16.083	16.090	16.096	2,800
2,810	16.096	16.103	16.110	16.116	16.123	16.130	16.136	16.143	16.150	16.156	16.163	2,810
2,820	16.163	16.170	16.176	16.183	16.189	16.196	16.203	16.209	16.216	16.223	16.229	2,820
2,830	16.229	16.236	16.243	16.249	16.256	16.262	16.269	16.276	16.282	16.289	16.296	2,830
2,840	16.296	16.302	16.309	16.315	16.322	16.329	16.335	16.342	16.349	16.355	16.362	2,840
2,850	16.362	16.368	16.375	16.382	16.388	16.395	16.402	16.408	16.415	16.421	16.428	2,850
2,860	16.428	16.435	16.441	16.448	16.454	16.461	16.468	16.474	16.481	16.488	16.494	2,860
2,870	16.494	16.501	16.507	16.514	16.521	16.527	16.534	16.540	16.547	16.554	16.560	2,870
2,880	16.560	16.567	16.573	16.580	16.587	16.593	16.600	16.606	16.613	16.620	16.626	2,880
2,890	16.626	16.633	16.639	16.646	16.653	16.659	16.666	16.672	16.679	16.686	16.692	2,890
2,900	16.692	16.699	16.705	16.712	16.719	16.725	16.732	16.738	16.745	16.751	16.758	2,900
2,910	16.758	16.765	16.771	16.778	16.784	16.791	16.797	16.804	16.811	16.817	16.824	2,910
2,920	16.824	16.830	16.837	16.844	16.850	16.857	16.863	16.870	16.876	16.883	16.890	2,920
2,930	16.890	16.896	16.903	16.909	16.916	16.922	16.929	16.935	16.942	16.949	16.955	2,930
2,940	16.955	16.962	16.968	16.975	16.981	16.988	16.995	17.001	17.008	17.014	17.021	2,940
2,950	17.021	17.027	17.034	17.040	17.047	17.053	17.060	17.067	17.073	17.080	17.086	2,950
2,960	17.086	17.093	17.099	17.106	17.112	17.119	17.125	17.132	17.139	17.145	17.152	2,960
2,970	17.152	17.158	17.165	17.171	17.178	17.184	17.191	17.197	17.204	17.210	17.217	2,970
2,980	17.217	17.223	17.230	17.237	17.243	17.250	17.256	17.263	17.269	17.276	17.282	2,980
2,990	17.282	17.289	17.295	17.302	17.308	17.315	17.321	17.328	17.334	17.341	17.347	2,990
3,000	17.347	17.354	17.360	17.367	17.373	17.380	17.386	17.393	17.399	17.406	17.412	3,000
3,010	17.412	17.419	17.425	17.432	17.438	17.445	17.451	17.458	17.464	17.471	17.477	3,010
3,020	17.477	17.484	17.490	17.497	17.503	17.510	17.516	17.523	17.529	17.536	17.542	3,020
3,030	17.542	17.549	17.555	17.562	17.568	17.575	17.581	17.588	17.594	17.601	17.607	3,030
3,040	17.607	17.614	17.620	17.627	17.633	17.639	17.646	17.652	17.659	17.665	17.672	3,040
3,050	17.672	17.678	17.685	17.691	17.698	17.704	17.711	17.717	17.723	17.730	17.736	3,050
3,060	17.736	17.743	17.749	17.756	17.762	17.769	17.775	17.781	17.788	17.794	17.801	3,060
3,070	17.801	17.807	17.814	17.820	17.826	17.833	17.839	17.846	17.852	17.859	17.865	3,070
3,080	17.865	17.871	17.878	17.884	17.891	17.897	17.903	17.910	17.916	17.923	17.929	3,080
3,090	17.929	17.935	17.942	17.948	17.954	17.961	17.967	17.974	17.980	17.986	17.993	3,090
3,100	17.993	17.999	18.005	18.012	18.018	18.024	18.031	18.037	18.043	18.050	18.056	3,100
3,110	18.056	18.063	18.069	18.075	18.081	18.088	18.094	18.100	18.107	18.113	18.119	3,110
3,120	18.119	18.126	18.132	18.138	18.145	18.151	18.157	18.163	18.170	18.176	18.182	3,120
3,130	18.182	18.189	18.195	18.201	18.207	18.214	18.220	18.226	18.232	18.239	18.245	3,130
3,140	18.245	18.251	18.257	18.264	18.270	18.276	18.282	18.289	18.295	18.301	18.307	3,140
DEG F	0	1	2	3	4	5	6	7	8	9	10	DEG F

[a] Converted from degrees Celsius (IPTS 1968).

TABLE 10.12—*Type S thermocouples (continued).*

Temperature in Degrees Fahrenheit[a]

EMF in Absolute Millivolts | Reference Junctions at 32 F

DEG F	0	1	2	3	4	5	6	7	8	9	10	DEG F
					THERMOELECTRIC VOLTAGE IN ABSOLUTE MILLIVOLTS							
3,150	18.307	18.313	18.320	18.326	18.332	18.338	18.344	18.351	18.357	18.363	18.369	3,150
3,160	18.369	18.375	18.381	18.388	18.394	18.400	18.406	18.412	18.418	18.424	18.431	3,160
3,170	18.431	18.437	18.443	18.449	18.455	18.461	18.467	18.473	18.479	18.486	18.492	3,170
3,180	18.492	18.498	18.504	18.510	18.516	18.522	18.528	18.534	18.540	18.546	18.552	3,180
3,190	18.552	18.558	18.564	18.570	18.576	18.582	18.588	18.594	18.600	18.606	18.612	3,190
3,200	18.612	18.618	18.624	18.630	18.636	18.642	18.648	18.654	18.660	18.666	18.672	3,200
3,210	18.672	18.678	18.684	18.690	18.696							3,210
DEG F	0	1	2	3	4	5	6	7	8	9	10	DEG F

[a] Converted from degrees Celsius (IPTS 1968).

TABLE 10.13—*Type S thermocouples (deg C-millivolts).*

Temperature in Degrees Celsius (IPTS 1968)

EMF in Absolute Millivolts | Reference Junctions at 0 C

DEG C	0	1	2	3	4	5	6	7	8	9	10	DEG C
	THERMOELECTRIC VOLTAGE IN ABSOLUTE MILLIVOLTS											
-50	-0.236											-50
-40	-0.194	-0.199	-0.203	-0.207	-0.211	-0.215	-0.220	-0.224	-0.228	-0.232	-0.236	-40
-30	-0.150	-0.155	-0.159	-0.164	-0.168	-0.173	-0.177	-0.181	-0.186	-0.190	-0.194	-30
-20	-0.103	-0.108	-0.112	-0.117	-0.122	-0.127	-0.132	-0.136	-0.141	-0.145	-0.150	-20
-10	-0.053	-0.058	-0.063	-0.068	-0.073	-0.078	-0.083	-0.088	-0.093	-0.098	-0.103	-10
- 0	0.000	-0.005	-0.011	-0.016	-0.021	-0.027	-0.032	-0.037	-0.042	-0.048	-0.053	- 0
0	0.000	0.005	0.011	0.016	0.022	0.027	0.033	0.038	0.044	0.050	0.055	0
10	0.055	0.061	0.067	0.072	0.078	0.084	0.090	0.095	0.101	0.107	0.113	10
20	0.113	0.119	0.125	0.131	0.137	0.142	0.148	0.154	0.161	0.167	0.173	20
30	0.173	0.179	0.185	0.191	0.197	0.203	0.210	0.216	0.222	0.228	0.235	30
40	0.235	0.241	0.247	0.254	0.260	0.266	0.273	0.279	0.286	0.292	0.299	40
50	0.299	0.305	0.312	0.318	0.325	0.331	0.338	0.345	0.351	0.358	0.365	50
60	0.365	0.371	0.378	0.385	0.391	0.398	0.405	0.412	0.419	0.425	0.432	60
70	0.432	0.439	0.446	0.453	0.460	0.467	0.474	0.481	0.488	0.495	0.502	70
80	0.502	0.509	0.516	0.523	0.530	0.537	0.544	0.551	0.558	0.566	0.573	80
90	0.573	0.580	0.587	0.594	0.602	0.609	0.616	0.623	0.631	0.638	0.645	90
100	0.645	0.653	0.660	0.667	0.675	0.682	0.690	0.697	0.704	0.712	0.719	100
110	0.719	0.727	0.734	0.742	0.749	0.757	0.764	0.772	0.780	0.787	0.795	110
120	0.795	0.802	0.810	0.818	0.825	0.833	0.841	0.848	0.856	0.864	0.872	120
130	0.872	0.879	0.887	0.895	0.903	0.910	0.918	0.926	0.934	0.942	0.950	130
140	0.950	0.957	0.965	0.973	0.981	0.989	0.997	1.005	1.013	1.021	1.029	140
150	1.029	1.037	1.045	1.053	1.061	1.069	1.077	1.085	1.093	1.101	1.109	150
160	1.109	1.117	1.125	1.133	1.141	1.149	1.158	1.166	1.174	1.182	1.190	160
170	1.190	1.198	1.207	1.215	1.223	1.231	1.240	1.248	1.256	1.264	1.273	170
180	1.273	1.281	1.289	1.297	1.306	1.314	1.322	1.331	1.339	1.347	1.356	180
190	1.356	1.364	1.373	1.381	1.389	1.398	1.406	1.415	1.423	1.432	1.440	190
200	1.440	1.448	1.457	1.465	1.474	1.482	1.491	1.499	1.508	1.516	1.525	200
210	1.525	1.534	1.542	1.551	1.559	1.568	1.576	1.585	1.594	1.602	1.611	210
220	1.611	1.620	1.628	1.637	1.645	1.654	1.663	1.671	1.680	1.689	1.698	220
230	1.698	1.706	1.715	1.724	1.732	1.741	1.750	1.759	1.767	1.776	1.785	230
240	1.785	1.794	1.802	1.811	1.820	1.829	1.838	1.846	1.855	1.864	1.873	240
250	1.873	1.882	1.891	1.899	1.908	1.917	1.926	1.935	1.944	1.953	1.962	250
260	1.962	1.971	1.979	1.988	1.997	2.006	2.015	2.024	2.033	2.042	2.051	260
270	2.051	2.060	2.069	2.078	2.087	2.096	2.105	2.114	2.123	2.132	2.141	270
280	2.141	2.150	2.159	2.168	2.177	2.186	2.195	2.204	2.213	2.222	2.232	280
290	2.232	2.241	2.250	2.259	2.268	2.277	2.286	2.295	2.304	2.314	2.323	290
300	2.323	2.332	2.341	2.350	2.359	2.368	2.378	2.387	2.396	2.405	2.414	300
310	2.414	2.424	2.433	2.442	2.451	2.460	2.470	2.479	2.488	2.497	2.506	310
320	2.506	2.516	2.525	2.534	2.543	2.553	2.562	2.571	2.581	2.590	2.599	320
330	2.599	2.608	2.618	2.627	2.636	2.646	2.655	2.664	2.674	2.683	2.692	330
340	2.692	2.702	2.711	2.720	2.730	2.739	2.748	2.758	2.767	2.776	2.786	340
350	2.786	2.795	2.805	2.814	2.823	2.833	2.842	2.852	2.861	2.870	2.880	350
360	2.880	2.889	2.899	2.908	2.917	2.927	2.936	2.946	2.955	2.965	2.974	360
370	2.974	2.984	2.993	3.003	3.012	3.022	3.031	3.041	3.050	3.059	3.069	370
380	3.069	3.078	3.088	3.097	3.107	3.117	3.126	3.136	3.145	3.155	3.164	380
390	3.164	3.174	3.183	3.193	3.202	3.212	3.221	3.231	3.241	3.250	3.260	390
400	3.260	3.269	3.279	3.288	3.298	3.308	3.317	3.327	3.336	3.346	3.356	400
410	3.356	3.365	3.375	3.384	3.394	3.404	3.413	3.423	3.433	3.442	3.452	410
420	3.452	3.462	3.471	3.481	3.491	3.500	3.510	3.520	3.529	3.539	3.549	420
430	3.549	3.558	3.568	3.578	3.587	3.597	3.607	3.616	3.626	3.636	3.645	430
440	3.645	3.655	3.665	3.675	3.684	3.694	3.704	3.714	3.723	3.733	3.743	440
450	3.743	3.752	3.762	3.772	3.782	3.791	3.801	3.811	3.821	3.831	3.840	450
460	3.840	3.850	3.860	3.870	3.879	3.889	3.899	3.909	3.919	3.928	3.938	460
470	3.938	3.948	3.958	3.968	3.977	3.987	3.997	4.007	4.017	4.027	4.036	470
480	4.036	4.046	4.056	4.066	4.076	4.086	4.095	4.105	4.115	4.125	4.135	480
490	4.135	4.145	4.155	4.164	4.174	4.184	4.194	4.204	4.214	4.224	4.234	490
500	4.234	4.243	4.253	4.263	4.273	4.283	4.293	4.303	4.313	4.323	4.333	500
510	4.333	4.343	4.352	4.362	4.372	4.382	4.392	4.402	4.412	4.422	4.432	510
520	4.432	4.442	4.452	4.462	4.472	4.482	4.492	4.502	4.512	4.522	4.532	520
530	4.532	4.542	4.552	4.562	4.572	4.582	4.592	4.602	4.612	4.622	4.632	530
540	4.632	4.642	4.652	4.662	4.672	4.682	4.692	4.702	4.712	4.722	4.732	540
550	4.732	4.742	4.752	4.762	4.772	4.782	4.792	4.802	4.812	4.822	4.832	550
560	4.832	4.842	4.852	4.862	4.873	4.883	4.893	4.903	4.913	4.923	4.933	560
570	4.933	4.943	4.953	4.963	4.973	4.984	4.994	5.004	5.014	5.024	5.034	570
580	5.034	5.044	5.054	5.065	5.075	5.085	5.095	5.105	5.115	5.125	5.136	580
590	5.136	5.146	5.156	5.166	5.176	5.186	5.197	5.207	5.217	5.227	5.237	590
DEG C	0	1	2	3	4	5	6	7	8	9	10	DEG C

TABLE 10.13—*Type S thermocouples (continued).*

Temperature in Degrees Celsius (IPTS 1968)

EMF in Absolute Millivolts **Reference Junctions at 0 C**

DEG C	0	1	2	3	4	5	6	7	8	9	10	DEG C
	THERMOELECTRIC VOLTAGE IN ABSOLUTE MILLIVOLTS											
600	5.237	5.247	5.258	5.268	5.278	5.288	5.298	5.309	5.319	5.329	5.339	600
610	5.339	5.350	5.360	5.370	5.380	5.391	5.401	5.411	5.421	5.431	5.442	610
620	5.442	5.452	5.462	5.473	5.483	5.493	5.503	5.514	5.524	5.534	5.544	620
630	5.544	5.555	5.565	5.575	5.586	5.596	5.606	5.617	5.627	5.637	5.648	630
640	5.648	5.658	5.668	5.679	5.689	5.700	5.710	5.720	5.731	5.741	5.751	640
650	5.751	5.762	5.772	5.782	5.793	5.803	5.814	5.824	5.834	5.845	5.855	650
660	5.855	5.866	5.876	5.887	5.897	5.907	5.918	5.928	5.939	5.949	5.960	660
670	5.960	5.970	5.980	5.991	6.001	6.012	6.022	6.033	6.043	6.054	6.064	670
680	6.064	6.075	6.085	6.096	6.106	6.117	6.127	6.138	6.148	6.159	6.169	680
690	6.169	6.180	6.190	6.201	6.211	6.222	6.232	6.243	6.253	6.264	6.274	690
700	6.274	6.285	6.295	6.306	6.316	6.327	6.338	6.348	6.359	6.369	6.380	700
710	6.380	6.390	6.401	6.412	6.422	6.433	6.443	6.454	6.465	6.475	6.486	710
720	6.486	6.496	6.507	6.518	6.528	6.539	6.549	6.560	6.571	6.581	6.592	720
730	6.592	6.603	6.613	6.624	6.635	6.645	6.656	6.667	6.677	6.688	6.699	730
740	6.699	6.709	6.720	6.731	6.741	6.752	6.763	6.773	6.784	6.795	6.805	740
750	6.805	6.816	6.827	6.838	6.848	6.859	6.870	6.880	6.891	6.902	6.913	750
760	6.913	6.923	6.934	6.945	6.956	6.966	6.977	6.988	6.999	7.009	7.020	760
770	7.020	7.031	7.042	7.053	7.063	7.074	7.085	7.096	7.107	7.117	7.128	770
780	7.128	7.139	7.150	7.161	7.171	7.182	7.193	7.204	7.215	7.225	7.236	780
790	7.236	7.247	7.258	7.269	7.280	7.291	7.301	7.312	7.323	7.334	7.345	790
800	7.345	7.356	7.367	7.377	7.388	7.399	7.410	7.421	7.432	7.443	7.454	800
810	7.454	7.465	7.476	7.486	7.497	7.508	7.519	7.530	7.541	7.552	7.563	810
820	7.563	7.574	7.585	7.596	7.607	7.618	7.629	7.640	7.651	7.661	7.672	820
830	7.672	7.683	7.694	7.705	7.716	7.727	7.738	7.749	7.760	7.771	7.782	830
840	7.782	7.793	7.804	7.815	7.826	7.837	7.848	7.859	7.870	7.881	7.892	840
850	7.892	7.904	7.915	7.926	7.937	7.948	7.959	7.970	7.981	7.992	8.003	850
860	8.003	8.014	8.025	8.036	8.047	8.058	8.069	8.081	8.092	8.103	8.114	860
870	8.114	8.125	8.136	8.147	8.158	8.169	8.180	8.192	8.203	8.214	8.225	870
880	8.225	8.236	8.247	8.258	8.270	8.281	8.292	8.303	8.314	8.325	8.336	880
890	8.336	8.348	8.359	8.370	8.381	8.392	8.404	8.415	8.426	8.437	8.448	890
900	8.448	8.460	8.471	8.482	8.493	8.504	8.516	8.527	8.538	8.549	8.560	900
910	8.560	8.572	8.583	8.594	8.605	8.617	8.628	8.639	8.650	8.662	8.673	910
920	8.673	8.684	8.695	8.707	8.718	8.729	8.741	8.752	8.763	8.774	8.786	920
930	8.786	8.797	8.808	8.820	8.831	8.842	8.854	8.865	8.876	8.888	8.899	930
940	8.899	8.910	8.922	8.933	8.944	8.956	8.967	8.978	8.990	9.001	9.012	940
950	9.012	9.024	9.035	9.047	9.058	9.069	9.081	9.092	9.103	9.115	9.126	950
960	9.126	9.138	9.149	9.160	9.172	9.183	9.195	9.206	9.217	9.229	9.240	960
970	9.240	9.252	9.263	9.275	9.286	9.298	9.309	9.320	9.332	9.343	9.355	970
980	9.355	9.366	9.378	9.389	9.401	9.412	9.424	9.435	9.447	9.458	9.470	980
990	9.470	9.481	9.493	9.504	9.516	9.527	9.539	9.550	9.562	9.573	9.585	990
1,000	9.585	9.596	9.608	9.619	9.631	9.642	9.654	9.665	9.677	9.689	9.700	1,000
1,010	9.700	9.712	9.723	9.735	9.746	9.758	9.770	9.781	9.793	9.804	9.816	1,010
1,020	9.816	9.828	9.839	9.851	9.862	9.874	9.886	9.897	9.909	9.920	9.932	1,020
1,030	9.932	9.944	9.955	9.967	9.979	9.990	10.002	10.013	10.025	10.037	10.048	1,030
1,040	10.048	10.060	10.072	10.083	10.095	10.107	10.118	10.130	10.142	10.154	10.165	1,040
1,050	10.165	10.177	10.189	10.200	10.212	10.224	10.235	10.247	10.259	10.271	10.282	1,050
1,060	10.282	10.294	10.306	10.318	10.329	10.341	10.353	10.364	10.376	10.388	10.400	1,060
1,070	10.400	10.411	10.423	10.435	10.447	10.459	10.470	10.482	10.494	10.506	10.517	1,070
1,080	10.517	10.529	10.541	10.553	10.565	10.576	10.588	10.600	10.612	10.624	10.635	1,080
1,090	10.635	10.647	10.659	10.671	10.683	10.694	10.706	10.718	10.730	10.742	10.754	1,090
1,100	10.754	10.765	10.777	10.789	10.801	10.813	10.825	10.836	10.848	10.860	10.872	1,100
1,110	10.872	10.884	10.896	10.908	10.919	10.931	10.943	10.955	10.967	10.979	10.991	1,110
1,120	10.991	11.003	11.014	11.026	11.038	11.050	11.062	11.074	11.086	11.098	11.110	1,120
1,130	11.110	11.121	11.133	11.145	11.157	11.169	11.181	11.193	11.205	11.217	11.229	1,130
1,140	11.229	11.241	11.252	11.264	11.276	11.288	11.300	11.312	11.324	11.336	11.348	1,140
1,150	11.348	11.360	11.372	11.384	11.396	11.408	11.420	11.432	11.443	11.455	11.467	1,150
1,160	11.467	11.479	11.491	11.503	11.515	11.527	11.539	11.551	11.563	11.575	11.587	1,160
1,170	11.587	11.599	11.611	11.623	11.635	11.647	11.659	11.671	11.683	11.695	11.707	1,170
1,180	11.707	11.719	11.731	11.743	11.755	11.767	11.779	11.791	11.803	11.815	11.827	1,180
1,190	11.827	11.839	11.851	11.863	11.875	11.887	11.899	11.911	11.923	11.935	11.947	1,190
1,200	11.947	11.959	11.971	11.983	11.995	12.007	12.019	12.031	12.043	12.055	12.067	1,200
1,210	12.067	12.079	12.091	12.103	12.116	12.128	12.140	12.152	12.164	12.176	12.188	1,210
1,220	12.188	12.200	12.212	12.224	12.236	12.248	12.260	12.272	12.284	12.296	12.308	1,220
1,230	12.308	12.320	12.332	12.345	12.357	12.369	12.381	12.393	12.405	12.417	12.429	1,230
1,240	12.429	12.441	12.453	12.465	12.477	12.489	12.501	12.514	12.526	12.538	12.550	1,240
DEG C	0	1	2	3	4	5	6	7	8	9	10	DEG C

TABLE 10.13—*Type S thermocouples (continued).*

Temperature in Degrees Celsius (IPTS 1968)

EMF in Absolute Millivolts | Reference Junctions at 0 C

DEG C	0	1	2	3	4	5	6	7	8	9	10	DEG C
THERMOELECTRIC VOLTAGE IN ABSOLUTE MILLIVOLTS												
1,250	12.550	12.562	12.574	12.586	12.598	12.610	12.622	12.634	12.647	12.659	12.671	1,250
1,260	12.671	12.683	12.695	12.707	12.719	12.731	12.743	12.755	12.767	12.780	12.792	1,260
1,270	12.792	12.804	12.816	12.828	12.840	12.852	12.864	12.876	12.888	12.901	12.913	1,270
1,280	12.913	12.925	12.937	12.949	12.961	12.973	12.985	12.997	13.010	13.022	13.034	1,280
1,290	13.034	13.046	13.058	13.070	13.082	13.094	13.107	13.119	13.131	13.143	13.155	1,290
1,300	13.155	13.167	13.179	13.191	13.203	13.216	13.228	13.240	13.252	13.264	13.276	1,300
1,310	13.276	13.288	13.300	13.313	13.325	13.337	13.349	13.361	13.373	13.385	13.397	1,310
1,320	13.397	13.410	13.422	13.434	13.446	13.458	13.470	13.482	13.495	13.507	13.519	1,320
1,330	13.519	13.531	13.543	13.555	13.567	13.579	13.592	13.604	13.616	13.628	13.640	1,330
1,340	13.640	13.652	13.664	13.677	13.689	13.701	13.713	13.725	13.737	13.749	13.761	1,340
1,350	13.761	13.774	13.786	13.798	13.810	13.822	13.834	13.846	13.859	13.871	13.883	1,350
1,360	13.883	13.895	13.907	13.919	13.931	13.943	13.956	13.968	13.980	13.992	14.004	1,360
1,370	14.004	14.016	14.028	14.040	14.053	14.065	14.077	14.089	14.101	14.113	14.125	1,370
1,380	14.125	14.138	14.150	14.162	14.174	14.186	14.198	14.210	14.222	14.235	14.247	1,380
1,390	14.247	14.259	14.271	14.283	14.295	14.307	14.319	14.332	14.344	14.356	14.368	1,390
1,400	14.368	14.380	14.392	14.404	14.416	14.429	14.441	14.453	14.465	14.477	14.489	1,400
1,410	14.489	14.501	14.513	14.526	14.538	14.550	14.562	14.574	14.586	14.598	14.610	1,410
1,420	14.610	14.622	14.635	14.647	14.659	14.671	14.683	14.695	14.707	14.719	14.731	1,420
1,430	14.731	14.744	14.756	14.768	14.780	14.792	14.804	14.816	14.828	14.840	14.852	1,430
1,440	14.852	14.865	14.877	14.889	14.901	14.913	14.925	14.937	14.949	14.961	14.973	1,440
1,450	14.973	14.985	14.998	15.010	15.022	15.034	15.046	15.058	15.070	15.082	15.094	1,450
1,460	15.094	15.106	15.118	15.130	15.143	15.155	15.167	15.179	15.191	15.203	15.215	1,460
1,470	15.215	15.227	15.239	15.251	15.263	15.275	15.287	15.299	15.311	15.324	15.336	1,470
1,480	15.336	15.348	15.360	15.372	15.384	15.396	15.408	15.420	15.432	15.444	15.456	1,480
1,490	15.456	15.468	15.480	15.492	15.504	15.516	15.528	15.540	15.552	15.564	15.576	1,490
1,500	15.576	15.589	15.601	15.613	15.625	15.637	15.649	15.661	15.673	15.685	15.697	1,500
1,510	15.697	15.709	15.721	15.733	15.745	15.757	15.769	15.781	15.793	15.805	15.817	1,510
1,520	15.817	15.829	15.841	15.853	15.865	15.877	15.889	15.901	15.913	15.925	15.937	1,520
1,530	15.937	15.949	15.961	15.973	15.985	15.997	16.009	16.021	16.033	16.045	16.057	1,530
1,540	16.057	16.069	16.080	16.092	16.104	16.116	16.128	16.140	16.152	16.164	16.176	1,540
1,550	16.176	16.188	16.200	16.212	16.224	16.236	16.248	16.260	16.272	16.284	16.296	1,550
1,560	16.296	16.308	16.319	16.331	16.343	16.355	16.367	16.379	16.391	16.403	16.415	1,560
1,570	16.415	16.427	16.439	16.451	16.462	16.474	16.486	16.498	16.510	16.522	16.534	1,570
1,580	16.534	16.546	16.558	16.569	16.581	16.593	16.605	16.617	16.629	16.641	16.653	1,580
1,590	16.653	16.664	16.676	16.688	16.700	16.712	16.724	16.736	16.747	16.759	16.771	1,590
1,600	16.771	16.783	16.795	16.807	16.819	16.830	16.842	16.854	16.866	16.878	16.890	1,600
1,610	16.890	16.901	16.913	16.925	16.937	16.949	16.960	16.972	16.984	16.996	17.008	1,610
1,620	17.008	17.019	17.031	17.043	17.055	17.067	17.078	17.090	17.102	17.114	17.125	1,620
1,630	17.125	17.137	17.149	17.161	17.173	17.184	17.196	17.208	17.220	17.231	17.243	1,630
1,640	17.243	17.255	17.267	17.278	17.290	17.302	17.313	17.325	17.337	17.349	17.360	1,640
1,650	17.360	17.372	17.384	17.396	17.407	17.419	17.431	17.442	17.454	17.466	17.477	1,650
1,660	17.477	17.489	17.501	17.512	17.524	17.536	17.548	17.559	17.571	17.583	17.594	1,660
1,670	17.594	17.606	17.617	17.629	17.641	17.652	17.664	17.676	17.687	17.699	17.711	1,670
1,680	17.711	17.722	17.734	17.745	17.757	17.769	17.780	17.792	17.803	17.815	17.826	1,680
1,690	17.826	17.838	17.850	17.861	17.873	17.884	17.896	17.907	17.919	17.930	17.942	1,690
1,700	17.942	17.953	17.965	17.976	17.988	17.999	18.010	18.022	18.033	18.045	18.056	1,700
1,710	18.056	18.068	18.079	18.090	18.102	18.113	18.124	18.136	18.147	18.158	18.170	1,710
1,720	18.170	18.181	18.192	18.204	18.215	18.226	18.237	18.249	18.260	18.271	18.282	1,720
1,730	18.282	18.293	18.305	18.316	18.327	18.338	18.349	18.360	18.372	18.383	18.394	1,730
1,740	18.394	18.405	18.416	18.427	18.438	18.449	18.460	18.471	18.482	18.493	18.504	1,740
1,750	18.504	18.515	18.526	18.536	18.547	18.558	18.569	18.580	18.591	18.602	18.612	1,750
1,760	18.612	18.623	18.634	18.645	18.655	18.666	18.677	18.687	18.698			1,760
DEG C	0	1	2	3	4	5	6	7	8	9	10	DEG C

TABLE 10.14—*Type T thermocouples (deg F-millivolts).*

Temperature in Degrees Fahrenheit[a]

EMF in Absolute Millivolts — Reference Junctions at 32 F

DEG F	0	1	2	3	4	5	6	7	8	9	10	DEG F
					THERMOELECTRIC VOLTAGE IN ABSOLUTE MILLIVOLTS							
-450	-6.254	-6.255	-6.256	-6.257	-6.258							-450
-440	-6.240	-6.242	-6.243	-6.245	-6.247	-6.248	-6.250	-6.251	-6.252	-6.253	-6.254	-440
-430	-6.217	-6.220	-6.223	-6.225	-6.227	-6.230	-6.232	-6.234	-6.236	-6.238	-6.240	-430
-420	-6.187	-6.191	-6.194	-6.197	-6.200	-6.203	-6.206	-6.209	-6.212	-6.215	-6.217	-420
-410	-6.150	-6.154	-6.158	-6.162	-6.166	-6.170	-6.173	-6.177	-6.181	-6.184	-6.187	-410
-400	-6.105	-6.110	-6.115	-6.119	-6.124	-6.128	-6.133	-6.137	-6.142	-6.146	-6.150	-400
-390	-6.053	-6.059	-6.064	-6.069	-6.075	-6.080	-6.085	-6.090	-6.095	-6.100	-6.105	-390
-380	-5.995	-6.001	-6.007	-6.013	-6.019	-6.025	-6.030	-6.036	-6.042	-6.048	-6.053	-380
-370	-5.930	-5.937	-5.943	-5.950	-5.957	-5.963	-5.969	-5.976	-5.982	-5.988	-5.995	-370
-360	-5.860	-5.867	-5.874	-5.881	-5.889	-5.896	-5.903	-5.910	-5.916	-5.923	-5.930	-360
-350	-5.785	-5.792	-5.800	-5.808	-5.815	-5.823	-5.830	-5.838	-5.845	-5.853	-5.860	-350
-340	-5.705	-5.713	-5.721	-5.729	-5.737	-5.745	-5.753	-5.761	-5.769	-5.777	-5.785	-340
-330	-5.620	-5.629	-5.638	-5.646	-5.655	-5.663	-5.672	-5.680	-5.688	-5.697	-5.705	-330
-320	-5.532	-5.541	-5.550	-5.559	-5.568	-5.576	-5.585	-5.594	-5.603	-5.612	-5.620	-320
-310	-5.439	-5.448	-5.457	-5.467	-5.476	-5.486	-5.495	-5.504	-5.513	-5.522	-5.532	-310
-300	-5.341	-5.351	-5.361	-5.371	-5.381	-5.390	-5.400	-5.410	-5.419	-5.429	-5.439	-300
-290	-5.240	-5.250	-5.261	-5.271	-5.281	-5.291	-5.301	-5.311	-5.321	-5.331	-5.341	-290
-280	-5.135	-5.145	-5.156	-5.167	-5.177	-5.188	-5.198	-5.209	-5.219	-5.230	-5.240	-280
-270	-5.025	-5.036	-5.047	-5.058	-5.069	-5.080	-5.091	-5.102	-5.113	-5.124	-5.135	-270
-260	-4.912	-4.923	-4.935	-4.946	-4.958	-4.969	-4.980	-4.992	-5.003	-5.014	-5.025	-260
-250	-4.794	-4.806	-4.818	-4.830	-4.842	-4.853	-4.865	-4.877	-4.889	-4.900	-4.912	-250
-240	-4.673	-4.685	-4.698	-4.710	-4.722	-4.734	-4.746	-4.758	-4.770	-4.782	-4.794	-240
-230	-4.548	-4.560	-4.573	-4.586	-4.598	-4.611	-4.623	-4.636	-4.648	-4.661	-4.673	-230
-220	-4.419	-4.432	-4.445	-4.458	-4.471	-4.484	-4.497	-4.509	-4.522	-4.535	-4.548	-220
-210	-4.286	-4.299	-4.313	-4.326	-4.339	-4.353	-4.366	-4.379	-4.392	-4.406	-4.419	-210
-200	-4.149	-4.163	-4.177	-4.191	-4.204	-4.218	-4.232	-4.245	-4.259	-4.272	-4.286	-200
-190	-4.009	-4.023	-4.037	-4.051	-4.065	-4.079	-4.093	-4.107	-4.121	-4.135	-4.149	-190
-180	-3.864	-3.879	-3.894	-3.908	-3.923	-3.937	-3.951	-3.966	-3.980	-3.994	-4.009	-180
-170	-3.717	-3.732	-3.746	-3.761	-3.776	-3.791	-3.806	-3.820	-3.835	-3.850	-3.864	-170
-160	-3.565	-3.580	-3.596	-3.611	-3.626	-3.641	-3.656	-3.671	-3.687	-3.702	-3.717	-160
-150	-3.410	-3.425	-3.441	-3.457	-3.472	-3.488	-3.503	-3.519	-3.534	-3.550	-3.565	-150
-140	-3.251	-3.267	-3.283	-3.299	-3.315	-3.331	-3.347	-3.362	-3.378	-3.394	-3.410	-140
-130	-3.089	-3.105	-3.121	-3.138	-3.154	-3.170	-3.186	-3.203	-3.219	-3.235	-3.251	-130
-120	-2.923	-2.939	-2.956	-2.973	-2.989	-3.006	-3.023	-3.039	-3.056	-3.072	-3.089	-120
-110	-2.753	-2.771	-2.788	-2.805	-2.822	-2.838	-2.855	-2.872	-2.889	-2.906	-2.923	-110
-100	-2.581	-2.598	-2.616	-2.633	-2.650	-2.667	-2.685	-2.702	-2.719	-2.736	-2.753	-100
-90	-2.405	-2.422	-2.440	-2.458	-2.475	-2.493	-2.511	-2.528	-2.546	-2.563	-2.581	-90
-80	-2.225	-2.243	-2.261	-2.279	-2.297	-2.315	-2.333	-2.351	-2.369	-2.387	-2.405	-80
-70	-2.042	-2.061	-2.079	-2.098	-2.116	-2.134	-2.152	-2.171	-2.189	-2.207	-2.225	-70
-60	-1.856	-1.875	-1.894	-1.912	-1.931	-1.950	-1.968	-1.987	-2.005	-2.024	-2.042	-60
-50	-1.667	-1.686	-1.705	-1.724	-1.743	-1.762	-1.781	-1.800	-1.819	-1.838	-1.856	-50
-40	-1.475	-1.494	-1.513	-1.533	-1.552	-1.571	-1.591	-1.610	-1.629	-1.648	-1.667	-40
-30	-1.279	-1.299	-1.319	-1.338	-1.358	-1.377	-1.397	-1.416	-1.436	-1.455	-1.475	-30
-20	-1.081	-1.101	-1.121	-1.141	-1.160	-1.180	-1.200	-1.220	-1.240	-1.260	-1.279	-20
-10	-0.879	-0.899	-0.920	-0.940	-0.960	-0.980	-1.000	-1.021	-1.041	-1.061	-1.081	-10
-0	-0.674	-0.695	-0.716	-0.736	-0.757	-0.777	-0.798	-0.818	-0.838	-0.859	-0.879	-0
0	-0.674	-0.654	-0.633	-0.613	-0.592	-0.571	-0.550	-0.529	-0.509	-0.488	-0.467	0
10	-0.467	-0.446	-0.425	-0.404	-0.383	-0.362	-0.341	-0.320	-0.299	-0.277	-0.256	10
20	-0.256	-0.235	-0.214	-0.193	-0.171	-0.150	-0.129	-0.107	-0.086	-0.064	-0.043	20
30	-0.043	-0.022	0.000	0.022	0.043	0.065	0.086	0.108	0.130	0.151	0.173	30
40	0.173	0.195	0.216	0.238	0.260	0.282	0.303	0.325	0.347	0.369	0.391	40
50	0.391	0.413	0.435	0.457	0.479	0.501	0.523	0.545	0.567	0.589	0.611	50
60	0.611	0.634	0.656	0.678	0.700	0.722	0.745	0.767	0.789	0.812	0.834	60
70	0.834	0.857	0.879	0.902	0.924	0.947	0.969	0.992	1.014	1.037	1.060	70
80	1.060	1.082	1.105	1.128	1.151	1.173	1.196	1.219	1.242	1.265	1.288	80
90	1.288	1.311	1.334	1.357	1.380	1.403	1.426	1.449	1.472	1.495	1.518	90
100	1.518	1.542	1.565	1.588	1.611	1.635	1.658	1.681	1.705	1.728	1.752	100
110	1.752	1.775	1.799	1.822	1.846	1.869	1.893	1.917	1.940	1.964	1.988	110
120	1.988	2.011	2.035	2.059	2.083	2.107	2.131	2.154	2.178	2.202	2.226	120
130	2.226	2.250	2.274	2.298	2.322	2.347	2.371	2.395	2.419	2.443	2.467	130
140	2.467	2.492	2.516	2.540	2.565	2.589	2.613	2.638	2.662	2.687	2.711	140
DEG F	0	1	2	3	4	5	6	7	8	9	10	DEG F

[a] Converted from degrees Celsius (IPTS 1968).

TABLE 10.14—*Type T thermocouples (continued).*

Temperature in Degrees Fahrenheit[a]

EMF in Absolute Millivolts

Reference Junctions at 32 F

DEG F	0	1	2	3	4	5	6	7	8	9	10	DEG F
					THERMOELECTRIC VOLTAGE IN ABSOLUTE MILLIVOLTS							
150	2.711	2.736	2.760	2.785	2.809	2.834	2.859	2.883	2.908	2.933	2.958	150
160	2.958	2.982	3.007	3.032	3.057	3.082	3.107	3.131	3.156	3.181	3.206	160
170	3.206	3.231	3.256	3.281	3.307	3.332	3.357	3.382	3.407	3.432	3.458	170
180	3.458	3.483	3.508	3.533	3.559	3.584	3.609	3.635	3.660	3.686	3.711	180
190	3.711	3.737	3.762	3.788	3.813	3.839	3.864	3.890	3.916	3.941	3.967	190
200	3.967	3.993	4.019	4.044	4.070	4.096	4.122	4.148	4.174	4.199	4.225	200
210	4.225	4.251	4.277	4.303	4.329	4.355	4.381	4.408	4.434	4.460	4.486	210
220	4.486	4.512	4.538	4.565	4.591	4.617	4.643	4.670	4.696	4.722	4.749	220
230	4.749	4.775	4.801	4.828	4.854	4.881	4.907	4.934	4.960	4.987	5.014	230
240	5.014	5.040	5.067	5.093	5.120	5.147	5.174	5.200	5.227	5.254	5.281	240
250	5.281	5.307	5.334	5.361	5.388	5.415	5.442	5.469	5.496	5.523	5.550	250
260	5.550	5.577	5.604	5.631	5.658	5.685	5.712	5.739	5.767	5.794	5.821	260
270	5.821	5.848	5.875	5.903	5.930	5.957	5.985	6.012	6.039	6.067	6.094	270
280	6.094	6.122	6.149	6.177	6.204	6.232	6.259	6.287	6.314	6.342	6.369	280
290	6.369	6.397	6.425	6.452	6.480	6.508	6.536	6.563	6.591	6.619	6.647	290
300	6.647	6.675	6.702	6.730	6.758	6.786	6.814	6.842	6.870	6.898	6.926	300
310	6.926	6.954	6.982	7.010	7.038	7.066	7.094	7.122	7.151	7.179	7.207	310
320	7.207	7.235	7.263	7.292	7.320	7.348	7.377	7.405	7.433	7.462	7.490	320
330	7.490	7.518	7.547	7.575	7.604	7.632	7.661	7.689	7.718	7.746	7.775	330
340	7.775	7.804	7.832	7.861	7.889	7.918	7.947	7.975	8.004	8.033	8.062	340
350	8.062	8.090	8.119	8.148	8.177	8.206	8.235	8.264	8.292	8.321	8.350	350
360	8.350	8.379	8.408	8.437	8.466	8.495	8.524	8.553	8.583	8.612	8.641	360
370	8.641	8.670	8.699	8.728	8.757	8.787	8.816	8.845	8.874	8.904	8.933	370
380	8.933	8.962	8.992	9.021	9.050	9.080	9.109	9.139	9.168	9.198	9.227	380
390	9.227	9.257	9.286	9.316	9.345	9.375	9.404	9.434	9.464	9.493	9.523	390
400	9.523	9.553	9.582	9.612	9.642	9.671	9.701	9.731	9.761	9.791	9.820	400
410	9.820	9.850	9.880	9.910	9.940	9.970	10.000	10.030	10.060	10.090	10.120	410
420	10.120	10.150	10.180	10.210	10.240	10.270	10.300	10.330	10.360	10.390	10.420	420
430	10.420	10.451	10.481	10.511	10.541	10.572	10.602	10.632	10.662	10.693	10.723	430
440	10.723	10.753	10.784	10.814	10.845	10.875	10.905	10.936	10.966	10.997	11.027	440
450	11.027	11.058	11.088	11.119	11.149	11.180	11.211	11.241	11.272	11.302	11.333	450
460	11.333	11.364	11.394	11.425	11.456	11.487	11.517	11.548	11.579	11.610	11.640	460
470	11.640	11.671	11.702	11.733	11.764	11.795	11.826	11.856	11.887	11.918	11.949	470
480	11.949	11.980	12.011	12.042	12.073	12.104	12.135	12.166	12.198	12.229	12.260	480
490	12.260	12.291	12.322	12.353	12.384	12.416	12.447	12.478	12.509	12.540	12.572	490
500	12.572	12.603	12.634	12.666	12.697	12.728	12.760	12.791	12.822	12.854	12.885	500
510	12.885	12.917	12.948	12.979	13.011	13.042	13.074	13.105	13.137	13.168	13.200	510
520	13.200	13.232	13.263	13.295	13.326	13.358	13.390	13.421	13.453	13.485	13.516	520
530	13.516	13.548	13.580	13.611	13.643	13.675	13.707	13.739	13.770	13.802	13.834	530
540	13.834	13.866	13.898	13.930	13.961	13.993	14.025	14.057	14.089	14.121	14.153	540
550	14.153	14.185	14.217	14.249	14.281	14.313	14.345	14.377	14.409	14.441	14.474	550
560	14.474	14.506	14.538	14.570	14.602	14.634	14.666	14.699	14.731	14.763	14.795	560
570	14.795	14.828	14.860	14.892	14.924	14.957	14.989	15.021	15.054	15.086	15.118	570
580	15.118	15.151	15.183	15.216	15.248	15.280	15.313	15.345	15.378	15.410	15.443	580
590	15.443	15.475	15.508	15.540	15.573	15.605	15.638	15.671	15.703	15.736	15.769	590
600	15.769	15.801	15.834	15.866	15.899	15.932	15.965	15.997	16.030	16.063	16.096	600
610	16.096	16.128	16.161	16.194	16.227	16.259	16.292	16.325	16.358	16.391	16.424	610
620	16.424	16.457	16.490	16.523	16.555	16.588	16.621	16.654	16.687	16.720	16.753	620
630	16.753	16.786	16.819	16.852	16.886	16.919	16.952	16.985	17.018	17.051	17.084	630
640	17.084	17.117	17.150	17.184	17.217	17.250	17.283	17.316	17.350	17.383	17.416	640
650	17.416	17.450	17.483	17.516	17.549	17.583	17.616	17.649	17.683	17.716	17.750	650
660	17.750	17.783	17.816	17.850	17.883	17.917	17.950	17.984	18.017	18.051	18.084	660
670	18.084	18.118	18.151	18.185	18.218	18.252	18.285	18.319	18.353	18.386	18.420	670
680	18.420	18.454	18.487	18.521	18.555	18.588	18.622	18.656	18.689	18.723	18.757	680
690	18.757	18.791	18.824	18.858	18.892	18.926	18.960	18.993	19.027	19.061	19.095	690
700	19.095	19.129	19.163	19.197	19.230	19.264	19.298	19.332	19.366	19.400	19.434	700
710	19.434	19.468	19.502	19.536	19.570	19.604	19.638	19.672	19.706	19.740	19.774	710
720	19.774	19.808	19.843	19.877	19.911	19.945	19.979	20.013	20.047	20.081	20.116	720
730	20.116	20.150	20.184	20.218	20.252	20.287	20.321	20.355	20.389	20.423	20.458	730
740	20.458	20.492	20.526	20.560	20.595	20.629	20.663	20.698	20.732	20.766	20.801	740
750	20.801	20.835	20.869									750
DEG F	0	1	2	3	4	5	6	7	8	9	10	DEG

[a] Converted from degrees Celsius (IPTS 1968).

TABLE 10.15—*Type T thermocouples (deg C-millivolts).*

Temperature in Degrees Celsius (IPTS 1968)

EMF in Absolute Millivolts Reference Junctions at 0 C

DEG C	0	1	2	3	4	5	6	7	8	9	10	DEG C
	THERMOELECTRIC VOLTAGE IN ABSOLUTE MILLIVOLTS											
-270	-6.258											-270
-260	-6.232	-6.236	-6.239	-6.242	-6.245	-6.248	-6.251	-6.253	-6.255	-6.256	-6.258	-260
-250	-6.181	-6.187	-6.193	-6.198	-6.204	-6.209	-6.214	-6.219	-6.224	-6.228	-6.232	-250
-240	-6.105	-6.114	-6.122	-6.130	-6.138	-6.146	-. 153	-6.160	-6.167	-6.174	-6.181	-240
-230	-6.007	-6.018	-6.028	-6.039	-6.049	-6.059	-6.068	-6.078	-6.087	-6.096	-6.105	-230
-220	-5.889	-5.901	-5.914	-5.926	-5.938	-5.950	-5.962	-5.973	-5.985	-5.996	-6.007	-220
-210	-5.753	-5.767	-5.782	-5.795	-5.809	-5.823	-5.836	-5.850	-5.863	-5.876	-5.889	-210
-200	-5.603	-5.619	-5.634	-5.650	-5.665	-5.680	-5.695	-5.710	-5.724	-5.739	-5.753	-200
-190	-5.439	-5.456	-5.473	-5.489	-5.506	-5.522	-5.539	-5.555	-5.571	-5.587	-5.603	-190
-180	-5.261	-5.279	-5.297	-5.315	-5.333	-5.351	-5.369	-5.387	-5.404	-5.421	-5.439	-180
-170	-5.069	-5.089	-5.109	-5.128	-5.147	-5.167	-5.186	-5.205	-5.223	-5.242	-5.261	-170
-160	-4.865	-4.886	-4.907	-4.928	-4.948	-4.969	-4.989	-5.010	-5.030	-5.050	-5.069	-160
-150	-4.648	-4.670	-4.693	-4.715	-4.737	-4.758	-4.780	-4.801	-4.823	-4.844	-4.865	-150
-140	-4.419	-4.442	-4.466	-4.489	-4.512	-4.535	-4.558	-4.581	-4.603	-4.626	-4.648	-140
-130	-4.177	-4.202	-4.226	-4.251	-4.275	-4.299	-4.323	-4.347	-4.371	-4.395	-4.419	-130
-120	-3.923	-3.949	-3.974	-4.000	-4.026	-4.051	-4.077	-4.102	-4.127	-4.152	-4.177	-120
-110	-3.656	-3.684	-3.711	-3.737	-3.764	-3.791	-3.818	-3.844	-3.870	-3.897	-3.923	-110
-100	-3.378	-3.407	-3.435	-3.463	-3.491	-3.519	-3.547	-3.574	-3.602	-3.629	-3.656	-100
-90	-3.089	-3.118	-3.147	-3.177	-3.206	-3.235	-3.264	-3.293	-3.321	-3.350	-3.378	-90
-80	-2.788	-2.818	-2.849	-2.879	-2.909	-2.939	-2.970	-2.999	-3.029	-3.059	-3.089	-80
-70	-2.475	-2.507	-2.539	-2.570	-2.602	-2.633	-2.664	-2.695	-2.726	-2.757	-2.788	-70
-60	-2.152	-2.185	-2.218	-2.250	-2.283	-2.315	-2.348	-2.380	-2.412	-2.444	-2.475	-60
-50	-1.819	-1.853	-1.886	-1.920	-1.953	-1.987	-2.020	-2.053	-2.087	-2.120	-2.152	-50
-40	-1.475	-1.510	-1.544	-1.579	-1.614	-1.648	-1.682	-1.717	-1.751	-1.785	-1.819	-40
-30	-1.121	-1.157	-1.192	-1.228	-1.263	-1.299	-1.334	-1.370	-1.405	-1.440	-1.475	-30
-20	-0.757	-0.794	-0.830	-0.867	-0.903	-0.940	-0.976	-1.013	-1.049	-1.085	-1.121	-20
-10	-0.383	-0.421	-0.458	-0.496	-0.534	-0.571	-0.608	-0.646	-0.683	-0.720	-0.757	-10
- 0	0.000	-0.039	-0.077	-0.116	-0.154	-0.193	-0.231	-0.269	-0.307	-0.345	-0.383	- 0
0	0.000	0.039	0.078	0.117	0.156	0.195	0.234	0.273	0.312	0.351	0.391	0
10	0.391	0.430	0.470	0.510	0.549	0.589	0.629	0.669	0.709	0.749	0.789	10
20	0.789	0.830	0.870	0.911	0.951	0.992	1.032	1.073	1.114	1.155	1.196	20
30	1.196	1.237	1.279	1.320	1.361	1.403	1.444	1.486	1.528	1.569	1.611	30
40	1.611	1.653	1.695	1.738	1.780	1.822	1.865	1.907	1.950	1.992	2.035	40
50	2.035	2.078	2.121	2.164	2.207	2.250	2.294	2.337	2.380	2.424	2.467	50
60	2.467	2.511	2.555	2.599	2.643	2.687	2.731	2.775	2.819	2.864	2.908	60
70	2.908	2.953	2.997	3.042	3.087	3.131	3.176	3.221	3.266	3.312	3.357	70
80	3.357	3.402	3.447	3.493	3.538	3.584	3.630	3.676	3.721	3.767	3.813	80
90	3.813	3.859	3.906	3.952	3.998	4.044	4.091	4.137	4.184	4.231	4.277	90
100	4.277	4.324	4.371	4.418	4.465	4.512	4.559	4.607	4.654	4.701	4.749	100
110	4.749	4.796	4.844	4.891	4.939	4.987	5.035	5.083	5.131	5.179	5.227	110
120	5.227	5.275	5.324	5.372	5.420	5.469	5.517	5.566	5.615	5.663	5.712	120
130	5.712	5.761	5.810	5.859	5.908	5.957	6.007	6.056	6.105	6.155	6.204	130
140	6.204	6.254	6.303	6.353	6.403	6.452	6.502	6.552	6.602	6.652	6.702	140
150	6.702	6.753	6.803	6.853	6.903	6.954	7.004	7.055	7.106	7.156	7.207	150
160	7.207	7.258	7.309	7.360	7.411	7.462	7.513	7.564	7.615	7.666	7.718	160
170	7.718	7.769	7.821	7.872	7.924	7.975	8.027	8.079	8.131	8.183	8.235	170
180	8.235	8.287	8.339	8.391	8.443	8.495	8.548	8.600	8.652	8.705	8.757	180
190	8.757	8.810	8.863	8.915	8.968	9.021	9.074	9.127	9.180	9.233	9.286	190
200	9.286	9.339	9.392	9.446	9.499	9.553	9.606	9.659	9.713	9.767	9.820	200
210	9.820	9.874	9.928	9.982	10.036	10.090	10.144	10.198	10.252	10.306	10.360	210
220	10.360	10.414	10.469	10.523	10.578	10.632	10.687	10.741	10.796	10.851	10.905	220
230	10.905	10.960	11.015	11.070	11.125	11.180	11.235	11.290	11.345	11.401	11.456	230
240	11.456	11.511	11.566	11.622	11.677	11.733	11.788	11.844	11.900	11.956	12.011	240
250	12.011	12.067	12.123	12.179	12.235	12.291	12.347	12.403	12.459	12.515	12.572	250
260	12.572	12.628	12.684	12.741	12.797	12.854	12.910	12.967	13.024	13.080	13.137	260
270	13.137	13.194	13.251	13.307	13.364	13.421	13.478	13.535	13.592	13.650	13.707	270
280	13.707	13.764	13.821	13.879	13.936	13.993	14.051	14.108	14.166	14.223	14.281	280
290	14.281	14.339	14.396	14.454	14.512	14.570	14.628	14.686	14.744	14.802	14.860	290
300	14.860	14.918	14.976	15.034	15.092	15.151	15.209	15.267	15.326	15.384	15.443	300
310	15.443	15.501	15.560	15.619	15.677	15.736	15.795	15.853	15.912	15.971	16.030	310
320	16.030	16.089	16.148	16.207	16.266	16.325	16.384	16.444	16.503	16.562	16.621	320
330	16.621	16.681	16.740	16.800	16.859	16.919	16.978	17.038	17.097	17.157	17.217	330
340	17.217	17.277	17.336	17.396	17.456	17.516	17.576	17.636	17.696	17.756	17.816	340
DEG C	0	1	2	3	4	5	6	7	8	9	10	DEG C

TABLE 10.15—*Type T thermocouples (continued).*

Temperature in Degrees Celsius (IPTS 1968)

EMF in Absolute Millivolts — Reference Junctions at 0 C

DEG C	0	1	2	3	4	5	6	7	8	9	10	DEG C
	THERMOELECTRIC VOLTAGE IN ABSOLUTE MILLIVOLTS											
350	17.816	17.877	17.937	17.997	18.057	18.118	18.178	18.238	18.299	18.359	18.420	350
360	18.420	18.480	18.541	18.602	18.662	18.723	18.784	18.845	18.905	18.966	19.027	360
370	19.027	19.088	19.149	19.210	19.271	19.332	19.393	19.455	19.516	19.577	19.638	370
380	19.638	19.699	19.761	19.822	19.883	19.945	20.006	20.068	20.129	20.191	20.252	380
390	20.252	20.314	20.376	20.437	20.499	20.560	20.622	20.684	20.746	20.807	20.869	390
400	20.869											400
DEG C	0	1	2	3	4	5	6	7	8	9	10	DEG C

TABLE 10.16—*Power series coefficients for Type T thermocouples.*

Temperature Range	Degree	Coefficients	Power of T Term
−270 to 0°C	14	3.8740773840 E + 1	1
		4.4123932482 E − 2	2
		1.1405238498 E − 4	3
		1.9974406568 E − 5	4
		9.0445401187 E − 7	5
		2.2766018504 E − 8	6
		3.6247409380 E − 10	7
		3.8648924201 E − 12	8
		2.8298678519 E − 14	9
		1.4281383349 E − 16	10
		4.8833254364 E − 19	11
		1.0803474683 E − 21	12
		1.3949291026 E − 24	13
		7.9795893156 E − 28	14
0 to 400°C	8	3.8740773840 E + 1	1
		3.3190198092 E − 2	2
		2.0714183645 E − 4	3
		−2.1945834823 E − 6	4
		1.1031900550 E − 8	5
		−3.0927581898 E − 11	6
		4.5653337165 E − 14	7
		−2.7616878040 E − 17	8

TABLE 10.17—*Power series coefficients for Type J thermocouples.*

Temperature Range	Degree	Coefficients	Power of T Term
−210 to 760°C	7	5.0372753027 E + 1	1
		3.0425491284 E − 2	2
		−8.5669750464 E − 5	3
		1.3348825735 E − 7	4
		−1.7022405966 E − 10	5
		1.9416091001 E − 13	6
		−9.6391844859 E − 17	7

TABLE 10.18—*Power series coefficients for Type E thermocouples.*

Temperature Range	Degree	Coefficients	Power of T Term
−270 to 0°C	13	5.8695857799 E + 1	1
		5.1667517705 E − 2	2
		−4.4652683347 E − 4	3
		−1.7346270905 E − 5	4
		−4.8719368427 E − 7	5
		−8.8896550447 E − 9	6
		−1.0930767375 E − 10	7
		−9.1784535039 E − 13	8
		−5.2575158521 E − 15	9
		−2.0169601996 E − 17	10
		−4.9502138782 E − 20	11
		−7.0177980633 E − 23	12
		−4.3671808488 E − 26	13
0 to 1000°C	9	5.8695857799 E + 1	1
		4.3110945462 E − 2	2
		5.7220358202 E − 5	3
		−5.4020668085 E − 7	4
		1.5425922111 E − 9	5
		−2.4850089136 E − 12	6
		2.3389721459 E − 15	7
		−1.1946296815 E − 18	8
		2.5561127497 E − 22	9

TABLE 10.19—*Power series coefficients for Type K thermocouples.*

Temperature Range	Degree	Coefficients	Power of T Term
−270 to 0°C	10	3.9475433139 E + 1	1
		2.7465251138 E − 2	2
		−1.6565406716 E − 4	3
		−1.5190912392 E − 6	4
		−2.4581670924 E − 8	5
		−2.4757917816 E − 10	6
		−1.5585276173 E − 12	7
		−5.9729921255 E − 15	8
		−1.2688801216 E − 17	9
		−1.1382797374 E − 20	10
0 to 1372°C	8 + exp	−1.8533063273 E + 1	0
		3.8918344612 E + 1	1
		1.6645154356 E − 2	2
		−7.8702374448 E − 5	3
		2.2835785557 E − 7	4
		−3.5700231258 E − 10	5
		2.9932909136 E − 13	6
		−1.2849848798 E − 16	7
		2.2239974336 E − 20	8
		$+125 \exp\left[-\frac{1}{2}\left(\frac{T-127}{65}\right)^2\right]$	

TABLE 10.20—*Power series coefficients for Type R thermocouples.*

Temperature Range	Degree	Coefficients	Power of T Term
−50 to 631°C	7	5.2891395059 E 0	1
		1.3911109947 E − 2	2
		−2.4005238430 E − 5	3
		3.6201410595 E − 8	4
		−4.4645019036 E − 11	5
		3.8497691865 E − 14	6
		−1.5372641559 E − 17	7
631 to 1065°C	3	−2.6418007025 E ± 2	0
		8.0468680747 E 0	1
		2.9892293723 E − 3	2
		−2.6876058617 E − 7	3
1065 to 1665°C	3	1.4901702702 E + 3	0
		2.8639867552 E 0	1
		8.0823631189 E − 3	2
		−1.9338477638 E − 6	3

TABLE 10.21—*Power series coefficients for Type S thermocouples.*

Temperature Range	Degree	Coefficients	Power of T Term
−50 to 631°C	6	5.3995782346 E 0	1
		1.2519770000 E − 2	2
		−2.2448217997 E − 5	3
		2.8452164949 E − 8	4
		−2.2440584544 E − 11	5
		8.5054166936 E − 15	6
631 to 1064°C	2	−2.9824481615 E + 2	0
		8.2375528221 E 0	1
		1.6453909942 E − 3	2
1064 to 1665°C	3	1.2766292175 E + 3	0
		3.4970908041 E 0	1
		6.3824648666 E − 3	2
		−1.5722424599 E − 6	3

TABLE 10.22—*Power series coefficients for Type B thermocouples.*

Temperature Range	Degree	Coefficients	Power of T Term
0 to 1820°C	8	−2.4674601620 E − 1	1
		5.9102111169 E − 3	2
		−1.4307123430 E − 6	3
		2.1509149750 E − 9	4
		−3.1757800720 E − 12	5
		2.4010367459 E − 15	6
		−9.0928148159 E − 19	7
		1.3299505137 E − 22	8

perature-emf relationships, to as many decimal places as have any useful significance, for the thermocouple types included in this standard. These data will be useful in all applications where discontinuities in the established reference values are objectionable, or where storage of complete tables is impractical.

10.3.2 Methods of Generation

The values of each of the seven thermocouple types given in Tables 10.2 to 10.15 are generated by power series equations according to Ref *1*. The coefficients of these equations for the appropriate ranges are given in Tables 10.16 to 10.22. Note that these coefficients yield values of emf in microvolts. This method permits easy generation of the temperature-emf values by a digital computer. It provides a smooth, continuous relationship, and the values will round off to the established temperature-emf values given in Tables 10.2 to 10.15 [*2*].[2]

Alternative generation schemes for temperature-emf relationships are described in Refs *3* and *4*.

10.4 References

[*1*] Powell, R. L., Hall, W. J., Hyink, C. H., Sparks, L. L., Burns, G. W., Scroger, M. G., and Plumb, H. H., "Thermocouple Reference Tables Based on the IPTS-68," NBS Monograph 125, National Bureau of Standards, 1973.

[*2*] ASTM Standard E 230 "Temperature-Electromotive Force (EMF) Tables for Thermocouples," *1980 Annual Book of ASTM Standards*, Part 44.

[*3*] British Standard, BS 4937, 1973, is in substantial agreement with Ref *1*.

[*4*] Alternative generation schemes in use prior to 1973 are described in the following references:

(*a*) Benedict, R. P. and Ashby, H. F., "Improved Reference Tables for Thermocouples," *Temperature, Its Measurement and Control in Science and Industry*, Part 2, Vol. 3, 1962, p. 51.

(*b*) Benedict, R. P., "The Generation of Thermocouple Reference Tables," *Electrotechnology*, Nov. 1963, p. 80.

(*c*) Adams, R. K. and Simpson, R. L., "Review of Techniques for Determining Thermocouple EMF—Temperature Characteristics," *Temperature, Its Measurement and Control in Science and Industry*, Part 3, Vol. 4, 1972, p. 1603.

(*d*) Bedford, R. E. et al., "New Reference Tables for Platinum 10% Rhodium/Platinum and Platinum 13% Rhodium/Platinum Thermocouples," *Temperature, Its Measurement and Control in Science and Industry*, Part 3, Vol. 4, 1972, p. 1585.

[2]The italic numbers in brackets refer to the list of references appended to this chapter.

Chapter 11—Cryogenics

11.1 General Remarks

Although there is some variation in the defined temperature range involved, cryogenics usually indicates concern with temperatures in the liquid oxygen range (about 90 K or −183°C) or lower. This temperature range will be discussed primarily in this chapter. Since a triple point (of water) or ice bath reference junction often is used, additional comments and values will be given for the entire subzero (0°C) range.

Most aspects of cryogenic thermometry are similar to those applicable at room or high temperatures. In particular, the measurement systems and thermoelectric theory are nearly identical. However, there are significant differences with respect to some materials, techniques or assembly, and fabrication, calibration schemes, and methods of practical usage. Fortunately the added difficulties with some details are offset by the removal of a few problems peculiar to high-temperature thermometry: chemical transformations are insignificant; oxidation, reduction, and impurity migration do not occur because of the low temperatures. Annealing of physical imperfections is also absent for the same reasons. Maintenance of fixed points and techniques of calibration are usually considerably easier and sometimes much more accurate. Thermal radiation is usually not important, at least if simple precautions are taken to account for it.

Several books have been written on thermometry and on the experimental techniques necessary for cryogenic research by Scott [*1*][1] and by Hust et al [*2*] on cryogenic engineering, and by White [*3*] and by Rose-Innes [*4*] on smaller, scientific systems.

11.2 Materials

Many thermocouple materials developed for high-temperature usage have too low a sensitivity at low temperatures to be practical; others have a reasonable sensitivity, but are very erratic with large inhomogeneities and lot-to-lot variations. Only three of the letter-designated type thermocouples have proven themselves for cryogenic use: Types E, T, and K. Type E is definitely recommended as the most valuable for general low-temperature use. It is better because of its higher sensitivity (coupled with only average

[1] The italic numbers in brackets refer to the list of references appended to this chapter.

inhomogeneity voltages) and lower thermal conductivity of both thermocouple materials. The latter property is particularly important for obtaining good thermal tempering of measuring junctions. Tables for the total voltage and Seebeck coefficient are given in Section 11.3 for each of the thermocouples just mentioned. These tables were taken from Ref *5*.

There are other, but uncommon, materials that have a higher sensitivity at very low temperatures, below 50 K: gold-cobalt and gold-iron. Both are negative thermoelectric materials and should be matched with KP or EP, "normal" silver, or less preferably, TP, to obtain a thermocouple pair. The older material, gold-cobalt (2.1 atomic percent cobalt) has been found to have significant instabilities caused by room-temperature annealing of the metastable solution of cobalt in gold. Because of those problems gold-cobalt should not be used any more in the future. The alloy gold-0.07 atomic percent iron has replaced gold-cobalt: it does not have alloying instabilities, it is more homogeneous, and it has a larger Seebeck coefficient.

A table giving the total voltage and Seebeck coefficient of KP or EP versus gold-0.07 atomic percent iron is given also in Section 11.3. This table is taken from data published by the National Bureau of Standards [*6*]. Reference tables for several other material combinations that are used as thermocouples in the cryogenic range are also given in this reference.

Figure 11.1 compares the Seebeck coefficients of various thermocouples used in the cryogenic range. The Seebeck coefficient of Type K is not shown, but it would be slightly above that of Type T.

11.3 Reference Tables (for use in the cryogenic range)

TABLE 11.1—*Type E-thermoelectric voltage,* E(T), *Seebeck coefficient,* S(T), *and derivative of the Seebeck coefficient,* dS/dT [5].

T, K	*E*, μV	*S*, μV/K	*dS/dT*, nV/K²	*T*, K	*E*, μV	*S*, μV/K	*dS/dT*, nV/K²
0	0.00	−0.203	604.4	45	413.20	17.149	318.9
1	0.09	0.384	571.8	46	430.51	17.467	316.3
2	0.76	0.941	543.0	47	448.14	17.782	313.6
3	1.97	1.472	517.7	48	466.07	18.094	311.0
4	3.69	1.978	495.6	49	484.32	18.404	308.3
5	5.92	2.464	476.2	50	502.88	18.711	305.6
6	8.61	2.931	459.4	51	521.74	19.015	303.0
7	11.77	3.383	444.7	52	540.91	19.317	300.3
8	15.38	3.821	432.1	53	560.38	19.616	297.7
9	19.41	4.248	421.1	54	580.14	19.912	295.1
10	23.87	4.664	411.7	55	600.20	20.206	292.5
11	28.74	5.072	403.6	56	620.55	20.497	289.9
12	34.01	5.472	396.7	57	641.19	20.786	287.4
13	39.68	5.865	390.8	58	662.12	21.072	284.9
14	45.74	6.254	385.8	59	683.33	21.355	282.5

TABLE 11.1—*(Continued)*.

T, K	E, μV	S, μV/K	dS/dT, nV/K^2	T, K	E, μV	S, μV/K	dS/dT, nV/K_2
15	52.18	6.637	381.5	60	704.83	21.637	280.0
16	59.01	7.017	377.8	61	726.61	21.915	277.7
17	66.22	7.393	374.6	62	748.66	22.192	275.4
18	73.80	7.766	371.9	63	770.99	22.466	273.1
19	81.75	8.137	369.5	64	793.59	22.738	270.9
20	90.07	8.505	367.4	65	816.47	23.008	268.8
21	98.76	8.872	365.6	66	839.61	23.276	266.7
22	107.81	9.237	363.9	67	863.02	23.541	264.6
23	117.23	9.600	362.3	68	886.69	23.805	262.6
24	127.01	9.961	360.8	69	910.63	24.067	260.7
25	137.15	10.321	359.3	70	934.82	24.326	258.8
26	147.65	10.680	357.9	71	959.28	24.584	256.9
27	158.51	11.037	356.4	72	983.99	24.840	255.1
28	169.73	11.393	354.9	73	1008.96	25.094	253.4
29	181.30	11.747	353.4	74	1034.18	25.347	251.7
30	193.22	12.099	351.8	75	1059.65	25.598	250.0
31	205.50	12.450	350.1	76	1085.37	25.847	248.4
32	218.12	12.800	348.4	77	1111.35	26.095	246.9
33	231.09	13.147	346.5	78	1137.56	26.341	245.4
34	244.41	13.493	344.6	79	1164.03	26.585	243.9
35	258.08	13.836	342.7	80	1190.73	26.829	242.4
36	272.09	14.178	340.6	81	1217.68	27.070	241.0
37	286.43	14.517	338.4	82	1244.87	27.311	239.7
38	301.12	14.855	336.2	83	1272.30	27.550	238.3
39	316.14	15.190	333.9	84	1299.97	27.787	237.0
40	331.50	15.523	331.5	85	1327.88	28.024	235.7
41	347.19	15.853	329.1	86	1356.02	28.259	234.5
42	363.20	16.181	326.6	87	1384.40	28.493	233.2
43	379.55	16.506	324.1	88	1413.01	28.725	232.0
44	396.21	16.829	321.5	89	1441.85	28.957	230.9
90	1470.92	29.187	229.7	135	3001.29	38.512	187.3
91	1500.22	29.416	228.6	136	3039.89	38.699	186.6
92	1529.75	29.644	227.4	137	3078.69	38.885	185.9
93	1559.51	29.871	226.3	138	3117.66	39.070	185.1
94	1589.49	30.097	225.2	139	3156.83	39.255	184.4
95	1619.70	30.321	224.1	140	3196.17	39.439	183.7
96	1650.13	30.545	223.1	141	3235.70	39.623	183.0
97	1680.79	30.768	222.0	142	3275.42	39.805	182.3
98	1711.67	30.989	221.0	143	3315.31	39.987	181.6
99	1742.77	31.210	219.9	144	3355.39	40.169	180.9
100	1774.09	31.429	218.9	145	3395.65	40.349	180.3
101	1805.63	31.647	217.9	146	3436.09	40.529	179.6
102	1837.38	31.865	216.9	147	3476.71	40.708	179.0
103	1869.36	32.081	215.9	148	3517.51	40.887	178.3
104	1901.54	32.297	214.9	149	3558.48	41.065	177.7
105	1933.95	32.511	213.9	150	3599.64	41.242	177.0
106	1966.57	32.724	212.9	151	3640.97	41.419	176.4
107	1999.40	32.937	211.9	152	3682.47	41.595	175.8

TABLE 11.1—*(Continued)*.

T, K	*E*, μV	*S*, μV/K	*dS/dT*, nV/K^2	*T*, K	*E*, μV	*S*, μV/K	*dS/dT*, nV/K^2
108	2032.44	33.148	210.9	153	3724.16	41.771	175.2
109	2065.69	33.359	310.0	154	3766.02	41.946	174.6
110	2099.16	33.568	209.0	155	3808.05	42.120	173.9
111	2132.83	33.777	208.1	156	3850.26	42.293	173.3
112	2166.71	33.984	207.1	157	3892.64	42.466	172.7
113	2200.80	34.191	206.2	158	3935.19	42.639	172.1
114	2235.09	34.397	205.2	159	3977.91	42.811	171.5
115	2269.59	34.601	204.3	160	4020.81	42.982	170.9
116	2304.29	34.805	203.4	161	4063.88	42.153	170.3
117	2339.20	35.008	202.4	162	4107.11	43.323	169.7
118	2374.31	35.210	201.5	163	4150.52	43.492	169.1
119	2409.62	35.411	200.6	164	4194.10	43.661	168.5
120	2445.13	35.611	199.7	165	4237.84	43.829	167.9
121	2480.84	35.811	198.8	166	4281.76	43.997	167.3
122	2516.75	36.009	198.0	167	4325.84	44.164	166.7
123	2552.86	36.207	197.1	168	4370.08	44.330	166.1
124	2589.17	36.403	196.2	169	4414.50	44.496	165.5
125	2625.67	36.599	195.4	170	4459.07	44.661	164.9
126	2662.36	36.794	194.5	171	4503.82	44.826	164.3
127	2699.25	36.988	193.7	172	4548.73	44.990	163.7
128	2736.34	37.181	192.9	173	4593.80	45.153	163.1
129	2773.62	37.374	192.0	174	4639.03	45.316	162.5
130	2811.09	37.565	191.2	175	4684.43	45.478	161.8
131	2848.75	32.756	190.4	176	4729.99	45.640	161.2
132	2886.60	32.946	189.6	177	4775.71	45.800	160.6
133	2924.64	38.136	188.9	178	4821.59	45.961	160.0
134	2962.87	38.324	188.1	179	4867.63	45.120	159.4
180	4913.83	46.279	158.7	230	7413.67	53.482	130.9
181	4960.19	46.438	158.1	231	7467.22	53.613	130.4
182	5006.70	46.596	157.5	232	7520.89	53.743	129.9
183	5053.38	46.753	156.9	233	7574.70	53.872	129.4
184	5100.21	46.909	156.2	234	7628.64	54.002	129.0
185	5147.20	47.065	155.6	235	7682.70	54.130	128.5
186	5194.34	47.221	155.0	236	7736.90	54.259	128.0
187	5241.64	47.375	154.4	237	7791.22	54.386	127.5
188	5289.09	47.529	153.7	238	7845.67	54.514	127.1
189	5336.70	47.683	153.1	239	7900.25	54.641	126.6
190	5384.46	47.835	152.5	240	7954.95	54.767	126.1
191	5432.37	47.988	151.9	241	8009.78	54.893	125.7
192	5480.43	48.139	151.2	242	8064.74	55.018	125.2
193	5528.65	48.290	150.6	243	8119.82	55.143	124.8
194	5577.01	48.440	150.0	244	8175.02	55.268	124.3
195	5625.53	48.590	149.4	245	8230.35	55.392	123.9
196	5674.19	48.739	148.8	246	8285.81	55.516	123.5
197	5723.00	48.888	148.2	247	8341.38	55.639	123.1
198	5771.97	49.036	147.6	248	8397.09	55.762	122.7
199	5821.08	49.183	147.0	249	8452.91	55.884	122.3

TABLE 11.1—*(Continued)*.

T, K	E, μV	S, μV/K	dS/dT, nV/K²	T, K	E, μV	S, μV/K	dS/dT, nV/K²
200	5870.33	49.330	146.4	250	8508.85	56.006	121.9
201	5919.74	49.476	145.9	251	8564.92	56.128	121.5
202	5969.28	49.622	145.3	252	8621.11	56.250	121.2
203	6018.98	49.767	144.7	253	8677.42	56.371	120.8
204	6068.82	49.911	144.1	254	8733.85	56.491	120.5
205	6118.80	50.055	143.6	255	8790.40	56.611	120.1
206	6168.93	50.198	143.0	256	8847.07	56.731	119.8
207	6219.20	50.341	142.5	257	8903.86	56.851	119.4
208	6269.61	50.483	141.9	258	8960.78	56.970	119.1
209	6320.16	50.625	141.4	259	9017.81	57.089	118.7
210	6370.86	50.766	140.9	260	9074.95	57.208	118.3
211	6421.69	50.907	140.3	261	9132.22	57.326	117.9
212	6472.67	51.047	139.8	262	9189.61	57.444	117.4
213	6523.79	51.186	139.3	263	9247.11	57.561	116.9
214	6575.04	51.325	138.8	264	9304.73	57.677	116.3
215	6626.44	51.464	138.3	265	9362.46	57.793	115.6
216	6677.97	51.602	137.8	266	9420.31	57.908	114.8
217	6729.64	51.739	137.3	267	9478.28	58.023	113.9
218	6781.45	51.876	136.8	268	9536.36	58.136	112.8
219	6833.39	52.013	136.3	269	9594.55	58.248	111.5
220	6885.47	52.149	135.8	270	9652.85	58.359	110.0
221	6937.69	52.284	135.3	271	9711.27	58.468	108.2
222	6990.04	52.419	134.8	272	9769.79	58.575	106.2
223	7042.53	52.554	134.3	273	9828.42	58.680	103.7
224	7095.15	52.688	133.8	274	9887.15	58.783	100.9
225	7147.90	52.821	133.3	275	9945.98	58.882	97.6
226	7200.79	52.955	132.8	276	10004.91	58.978	93.8
227	7253.81	53.087	132.3	277	10063.94	59.069	89.3
228	7306.97	53.219	131.9	278	10123.05	59.156	84.2
229	7360.25	53.351	131.4	279	10182.25	59.237	78.2
				280	10241.52	59.312	71.4

TABLE 11.2—*Type T-thermoelectric voltage,* E(T), *Seebeck coefficient,* S(T), *and derivative of the Seebeck coefficient,* dS/dT [5].

T, K	E, μV	S, μV/K	dS/dT, nV/K²	T, K	E, μV	S, μV/K	dS/dT, nV/K²
0	0.00	−0.400	526.6	45	272.34	11.245	193.5
1	−0.15	0.099	473.2	46	283.68	11.437	190.3
2	0.18	0.549	428.0	47	295.21	11.625	187.1
3	0.94	0.958	390.1	48	306.93	11.811	184.0
4	2.09	1.332	358.4	49	318.83	11.993	181.0
5	3.59	1.677	332.3	50	330.92	12.173	178.0
6	5.43	1.998	310.9	51	343.18	12.349	175.2
7	7.58	2.300	293.5	52	345.62	12.523	172.5

TABLE 11.2—*(Continued)*.

T, K	E, μV	S, μV/K	dS/dT, nV/K^2	T, K	E, μV	S, μV/K	dS/dT, nV/K^2
8	10.03	2.586	279.7	53	368.23	12.694	169.8
9	12.75	2.860	268.8	54	381.00	12.863	167.3
10	15.74	3.124	260.3	55	393.95	13.029	164.9
11	19.00	3.381	254.0	56	407.06	13.193	162.6
12	22.50	3.633	249.3	57	420.33	13.354	164.4
13	26.26	3.880	246.0	58	433.77	13.514	158.4
14	30.26	4.125	243.8	59	447.36	13.671	156.4
15	34.51	4.368	242.5	60	461.11	13.826	154.6
16	39.00	4.610	241.8	61	475.01	13.980	152.9
17	43.73	4.852	241.5	62	489.07	14.132	151.3
18	48.70	5.094	241.6	63	503.28	14.283	149.8
19	53.92	5.335	241.8	64	517.64	14.432	148.4
20	59.37	5.577	242.2	65	532.14	14.580	147.1
21	65.07	5.820	242.5	66	546.79	14.726	146.0
22	71.01	6.062	242.7	67	561.59	14.872	144.9
23	77.20	6.305	242.7	68	578.54	15.016	143.9
24	83.62	6.548	242.6	69	591.62	15.159	142.9
25	90.29	6.790	242.2	70	606.86	15.302	142.1
26	97.20	7.032	241.6	71	622.23	15.444	141.3
27	104.36	7.273	240.8	72	637.74	15.584	140.6
28	111.75	7.513	239.7	73	653.40	15.725	139.9
29	119.38	7.752	238.3	74	669.19	15.864	139.3
30	127.25	7.990	236.7	75	685.13	16.003	138.8
31	135.36	8.226	234.8	76	701.20	16.142	138.3
32	143.70	8.459	232.7	77	717.41	16.280	137.8
33	152.28	8.691	230.4	78	733.76	16.417	137.4
34	161.08	8.920	227.9	79	750.24	16.555	137.0
35	170.12	9.147	225.2	80	766.87	16.691	136.6
36	179.38	9.371	222.4	81	783.63	16.828	136.2
37	188.86	9.592	219.4	82	800.52	16.964	135.9
38	198.56	9.809	216.4	83	817.55	17.100	135.6
39	208.48	10.024	213.2	84	834.72	17.235	135.3
40	218.61	10.236	210.0	85	852.02	17.370	135.0
41	228.95	10.444	206.7	86	869.46	17.505	134.7
42	239.49	10.649	203.4	87	887.03	17.640	134.5
43	250.24	10.851	200.1	88	904.74	17.774	134.2
44	261.19	11.049	196.8	89	922.58	17.908	133.9
90	940.56	18.042	133.7	140	2004.33	24.420	123.2
91	958.66	18.175	133.4	141	2028.82	24.543	123.0
92	976.91	18.309	133.1	142	2053.42	24.666	122.9
93	995.28	18.442	132.9	143	2078.15	24.789	122.7
94	1013.79	18.574	132.6	144	2103.00	24.911	122.6
95	1032.43	18.707	132.3	145	2127.97	25.034	122.4
96	1051.20	18.839	132.1	146	2153.07	25.156	122.3
97	1070.11	18.971	131.8	147	2178.28	25.279	122.1
98	1089.14	19.102	131.5	148	2203.62	25.401	121.9
99	1108.31	19.234	131.2	149	2229.08	25.522	121.8

TABLE 11.2—*(Continued)*.

T, K	E, μV	S, μV/K	dS/dT, nV/K²	T, K	E, μV	S, μV/K	dS/dT, nV/K²
100	1127.61	19.365	131.0	150	2254.67	25.644	121.6
101	1147.04	19.496	130.7	151	2280.37	25.766	121.4
102	1166.60	19.626	130.4	152	2306.20	25.887	121.2
103	1186.29	19.757	130.1	153	2332.15	26.008	121.0
104	1206.12	19.886	129.8	154	2358.21	26.129	120.7
105	1226.07	20.016	129.6	155	2384.40	26.249	120.5
106	1246.15	20.146	129.3	156	2410.71	26.370	120.3
107	1266.36	20.275	129.0	157	2437.14	26.490	120.1
108	1286.70	20.404	128.7	158	2463.69	26.610	119.8
109	1307.17	20.532	128.5	159	2490.36	26.730	119.6
110	1327.76	20.661	128.2	160	2517.15	26.849	119.3
111	1348.49	20.789	128.0	161	2544.06	26.968	119.0
112	1369.34	20.916	127.7	162	2571.09	27.087	118.8
113	1390.32	21.044	127.5	163	2598.24	27.206	118.5
114	1411.43	21.171	127.2	164	2625.50	27.324	118.2
115	1432.66	21.298	127.0	165	2652.88	27.442	118.0
116	1454.02	21.425	126.8	166	2680.39	27.560	117.7
117	1475.51	21.552	126.6	167	2708.00	27.678	117.4
118	1497.13	21.678	126.3	168	2735.74	27.795	117.1
119	1518.87	21.805	126.1	169	2763.59	27.912	116.8
120	1540.74	21.931	126.0	170	2791.56	28.029	116.6
121	1562.73	22.057	125.8	171	2819.65	28.145	116.3
122	1584.85	22.182	125.6	172	2847.85	28.261	116.0
123	1607.10	22.308	125.4	173	2876.17	28.377	115.7
124	1629.47	22.433	125.3	174	2904.61	28.493	115.5
125	1651.96	22.558	125.1	175	2933.16	28.608	115.2
126	1674.58	22.683	125.0	176	2961.82	28.723	114.9
127	1697.33	22.808	124.8	177	2990.60	28.838	114.7
128	1720.20	22.933	124.7	178	3019.50	28.952	114.4
129	1743.20	23.058	124.5	179	3048.51	29.067	114.2
130	1766.31	23.182	124.4	180	3077.63	29.181	113.9
131	1789.56	23.306	124.3	181	3106.87	29.294	113.7
132	1812.93	23.431	124.2	182	3136.22	29.408	113.4
133	1836.42	23.555	124.0	183	3165.68	29.521	113.2
134	1860.04	23.679	123.9	184	3195.28	29.634	113.0
135	1883.78	23.803	123.8	185	3224.95	29.747	112.8
136	1907.64	23.926	123.7	186	3254.76	29.860	112.5
137	1931.63	24.050	123.5	187	3284.67	29.972	112.3
138	1955.74	24.173	123.4	188	3314.70	30.085	112.1
139	1979.98	24.297	123.3	189	3344.84	30.197	111.9
190	3375.09	30.308	111.7	235	4848.69	35.093	100.0
191	3405.46	30.420	111.5	236	4883.84	35.193	99.9
192	3435.93	30.531	111.3	237	4919.08	35.293	99.7
193	3466.52	30.643	111.1	238	4954.42	35.392	99.5
194	3497.22	30.754	111.0	239	4989.86	35.492	99.4
195	3528.03	30.865	110.8	240	5025.40	35.591	99.3
196	3558.95	30.975	110.6	241	5061.04	35.690	99.2
197	3589.98	31.086	110.4	242	5096.78	35.789	99.1

TABLE 11.2—*(Continued).*

T, K	E, μV	S, μV/K	dS/dT, nV/K^2	T, K	E, μV	S, μV/K	dS/dT, nV/K^2
198	3621.12	31.196	110.2	243	5132.62	35.888	99.0
199	3652.37	31.306	110.0	244	5168.56	35.987	98.9
200	3683.73	31.416	109.8	245	5204.60	36.086	98.8
201	3715.20	31.526	109.6	246	5240.73	36.185	98.8
202	3746.78	31.635	109.4	247	5276.97	36.284	98.7
203	3778.47	31.744	109.2	248	5313.30	36.382	98.6
204	3810.27	31.853	108.9	249	5349.73	36.481	98.5
205	3842.18	31.962	108.7	250	5386.26	36.579	98.4
206	3874.20	32.071	108.5	251	5422.89	36.678	98.3
207	3906.32	32.179	108.2	252	5459.62	36.776	98.1
208	3938.55	32.287	108.0	253	5496.44	36.874	97.9
209	3970.90	32.395	107.7	254	5533.37	36.972	97.6
210	4003.34	32.503	107.4	255	5570.39	37.069	97.3
211	4035.90	32.610	107.2	256	5607.50	37.166	97.0
212	4068.56	32.717	106.9	257	5644.72	37.263	96.6
213	4101.34	32.824	106.6	258	5682.03	37.359	96.1
214	4134.21	32.930	106.3	259	5719.44	37.455	95.6
215	4167.20	33.036	106.0	260	5756.94	37.551	95.0
216	4200.29	33.142	105.7	261	5794.54	37.645	94.3
217	4233.48	33.248	105.4	262	5832.23	37.739	93.6
218	4266.78	33.353	105.0	263	5870.02	37.833	92.9
219	4300.19	33.458	104.7	264	5907.90	37.925	92.1
220	4333.70	33.562	104.4	265	5945.87	38.017	91.3
221	4367.31	33.667	104.0	266	5983.93	38.108	90.5
222	4401.03	33.770	103.7	267	6022.08	38.198	89.8
223	4434.85	33.874	103.4	268	6060.32	38.287	89.0
224	4468.78	33.977	103.1	269	6098.66	38.376	88.4
225	4502.81	34.080	102.7	270	6137.08	38.464	88.0
226	4536.94	34.183	102.4	271	6175.58	38.552	87.7
227	4571.17	34.285	102.1	272	6214.18	38.640	87.8
228	4605.51	34.387	101.8	273	6252.86	38.728	88.2
229	4639.94	34.488	101.5	274	6291.64	38.816	89.0
230	4674.48	34.590	101.2	275	6330.50	38.906	90.5
231	4709.12	34.691	101.0	276	6369.45	38.997	92.6
232	4743.87	34.792	100.7	277	6408.49	39.091	95.6
233	4778.71	34.892	100.5	278	6447.63	39.189	99.7
234	4813.65	34.993	100.2	279	6486.87	39.291	105.0
				280	6526.22	39.399	111.8

TABLE 11.3—*Type K-thermoelectric voltage,* E(T), *Seebeck coefficient,* S(T), *and derivative of Seebeck coefficient,* dS/dT [5].

T, K	*E*, μV	*S*, μV/K	*dS/dT*, nV/K²	*T*, K	*E*, μV	*S*, μV/K	*dS/dT*, nV/K²
0	0.00	0.241	146.9	45	214.95	9.683	212.9
1	0.32	0.391	154.3	46	224.74	9.896	211.8
2	0.78	0.549	161.3	47	234.75	10.107	210.8
3	1.42	0.714	167.7	48	244.96	10.317	209.7
4	2.21	0.884	173.7	49	255.38	10.526	208.7
5	3.19	1.061	179.2	50	266.01	10.735	207.6
6	4.34	1.243	184.3	51	276.85	10.942	206.5
7	5.67	1.429	189.0	52	287.89	11.148	205.5
8	7.20	1.621	193.4	53	299.14	11.353	204.4
9	8.92	1.816	197.3	54	310.60	11.556	203.3
10	10.83	2.015	200.9	55	322.26	11.759	202.3
11	12.95	2.218	204.2	56	334.12	11.961	201.2
12	15.27	2.424	207.2	57	346.18	12.162	200.1
13	17.80	2.632	209.9	58	358.44	12.361	199.1
14	20.53	2.843	212.3	59	370.90	12.560	198.1
15	23.48	3.057	214.5	60	383.56	12.757	197.0
16	26.65	3.272	216.4	61	396.41	12.954	196.0
17	30.03	3.489	218.1	62	409.47	13.149	195.0
18	33.63	3.708	219.5	63	422.71	13.344	194.0
19	37.45	3.928	220.8	64	436.15	13.537	193.0
20	41.48	4.150	221.9	65	449.79	13.730	192.1
21	45.75	4.372	222.8	66	463.61	13.922	191.1
22	50.23	4.595	223.5	67	477.63	14.112	190.2
23	54.94	4.819	224.0	68	491.84	14.302	189.2
24	59.87	5.043	224.4	69	506.23	14.491	188.3
25	65.02	5.268	224.7	70	520.82	14.678	187.4
26	70.40	5.493	224.9	71	535.59	14.865	186.5
27	76.01	5.718	224.9	72	550.55	15.051	185.6
28	81.84	5.942	224.8	73	565.69	15.237	184.8
29	87.89	6.167	224.6	74	581.02	15.421	183.9
30	94.17	6.392	224.3	75	596.53	15.604	183.1
31	100.68	6.616	224.0	76	612.23	15.787	182.2
32	107.40	6.840	223.5	77	628.11	15.969	181.4
33	114.36	7.063	223.0	78	644.17	16.150	180.6
34	121.53	7.285	222.4	79	660.41	16.330	179.8
35	128.93	7.508	221.7	80	676.83	16.510	179.1
36	136.54	7.729	221.0	81	693.43	16.688	178.3
37	144.38	7.950	220.3	82	710.20	16.866	177.5
38	152.44	8.169	219.4	83	727.16	17.043	176.8
39	160.72	8.388	218.6	84	744.29	17.220	176.1
40	169.22	8.607	217.7	85	761.60	17.396	175.3
41	177.94	8.824	216.8	86	779.08	17.571	174.6
42	186.87	9.040	215.8	87	796.74	17.745	173.9
43	196.02	9.256	214.9	88	814.57	17.918	173.2
44	205.38	9.470	213.9	89	832.58	18.091	172.6
90	850.75	18.264	171.9	140	1966.10	26.107	142.3
91	869.10	18.435	171.2	141	1992.28	26.249	141.7

TABLE 11.3—*(Continued).*

T, K	*E*, μV	*S*, μV/K	*dS/dT*, nV/K²	*T*, K	*E*, μV	*S*, μV/K	*dS/dT*, nV/K₂
92	887.62	18.606	170.6	142	2018.60	26.391	141.1
93	906.31	18.776	169.9	143	2045.06	26.532	140.6
94	925.18	18.946	169.3	144	2071.66	26.672	140.0
95	944.21	19.115	168.6	145	2098.40	26.811	139.4
96	963.40	19.283	168.0	146	2125.28	26.950	138.8
97	982.77	19.451	167.4	147	2152.30	27.089	138.2
98	1002.31	19.618	166.8	148	2179.46	27.227	137.6
99	1022.01	19.784	166.2	149	2206.75	27.364	137.0
100	1041.87	19.950	165.5	150	2234.19	27.501	136.4
101	1061.91	20.115	164.9	151	2261.76	27.637	135.7
102	1082.11	20.280	164.3	152	2289.46	27.772	135.1
103	1102.47	20.444	163.7	153	2317.30	27.907	134.5
104	1122.99	20.608	163.2	154	2345.27	28.041	133.9
105	1143.68	20.770	162.6	155	2373.38	28.175	133.3
106	1164.53	20.933	162.0	156	2401.62	28.308	132.7
107	1185.55	21.094	161.4	157	2430.00	28.440	132.1
108	1206.72	21.255	160.8	158	2458.50	28.572	131.5
109	1228.06	21.416	160.2	159	2487.14	28.703	130.9
110	1249.55	21.576	159.7	160	2515.91	28.834	130.3
111	1271.21	21.735	159.1	161	2544.81	28.964	129.7
112	1293.02	21.894	158.5	162	2573.84	29.093	129.0
113	1315.00	22.052	157.9	163	2603.00	29.222	128.4
114	1337.13	22.210	157.4	164	2632.28	29.350	127.8
115	1359.42	22.367	156.8	165	2661.70	29.478	127.2
116	1381.86	22.524	156.2	166	2691.24	29.604	126.6
117	1404.46	22.679	155.6	167	2720.90	29.731	126.0
118	1427.22	22.835	155.1	168	2750.70	29.856	125.3
119	1450.13	22.990	154.5	169	2780.62	29.981	124.7
120	1473.20	23.144	153.9	170	2810.66	30.106	124.1
121	1496.42	23.297	153.4	171	2840.83	30.230	123.5
122	1519.80	23.451	152.8	172	2871.12	30.353	122.9
123	1543.32	23.603	152.2	173	2901.53	30.475	122.2
124	1567.00	23.755	151.6	174	2932.07	30.597	121.6
125	1590.83	23.906	151.1	175	2962.73	30.718	121.0
126	1614.81	24.057	150.5	176	2993.51	30.839	120.4
127	1638.95	24.207	149.9	177	3024.41	30.959	119.7
128	1663.23	24.357	149.3	178	3055.42	31.079	119.1
129	1687.66	24.506	148.8	179	3086.56	31.197	118.5
130	1712.24	24.654	148.2	180	3117.82	31.316	117.9
131	1736.97	24.802	147.6	181	3149.19	31.433	117.2
132	1761.84	24.950	147.0	182	3180.68	31.550	116.6
133	1786.87	25.096	146.4	183	3212.29	31.666	116.0
134	1812.04	25.243	145.9	184	3244.02	31.782	115.3
135	1837.35	25.388	145.3	185	3275.86	31.897	114.7
136	1862.81	25.533	144.7	186	3307.81	32.011	114.1
137	1888.42	25.678	144.1	187	3339.88	32.125	113.4
138	1914.17	25.821	143.5	188	3372.06	32.238	112.8
139	1940.06	25.965	142.9	189	3404.36	32.351	112.2

TABLE 11.3—*(Continued).*

T, K	*E*, μV	*S*, μV/K	*dS/dT*, nV/K²	*T*, K	*E*, μV	*S*, μV/K	*dS/dT*, nV/K²
190	3436.76	32.463	111.5	235	5000.69	36.825	82.2
191	3469.28	32.574	110.9	236	5037.55	36.907	81.6
192	3501.91	32.684	110.3	237	5074.50	36.988	80.9
193	3534.65	32.794	109.6	238	5111.53	37.069	80.3
194	3567.50	32.904	109.0	239	5148.64	37.149	79.6
195	3600.46	33.012	108.3	240	5185.83	37.228	79.0
196	3633.52	33.120	107.7	241	5223.10	37.307	78.4
197	3666.70	33.228	107.1	242	5260.44	37.385	77.7
198	3699.98	33.334	106.4	243	5297.87	37.462	77.1
199	3733.36	33.440	105.8	244	5335.37	37.539	76.4
200	3766.86	33.546	105.1	245	5372.94	37.615	75.8
201	3800.46	33.651	104.5	246	5410.60	37.691	75.1
202	3834.16	33.755	103.8	247	5448.32	37.765	74.5
203	3867.96	33.858	103.2	248	5486.13	37.840	73.8
204	3901.87	33.961	102.5	249	5524.00	37.913	73.2
205	3935.89	34.063	101.9	250	5561.95	37.986	72.5
206	3970.00	34.165	101.2	251	5599.98	38.058	71.9
207	4004.22	34.266	100.6	252	5638.07	38.130	71.2
208	4038.53	34.366	99.9	253	5676.23	38.201	70.6
209	4072.95	34.466	99.3	254	5714.47	38.271	69.9
210	4107.46	34.565	98.6	255	5752.78	38.340	69.2
211	4142.08	34.663	98.0	256	5791.15	38.409	68.5
212	4176.79	34.761	97.3	257	5829.59	38.477	67.9
213	4211.60	34.857	96.7	258	5868.10	38.545	67.2
214	4246.50	34.954	96.0	259	5906.68	38.612	66.5
215	4281.51	35.049	95.3	260	5945.33	38.678	65.8
216	4316.60	35.144	94.7	261	5984.04	38.743	65.1
217	4351.79	35.239	94.0	262	6022.81	38.808	64.3
218	4387.08	35.333	93.4	263	6061.65	38.872	63.6
219	4422.46	35.426	92.7	264	6100.56	38.935	62.8
220	4457.93	35.518	92.1	265	6139.52	38.998	62.1
221	4493.49	35.610	91.4	266	6178.55	39.059	61.3
222	4529.15	35.701	90.7	267	6217.64	39.120	60.5
223	4564.90	35.791	90.1	268	6256.79	39.180	59.6
224	4600.73	35.881	89.4	269	6296.00	39.239	58.8
225	4636.66	35.970	88.8	270	6335.27	39.298	57.9
226	4672.67	36.059	88.1	271	6374.60	39.355	57.0
227	4708.77	36.146	87.5	272	6413.98	39.412	56.1
228	4744.96	36.233	86.8	273	6453.42	39.467	55.1
229	4781.24	36.320	86.2	274	6492.91	39.522	54.1
230	4817.60	36.406	85.5	275	6532.46	39.575	53.0
231	4854.05	36.491	84.8	276	6572.06	39.628	51.9
232	4890.59	36.575	84.2	277	6611.72	39.679	50.8
233	4927.20	36.659	83.5	278	6651.42	39.729	49.6
234	4963.90	36.743	82.9	279	6691.18	39.778	48.4
				280	6730.98	39.826	47.1

TABLE 11.4—*Thermocouple, KP or EP versus gold-0.07 atomic percent iron-thermoelectric voltage, Seebeck coefficient, and derivative of the Seebeck coefficient;* E = f (T) [6].

T, K	*E*, μV	*S*, μV/K	*dS/dT*, nV/K^2	*T*, K	*E*, μV	*S*, μV/K	*dS/dT*, nV/K^2
0	0.00	0.000	0.0	45	710.22	16.569	26.9
1	7.85	8.673	1565.8	46	726.81	16.597	29.3
2	17.27	10.127	1346.7	47	743.42	16.628	31.4
3	28.04	11.375	1152.4	48	760.06	16.660	33.3
4	39.96	12.439	980.4	49	776.74	16.694	35.0
5	52.86	13.342	828.8	50	793.45	16.730	36.5
6	66.59	14.103	695.4	51	810.20	16.767	37.8
7	81.03	14.739	578.6	52	826.99	16.806	38.9
8	96.04	15.265	476.7	53	843.81	16.845	39.9
9	111.52	15.697	388.1	54	860.68	16.885	40.7
10	127.40	16.045	311.5	55	877.58	16.926	41.4
11	143.59	16.323	245.6	56	894.53	16.968	42.0
12	160.03	16.540	189.2	57	911.52	17.010	42.5
13	176.65	16.704	141.4	58	928.55	17.053	42.9
14	193.42	16.825	101.0	59	945.63	17.096	43.2
15	210.29	16.909	67.3	60	962.74	17.139	43.4
16	227.23	16.962	39.5	61	979.90	17.183	43.6
17	244.21	16.989	16.8	62	997.11	17.226	43.7
18	261.20	16.997	−1.4	63	1014.36	17.270	43.7
19	278.19	16.988	−15.7	64	1031.65	17.314	43.7
20	295.17	16.966	−26.6	65	1048.99	17.358	43.7
21	312.12	16.935	−34.6	66	1066.36	17.401	43.7
22	329.04	16.898	−40.1	67	1083.79	17.445	43.6
23	345.92	16.856	−43.5	68	1101.25	17.489	43.5
24	362.75	16.811	−45.1	69	1118.76	17.532	43.4
25	379.54	16.766	−45.3	70	1136.32	17.575	43.3
26	396.28	16.721	−44.2	71	1153.92	17.619	43.2
27	412.98	16.678	−42.1	72	1171.56	17.662	43.1
28	429.64	16.637	−39.2	73	1189.24	17.705	42.9
29	446.26	16.600	−35.7	74	1206.96	17.748	42.8
30	462.84	16.566	−31.8	75	1224.73	17.790	42.7
31	479.39	16.536	−27.5	76	1242.55	17.833	42.6
32	495.92	16.511	−23.0	77	1260.40	17.875	42.4
33	512.42	16.490	−18.4	78	1278.30	17.918	42.3
34	528.90	16.474	−13.8	79	1296.24	17.960	42.2
35	545.37	16.463	−9.2	80	1314.22	18.002	42.0
36	561.83	16.456	−4.7	81	1332.24	18.044	41.9
37	578.28	16.453	−0.4	82	1350.30	18.086	41.8
38	594.73	16.455	3.8	83	1368.41	18.128	41.6
39	611.19	16.461	7.8	84	1386.56	18.169	41.5
40	627.66	16.471	11.6	85	1404.75	18.211	41.3
41	644.13	16.484	15.2	86	1422.98	18.252	41.2
42	660.63	16.501	18.5	87	1441.25	18.293	41.0
43	677.14	16.521	21.5	88	1459.57	18.334	40.9
44	693.67	16.544	24.3	89	1477.92	18.375	40.7
90	1496.32	18.415	40.5	140	2462.15	20.094	26.7
91	1514.75	18.456	40.3	141	2482.25	20.120	26.5

TABLE 11.4—*(Continued).*

T, K	E, μV	S, μV/K	dS/dT, nV/K^2	T, K	E, μV	S, μV/K	dS/dT, nV/K^2
92	1533.23	18.496	40.2	142	2502.39	20.147	26.3
93	1551.74	18.536	40.0	143	2522.55	20.173	26.1
94	1570.30	18.576	39.7	144	2542.73	20.199	26.0
95	1588.89	18.615	39.5	145	2562.94	20.225	25.8
96	1607.53	18.655	39.3	146	2583.18	20.250	25.6
97	1626.20	18.694	39.1	147	2603.45	20.276	25.4
98	1644.92	18.733	38.8	148	2623.73	20.301	25.3
99	1663.67	18.772	38.5	149	2644.05	20.327	25.1
100	1682.46	18.810	38.3	150	2664.39	20.352	24.9
101	1701.29	18.848	38.0	151	2684.75	20.376	24.7
102	1720.16	18.886	37.7	152	2705.14	20.401	24.6
103	1739.06	18.924	37.4	153	2725.55	20.426	24.4
104	1758.00	18.961	37.1	154	2745.99	20.450	24.2
105	1776.98	18.998	36.8	155	2766.45	20.474	24.1
106	1796.00	19.035	36.5	156	2786.94	20.498	23.9
107	1815.05	19.071	36.2	157	2807.45	20.522	23.7
108	1834.14	19.107	35.8	158	2827.98	20.545	23.5
109	1853.27	19.143	35.5	159	2848.54	20.569	23.4
110	1872.43	19.178	35.2	160	2869.12	20.592	23.2
111	1891.62	19.213	34.8	161	2889.72	20.615	23.0
112	1910.85	19.248	34.5	162	2910.35	20.638	22.8
113	1930.12	19.282	34.2	163	2931.00	20.661	22.6
114	1949.42	19.316	33.8	164	2951.67	20.683	22.5
115	1968.75	19.350	33.5	165	2972.37	20.706	22.3
116	1988.12	19.383	33.2	166	2993.08	20.728	22.1
117	2007.52	19.416	32.8	167	3013.82	20.750	21.9
118	2026.95	19.449	32.5	168	3034.58	20.772	21.7
119	2046.41	19.481	32.2	169	3055.37	20.793	21.5
120	2065.91	19.513	31.8	170	3076.17	20.815	21.3
121	2085.44	19.545	31.5	171	3096.99	20.836	21.1
122	2105.00	19.576	31.2	172	3117.84	20.857	21.0
123	2124.59	19.607	30.9	173	3138.71	20.878	20.8
124	2144.21	19.638	30.6	174	3159.60	20.899	20.6
125	2163.87	19.668	30.3	175	3180.51	20.919	20.4
126	2183.55	19.698	30.0	176	3201.44	20.939	20.2
127	2203.26	19.728	29.7	177	3222.38	20.960	20.1
128	2223.00	19.758	29.4	178	3243.35	20.980	19.9
129	2242.78	19.787	29.2	179	3264.34	20.999	19.7
130	2262.58	19.816	28.9	180	3285.35	21.019	19.6
131	2282.41	19.845	28.7	181	3306.38	21.038	19.4
132	2302.27	19.873	28.4	182	3327.43	21.058	19.2
133	2322.16	19.902	28.2	183	3348.50	21.077	19.1
134	2342.07	19.930	28.0	184	3369.58	21.096	18.9
135	2362.02	19.958	27.7	185	3390.69	21.115	18.8
136	2381.99	19.985	27.5	186	3411.81	21.133	18.6
137	2401.99	20.013	27.3	187	3432.96	21.152	18.5
138	2422.01	20.040	27.1	188	3454.12	21.171	18.4
139	2442.07	20.067	26.9	189	3475.30	21.189	18.3

TABLE 11.4—*(Continued).*

T, K	*E*, μV	*S*, μV/K	*dS/dT*, nV/K²	*T*, K	*E*, μV	*S*, μV/K	*dS/dT*, nV/K²
190	3496.49	21.207	18.1	235	4467.28	21.881	10.3
191	3517.71	21.225	18.0	236	4489.17	21.891	10.1
192	3538.94	21.243	17.9	237	4511.06	21.901	9.9
193	3560.20	21.261	17.8	238	4532.97	21.911	9.8
194	3581.47	21.279	17.7	239	4554.88	21.921	9.6
195	3602.75	21.296	17.6	240	4576.81	21.930	9.5
196	3624.06	21.314	17.5	241	4598.74	21.940	9.4
197	3645.38	21.331	17.4	242	4620.69	21.949	9.3
198	3666.72	21.348	17.3	243	4642.64	21.958	9.3
199	3688.08	21.366	17.2	244	4664.61	21.968	9.2
200	3709.45	21.383	17.1	245	4686.58	21.977	9.2
201	3730.84	21.400	17.0	246	4708.56	21.986	9.3
202	3752.25	21.417	16.9	247	4730.55	21.995	9.3
203	3773.68	21.434	16.8	248	4752.55	22.005	9.4
204	3795.12	21.450	16.7	249	4774.56	22.014	9.5
205	3816.58	21.467	16.5	250	4796.58	22.024	9.6
206	3838.05	21.483	16.4	251	4818.61	22.034	9.8
207	3859.54	21.500	16.3	252	4840.64	22.043	10.0
208	3881.05	21.516	16.2	253	4862.69	22.053	10.2
209	3902.58	21.532	16.0	254	4884.75	22.064	10.3
210	3924.12	21.548	15.9	255	4906.82	22.074	10.6
211	3945.67	21.564	15.8	256	4928.90	22.085	10.8
212	3967.24	21.580	15.6	257	4950.99	22.096	11.0
213	3988.83	21.595	15.4	258	4973.09	22.107	11.1
214	4010.43	21.610	15.3	259	4995.20	22.118	11.3
215	4032.05	21.626	15.1	260	5017.33	22.129	11.5
216	4053.69	21.641	14.9	261	5039.46	22.141	11.6
217	4075.33	21.655	14.7	262	5061.61	22.152	11.6
218	4097.00	21.670	14.5	263	5083.77	22.164	11.7
219	4118.67	21.684	14.3	264	5105.94	22.176	11.6
220	4140.36	21.698	14.0	265	5128.12	22.187	11.5
221	4162.07	21.712	13.8	266	5150.31	22.199	11.3
222	4183.79	21.726	13.6	267	5172.52	22.210	11.1
223	4205.52	21.739	13.3	268	5194.73	22.221	10.7
224	4227.27	21.753	13.1	269	5216.96	22.231	10.3
225	4249.03	21.766	12.8	270	5239.19	22.241	9.7
226	4270.80	21.778	12.5	271	5261.44	22.251	9.0
227	4292.58	21.791	12.3	272	5283.70	22.259	8.2
228	4314.38	21.803	12.0	273	5305.96	22.267	7.3
229	4336.19	21.815	11.8	274	5328.23	22.274	6.3
230	4358.01	21.826	11.5	275	5350.51	22.280	5.2
231	4379.84	21.838	11.3	276	5372.79	22.284	4.0
232	4401.68	21.849	11.0	277	5395.08	22.288	2.7
233	4423.54	21.860	10.8	278	5417.36	22.290	1.3
234	4445.40	21.870	10.5	279	5439.65	22.290	−0.1

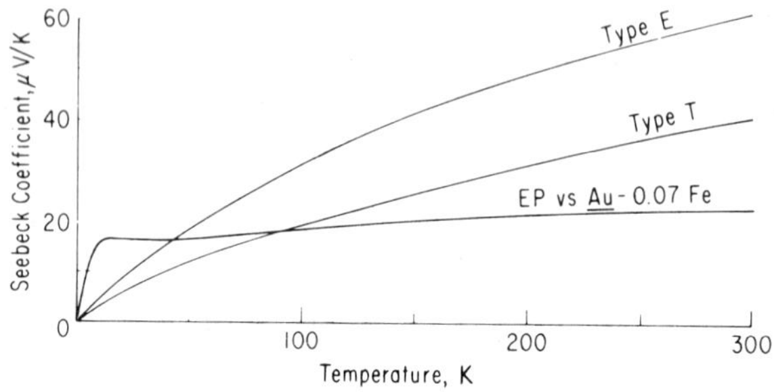

FIG. 11.1—*Seebeck coefficients for Types E, T, and EP versus Au-0.07 Fe.*

11.4 References

[1] Scott, R. B., *Cryogenic Engineering*, Van Nostrand, Princeton, N.J., 1959.

[2] Hust, J. G., Powell, R. L., and Sparks, L. L., "Methods for Cryogenic Thermocouple Thermometry," *Temperature, Its Measurement and Control in Science and Industry*, Part 3, Vol. 4, 1972, p. 1525.

[3] White, G. K., *Experimental Techniques in Low Temperature Physics*, Oxford University Press, London, 1959.

[4] Rose-Innes, A. C., *Low Temperature Techniques*, Van Nostrand, Princeton, N.J., 1964.

[5] Sparks, L. L., Powell, R. L., and Hall, W. J., "Reference Tables for Low-Temperature Thermocouples," NBS Monograph 124, National Bureau of Standards, 1972.

[6] Sparks, L. L. and Powell, R. L., "Low Temperature Thermocouples: KP, "Normal" Silver, and Copper Versus Au-0.02 at % Fe and Au-0.07 at % Fe," *Journal of Research*, National Bureau of Standards, Vol. 76A, No. 3, May-June 1972.

Chapter 12—Temperature Measurement Uncertainty

12.1 The General Problem

Every measurement has associated with it an error. This is a real certain value which really exists but cannot be known. It is different from the uncertainty of the measurement, which reflects only our understanding of the error. This chapter will deal primarily with the uncertainty.

An error may be caused by a mistake or by a broken instrument as well as by the usual variation in parameters affecting the temperature measurement. A mistake, like connecting a Type E thermocouple to a Type K instrument or using a broken instrument, in which perhaps a reference diode has failed, is regrettable but cannot be analyzed by usual statistical methods. Hence, these events are not treated herein.

Aside from mistakes and broken instruments, errors due to imprecision are apparent in a simple example: Suppose you want to heat a pot of water to 135°F. You insert a proper thermocouple and take a reading. If you take another reading, they will not be the same unless your experimental technique is too poor to resolve the difference. If you further check with another thermocouple or at a different point in the pot, or call an assistant to check you, or try another instrument, you probably will get as many different answers as the number of measurements made. This variation is to be expected—it reflects the uncertainty of the measurement. It may not present a problem. If you just want the water to wash your dishes, and will be happy with any temperature 125 to 145°F you probably need read no further. But if you are using this as a reagent where ±0.5 degree is important, you need to estimate the error, and you need to know how well you know the error.

As in all statistical analysis, the credibility of the work is enhanced by its success in predicting. When the analyst predicts a variability which is grossly different from that observed in a test, his judgment is flawed. Either his model or his analysis is wrong. Statistics do not lie, but they can be misleading by misapplication or by intent.

In the discussion which follows, no attempt will be made to teach statistics. This topic has been covered more than adequately by Benedict [*1*],[1] Chatfield [*2*], Spiegel [*3*], Abernethy [*4*], and the ASTM Committee E-11 on Statistical Techniques [*5*]. Certain terms developed by these authors are redefined for

[1]The italic numbers in brackets refer to the list of references appended to this paper.

convenience in paragraph 12.2 in the context of temperature measurements, and examples of their use are given in paragraph 12.3. The reader who wants to apply these tools will probably need to study some of the references if he is not familiar with their use.

12.2 Tools of the Trade

The language of measurement uncertainty is largely the language of statistics, and, like statistics, it is often misunderstood. Statistical words are very precise in their meaning, like legal terms, and he who uses them carelessly does so not only at his own risk but also at that of the reader. For convenience, the following terms are defined here in the context of thermocouple measurements.

12.2.1 Average and Mean

Average and *mean* are synonymous and are best defined by the equation

$$\text{average } X = \Sigma x_i/n,\text{, denoted } \overline{X}$$

where n is the number of measurements X_i. The definition "$\overline{X} = (X\text{max} + X\text{ min})/2$," though commonly used in other applications, has no place in measurement uncertainty. The symbol $\overline{X}$ is used for the average of a set of measurements; μ is the average of the total population.

The term *across* is introduced at this point. Like many statistical terms, an average is always across a specific group of data. The average across a series of readings (subset) will differ from the average across another subset, and from the average of the total population, which is seldom known. We must always know what we are averaging across.

12.2.2 Normal or Gaussian Distribution

Normal or *Gaussian distribution* is defined by the formula

$$f(x) = \frac{1}{\sigma\sqrt{2\pi}} e^{-1/2(x-\mu)^2/\sigma^2}$$

where σ is the standard deviation of the total population. In the analysis of measurement uncertainty, this formula is seldom used directly, but it defines the dispersion that usually exists in real physical data. The careful investigator will always check the normality of his test data before applying normal statistics to it. A visual check is often sufficient, especially with the aid of probability-plot paper (See 12.3.4).

12.2.3 Standard Deviation and Variance

The *standard deviation* and *variance* are measures of the precision or scatter of normally distributed data, that is, the extent to which such data departs from the mean. The variance is the sum of the squares of the deviations of the individual measurements from the mean, divided by the degrees of freedom. The standard deviation is the square root of the variance. An estimate of the standard deviation is s defined as

$$s = \sqrt{\frac{\Sigma(X_i - \overline{X})^2}{n - 1}}$$

Obviously, the variance, s^2, is the same equation without the radical. The impact of moderately priced personal calculators on these evaluations is clear. Statistical calculations can be performed easily which would probably be too tedious to perform longhand.

12.2.4 Bias, Precision, and Uncertainty

Bias, precision, and *uncertainty* are the key terms for describing defects in measurement, confusing at times because of careless use. *Precision* is the scatter, usually random, between measurements in the same set. *Bias* is the systematic difference between the means of two or more sets. *Uncertainty* is a less rigorous term which tries to combine bias and precision into a single term for talking purposes. One definition, by Abernethy [*4*], is $\pm(b + t_{95}s)$, where b is the best estimate of the limit of uncorrected bias, s is the precision, and t_{95} is the 95th percentile value of the two tailed Student t distribution. In words, it is the largest error which can reasonably be expected. An important rule is that an uncertainty statement is never made without including: (1) bias, (2) precision, (3) degrees of freedom, and (4) confidence interval. The opportunity for confusion lies in the fact that a component of uncertainty can move between bias and precision depending on the uncertainty being assessed; for example, the bias between reels of thermocouple wire becomes reel-to-reel precision when evaluating the uncertainty of K thermocouple measurements in general. The extent to which it can be "corrected out" of the uncertainty of the smaller set depends on the extent of calibration activities. The importance of identifying the restraints of the uncertainty statement is evident.

Degrees of freedom (df) is a term which is used in many of the equations of uncertainty, and relates to the size of the data sample on which the calculation is based. Chatfield [*2*] defines it as the number of independent comparisons available. For example in the formula for the average, df $= n$. In the formula for standard deviation it is $(n - 1)$, since there is only one difference between two values, one degree of freedom is lost. Similarly the constants in a regression equation each consume one degree of freedom as shown in paragraph 12.3.5.

12.2.5 Precision of the Mean

The *precision of the mean* is an important concept in temperature measurement. As the number of independent measurements of the same temperature is increased, the uncertainty of our knowledge of the true mean decreases, specifically by the square root of the number of measurements.

$$s_{\text{mean}} = \frac{s}{\sqrt{n}}$$

where n is the number of independent measurements. This can be applied to the time average, space average, production norm, or to any other set. For discussion see paragraph 12.3.3.

12.2.6 Regression Line or Least-Square Line

A *regression line* or *least-squares line* represents the equation which minimizes the sum of the squares of the displacements of the individual data points in a set from the line. The line may be a straight line or a polynomial, exponential, or any other form.

12.3 Typical Applications

12.3.1 General Considerations

The discussions herein will be confined to Type K, because of the availability of a statistically significant quantity of data for this thermocouple type. The same statistical concepts can be applied to other types.

ANSI/ASTM E 230 describes the temperature-emf characteristics of Type K wire. The "limits of error" stated in this standard are definitive, not statistical. Wire which does not conform to the stated limits is simply not Type K. Further, the limits of error are defined as referring only to new wire as delivered, and not to include the effects of insulating, sheathing, or exposure to operating conditions of the application. The statistical treatment discussed herein is, in a large part, similarly limited.

Topics to be discussed include: (1) the improvement in uncertainty which can be realized, especially at low temperatures, by calibration; (2) the use of the precision of the mean of several thermocouples; (3) probability plots; and (4) the use of regression analysis to eliminate variations for cause from random variations.

12.3.2 Wire Calibration

The Type K temperature-emf relationship, described by algorithm in NBS MN 125 [*6*], and tables presented in NBS 125 and ASTM E 230, are the results of empirical data developed by NBS from experimental data on real Type K thermocouples from several sources and including several sizes of wire. In other words, these documents present statistical information believed to represent the typical Type K thermocouple. Whether or not the deviations of the individual thermocouples tested from the algorithm were normally distributed is not known. We also do not know the scatter of these data about the K-curve.

Sanders [*7*] presents observations on incoming inspection calibrations performed by a large user of 20-28 B&S gage Type K wire with "soft" insulation. A small quantity of sheathed wire was considered separately. An analysis of variance was made to assess the effect of various insulations, cable makeup, vendors, and wire gages on the precision or bias of the calibration. The only significant influence was that of wire gage. The data are shown in Table 12.1. Precision is stated as two-standard deviations.

From these data it is clear that, first, the typical wire received by this user is systematically different from the K-curve as shown in Fig. 12.1, probably because of the difference in the range of sizes tested by the National Bureau of Standards (NBS) and this user. Thus, if he plans to use thermocouples made from this wire without further calibration, he would be advised to modify the K-curve by the bias curve of Table 12.1. Secondly, if he uses this modified curve, he can expect that such couples individually will vary from

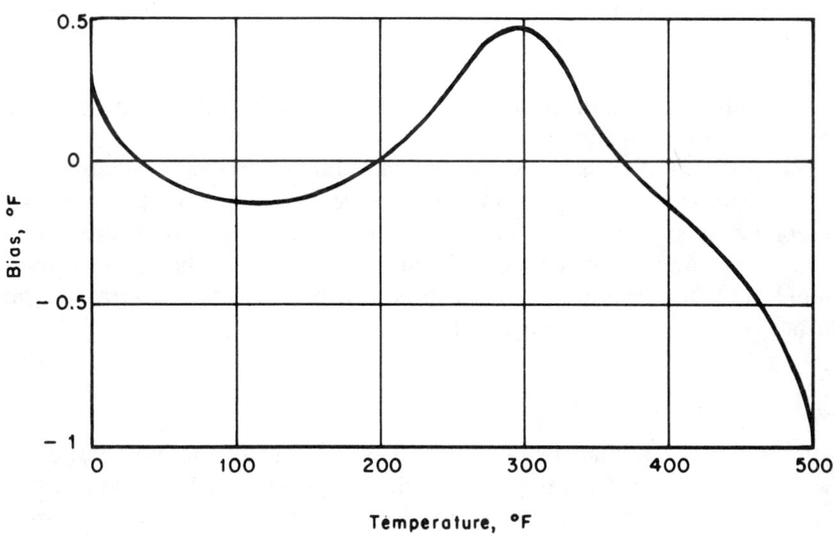

FIG. 12.1—*Bias of a typical Type K wire.*

TABLE 12.1—*Accuracy of unsheathed thermocouples.*

Temperature °F, T	Bias °F, b	Overall Precision °F, $t_{95}s$	In-Reel Precision °F, $t_{95}s$	Uncertainty °F, U
0	0.28	±0.09	±0.08	0.37
32	0.00	±0.06	±0.06	0.06
65	−0.16	±0.18	±0.06	0.34
100	−0.14	±0.29	±0.08	0.43
150	−0.10	±0.51	±0.11	0.61
200	0.00	±0.68	±0.11	0.68
250	0.26	±0.96	±0.15	1.22
300	0.47	±0.09	±0.18	1.56
350	0.13	±1.30	±0.20	1.43
400	−0.15	±1.40	±0.24	1.55
450	−0.41	±1.48	±0.28	1.89
500	−0.91	±1.33	±0.38	2.24

the curve by the precision shown in the table, which is substantially less than the ±2 deg permitted by ASTM E 230. If he does not correct for the bias, he can expect the true value to lie between the bounds of total uncertainty listed in the table, which is slightly in excess of the specification.

Sanders further concludes that the reel-to-reel precision, as expected, is worse than the precision within a reel; that is, that thermocouples whose readings are to be compared should be made, if possible, from the same reel of wire, especially if individual calibrations are not performed. If a piece of thermocouple wire is calibrated and then the junction is cut off in order to fabricate a probe, the calibration will be valid within 0.05°F for that probe if the new junction is within a few inches of the calibration junction. If it is more than a few inches away, the full in-reel precision will be developed. Similar calibration data for a limited sample of sheathed wire are given in Table 12.2, for comparison only.

The material discussed in the previous table represents a large, but specific, subset of small gage Type K wire below 500°F. Similar tests on a different data subset should be expected to vary to some extent, depending upon sample and measurement differences. However, the basic conclusions should hold, that the uncertainty can be reduced by batch calibration and further reduced by specific calibrations.

12.3.3 Means and Profiles

In all areas of human knowledge, iteration is one of the most respected ways to stress or reinforce impressions. If you are surprised at what you see, you take a second look. If you want to be sure what time your plane leaves, you check again. If you want confidence in the results of an experiment, you repeat it. It is then not surprising that the mean of several readings of a

TABLE 12.2—*Accuracy of sheathed thermocouples.*

Temperature °F, T	Bias °F, b	Overall Precision °F, $t_{95}s$	In-Reel Precision °F, $t_{95}s$	Uncertainty °F, U
100	−0.13	±0.37	±0.10	0.50
150	+0.01	±0.50	±0.14	0.51
200	+0.35	±0.70	±0.30	1.05
250	+0.82	±1.12	±0.42	1.94
300	+1.09	±1.36	±0.52	2.45
350	+0.78	±1.45	±0.61	2.23
400	+0.65	±1.51	±0.68	2.16
450	+0.74	±1.57	±0.90	2.31
500	+0.69	±1.40	±0.96	2.09

temperature, or of the readings of several sensors, or of the temperature-emf characteristics of a pair of alloy ingots lends more confidence than a single reading. Statistics provides the means to quantify this increase in confidence. Like all statistical tools, it must be properly used.

In the development of a turbine engine, the gas temperature at a particular axial section may be measured by one hundred or more sensors. For component efficiency studies the mean temperature at this section is a needed parameter. To assess incremental improvement in performance, the accuracy of this temperature can be quite critical. When a single probe is used, several error components exist. First, and usually dominant, is the temperature variation in space and time at different points across the section. Another error component is the degree to which the kinetic energy of the gas stream is not converted to junction temperature (aerodynamic recovery) (see 9.1.4). Others include manufacturing tolerances, conduction errors, and wire calibration. The latter is usually corrected out.

If we use 100 sensors the space profile, the manufacturing tolerances and the random errors in calibrating are reduced by $\sqrt{n} = 10$. Now when we compare the mean temperature before and after a change in configuration, the minimum significant change in performance is reduced by an order of magnitude. Note that the same improvement does not apply to the individual measurements in the set.

Further information may be derived from the same data for different purposes. The performance of the engine may be affected by the uniformity of the gas temperature, which is quantified by the standard deviation of the individual temperatures. This will also predict the maximum temperatures, which will affect the life of metal parts. A check of the normality of the data may show that a pattern exists.

It is important to realize the significance of the precision of the mean. It does us no good to improve the precision of measurement of the mean unless the mean is what we are going to use. For example, in calibration we frequently use an array of thermocouples to establish the degree of uniformity of temperatures of a liquid bath in which we will calibrate thermocouples. We can use the standard deviation of the measurements as a measure of uniformity, but we cannot claim knowledge of the local bath temperature to within the precision of the mean. Specifically

$$T_i = \overline{T} \pm ts, \text{ not } \overline{T} \pm ts/\sqrt{n}$$

where

T_i = local temperature at the location of the test sensor,
$\overline{T}$ = mean of temperature measurements,
s = standard deviation of temperature measurements, and
t = Student's t.

12.3.4 Probability Paper

Paragraph 12.2.2 refers to the normality of data and the need to assure yourself that data are normally distributed before applying the mathematics of classical statistics. Actually, the determination of this property can directly answer some questions and lead to accuracy improvement.

The normal or Gaussian distribution, plotted as error versus frequency of occurrence, yields the familiar "bell-shaped" curve. The integral of this curve, plotting error versus cumulative frequency, is an ogive. If the frequency axis of the ogive is properly warped, a special "probability paper" is created, on which normally distributed data are plotted as a straight line or "prob-plot" which is easily evaluated as to its straightness. Figure 12.2 reproduces this probability plot, which is commercially available from several sources. King [*8*], describes the mechanics of using this paper.

When the data are properly ranked and plotted on prob-plot paper, the nature of existing nonnormalities is often revealed. Data which are skewed to the left or right will plot concave upward or downward, respectively. One of the more common revelations is that the data are bimodal or polymodal. Such data will appear as a series of line segments of different slopes. This indicates that the data come from more than one subset, having different standard deviations or means or both. Data which are flatter or more peaked will yield plots which are similar to polymodal.

Figure 12.3 shows a straight line of finite slope, terminated at the top by a segment of zero slope. In real data the transition may be less sharp. Let us look at what we can conclude from this plot. First, unless there is an *a priori* reason to expect nonnormal temperature distribution, the highest temperatures which occurred were not measured.

If the data were taken from 1000 thermocouples in an engine or furnace, the mean temperature is shown to be 2500 deg (the 50th percentile), and there should have been 10 readings in excess of 2900 and one in excess of

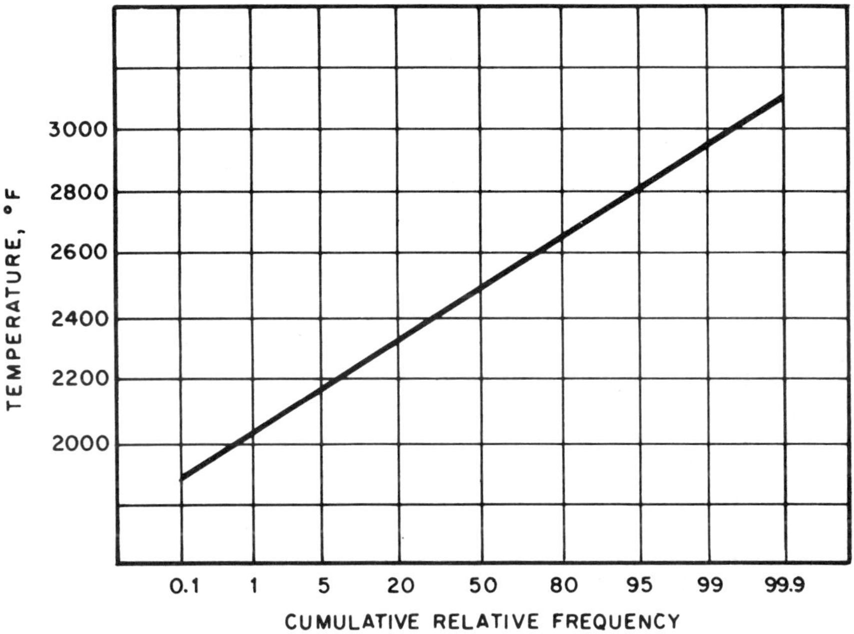

FIG. 12.2—*Typical probability plot.*

3000. Perhaps some of them burned out (truncation). Perhaps the experimenter thought he had reason to doubt any readings as high as 2900 (editing or false outlier rejection). Or perhaps there were no thermocouples located at the hottest spots (sampling error). But the evidence is clear, the probability of existence of temperatures above 3000 is high and cannot be rejected lightly. If the data are taken from 1000 successive readings of a single thermocouple, the reasoning is the same and the conclusions equally valid in the time domain. You do not have to have seen a temperature in excess of 3000 to have good reason to believe it occurred.

12.3.5 Regression Analyses

Regression analysis is used statistically to express a set of data in an analytical relationship. This relationship can be used to predict values of the dependent variable at values of the independent variable between those for which test data exists. The expression can be also used to smooth the curve of test data, based on physical knowledge beyond the mere statistics, for example, we may know from physical facts that the true relation is linear, polynomial, or exponential, and need only to determine a few constants to define it. Redundant data points beyond the number of constants to be de-

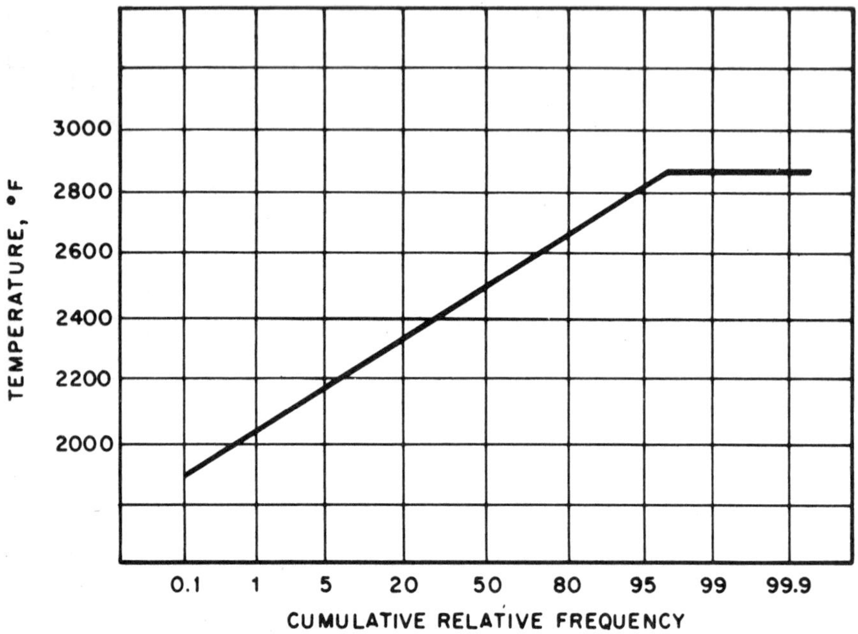

FIG. 12.3—*Typical probability plot—truncated data.*

rived are required to provide degrees of freedom in order to establish confidence in the constants. This confidence is an expression of the "goodness of fit" of the curve and is called the standard error of estimate (SEE) expressed by

$$\text{SEE} = \sqrt{\frac{\Sigma(Y_i - Y_{ci})^2}{n - (1 + q)}}$$

where

Y_{ci} = predicted value of $Y_i = f(X_i)$,
Y_i = measured value of Y_i at X_i,
n = number of data points, and
q = order of the derived equation.

The same SEE is used to assess the scatter of the data points around the curve. Obviously, if the scatter of the original data is large, this fact will be reflected in a large SEE of the curve, which implies a large uncertainty in the constants. Conversely, experimental data closely grouped around a simple, well-defined regression curve will produce a low value of SEE which indicates the precision of both the coefficients and the test data.

Care must be taken to avoid over fitting a curve to experimental data. Most physical relationships can be expressed with relatively few constants, for example, a low order polynomial. Any set of data, on the other hand, can be well described by a polynomial with a number of constants equal to the number of data points. Such a fit will generally be meaningless. The SEE will have increased because of the decrease in the degrees of freedom (denominator), and the equation will perfectly describe what you already know, while becoming worthless to predict interpolated or extrapolated values. A very interesting account of the regression of the temperature-emf tables is given in NBS-125 [6] and the references contained therein. In this unusual case the large quantity of data (degrees of freedom) permits regression to as much as a 14th order equation to reduce the SEE to the order of one microvolt, while retaining sufficient degrees of freedom.

Regression analysis is one of several techniques which can be used to identify a causal relationship between two variables. Other techniques such as correlation and analysis of variance (ANOVA) are somewhat more sophisticated and are discussed by Chatfield [2].

Thermocouples often are accused of drifting with time in use. If we have a series of data points representing time versus error determined by reference to noble metal thermocouples, resistance thermometer, optical pyrometer, or other alternate technique, there will be a data scatter for one or more experimental reasons between the points. One way to decide whether the error is a linear function of time is to determine the coefficients of a linear regression of the data. If there is no linear relationship, the SEE will approximate the standard deviation; if the points all fall on a straight line, it will be zero.

12.4 References

[1] Benedict, R. P., *Fundamentals of Temperature, Pressure, and Flow Measurements*, 2nd edition, Wiley, New York, 1977, Chapter 10.

[2] Chatfield, C., *Statistics for Technology*, Wiley, New York, 1978.

[3] Spiegel, M. R., *Theory and Problems of Statistics*, Scham's Outline Series, McGraw-Hill, New York, 1961.

[4] Abernethy, R. B. et al, "Uncertainty in Gas Turbine Measurements," Revised 1980, Instrument Society of America, 1483-3.

[5] "ASTM Standards on Precision and Accuracy for Various Applications," American Society for Testing and Materials, 1977.

[6] Powell, R. L. et al, "Thermocouple Reference Tables Based on the IPTS-68," National Bureau of Standards MN-125, Department of Commerce, Washington, D.C., 1974.

[7] Sanders, D. G., "Accuracy of Type K Thermocouples Below 500°: A Statistical Analysis," *Transactions*, Vol. 1, No. 4, Instrument Society of America, 1974.

[8] King, J. R., "Probability Charts for Decision Making," Industrial Press, Inc., New York, 1971.

[9] Ku, H. H., Ed., "Precision Measurement and Calibration: Statistical Concepts and Procedures," *NBS Special Publication 300*, Vol. 1, U.S. Government Printing Offices, 1969. (Paper 5.2 Statistical Concepts in Metrology).

Chapter 13—Terminology

calibrate, *v.*: 1. *general*—to determine the indication or output of a measuring device with respect to that of a standard.
2. *thermocouple*—to determine the emf developed by a thermocouple with respect to temperature established by a standard.

calibration point, *n.*: 1. *general*—a specific value, established by a standard, at which the indication or output of a measuring device is determined.
2. *thermocouple*—a temperature, established by a standard, at which the emf developed by a thermocouple is determined.

Celsius, *n.*—the designation of the degree on the International Practical Temperature Scale. Also used for the name of the Scale, as "Celsius temperature scale." Formerly (prior to 1948) called "centigrade."

centigrade, *n.*—the designation of the degree on the International Temperature Scale prior to 1948. (See *Celsius*.)

coaxial thermocouple element, *n.*—a thermocouple element consisting of a thermoelement in wire form, within a thermoelement in tube form with the two thermoelements insulated from each other and from the tube except at the measuring junction.

connection head, *n.*—a housing enclosing a terminal block for an electrical temperature-sensing device and usually provided with threaded openings for attachment to a protecting tube and for attachment of conduit.

defining fixed points, *n.*—the reproducible temperatures upon which the International Practical Temperature Scale is based.

degree, *n.*—the unit of a temperature scale. See Celsius, centigrade, Fahrenheit.

electromotive force (emf), *n.*—the electrical potential difference which produces or tends to produce an electric current.

extension wire, *n.*—a pair of wires having such temperature-emf characteristics relative to the thermocouple with which the wires are intended to be used that, when properly connected to the thermocouple, the reference junction is transferred to the other end of the wires.

Fahrenheit, *n.*—the designation of the degree and the temperature scale used commonly in public life and engineering circles in English-speaking countries. Related to the International Practical Temperature Scale by means of the equation

$$t_F = 9/5\, t_c + 32$$

fixed point, *n.*—a reproducible temperature of equilibrium between different phases of a material. (See *defining fixed points* and *secondary reference points.*)

freezing point, *n.*—the fixed point between the solid and liquid phases of a material when approached from the liquid phase under a pressure of 1 standard atm (101325 N/m^2). For a pure material this is also the melting point.

ice point, *n.*—the fixed point between ice and air-saturated water under a pressure of 1 standard atm (101325 N/m^2). This temperature is 0°C on the International Practical Temperature Scale.

International Practical Temperature Scale of 1948 (IPTS-48), *n.*—the temperature scale adopted by the 11th General Conference on Weights and Measures in 1960. Replaced in 1968 by the International Practical Temperature Scale of 1968.

International Practical Temperature Scale of 1968 (IPTS-68), *n.*—the temperature scale, which through adoption by the 13th General Conference on Weights and Measures in 1968, is defined in terms of fixed and reproducible equilibrium temperatures (defining fixed points) to which numerical values have been assigned, and equations establishing the relation between temperature and the indications of sensing instruments calibrated by means of the values assigned to the defining fixed points.

kelvin, *n.*—the designation of the thermodynamic temperature scale and the interval on this scale. This kelvin scale was defined by the Tenth General Conference on Weights and Measures in 1954 by assigning the temperature of 273.16 K to the triple point of water. Also the interval on the International Practical Kelvin Temperature Scale.

liquid-in-glass thermometer, *n.*—a temperature-measuring instrument whose indications are based on the temperature coefficient of expansion of a liquid relative to that of its containing glass envelope.

lower range-value, *n.*—the lowest quantity that an instrument is adjusted to measure.

measuring junction, *n.*—that junction of a thermocouple which is subjected to the temperature to be measured.

melting point, n.—the fixed point between the solid and liquid phases of a material when approached from the solid phase under a pressure of 1 standard atm (101325 N/m^2). For a pure material this is also the freezing point.

Peltier coefficient, *n.*—the reversible heat which is absorbed or evolved at a thermocouple junction when unit current passes in unit time. Synonymous with *Peltier emf.*

Peltier emf, *n.*—synonymous with *Peltier coefficient.*

platinum 27 (Pt-27), *n.*—the platinum standard to which the National Bureau of Standards referred thermoelectric measurements prior to 1973.

platinum 67 (Pt-67), *n.*—the platinum standard used by the National Bureau of Standards after 1972 as the reference to which thermoelectric measurements are referred.

potentiometer, Group A, *n.*—a laboratory high-precision type potentiometer having limits of error of approximately 0.2 μV at 1000 μV, and 5 μV or less at 50 000 μV.

potentiometer, Group B, *n.*—a laboratory precision type potentiometer having limits of error of approximately 1 μV at 1000 μV and 12 μV or less at 50 000 μV.

primary standard thermocouple, *n.*—a thermocouple that has had its temperature-emf relationship determined in accordance with methods described in the text establishing the International Practical Temperature Scale.

protecting tube, *n.*—a tube designed to enclose a temperature-sensing device and protect it from the deleterious effects of the environment. It may provide for attachment to a connection head but is not primarily designed for pressure-tight attachment to a vessel.

range, *n.*—the region between the limits within which a quantity is measured. It is expressed by stating the lower and upper range-values.

reference junction, *n.*—that junction of a thermocouple which is at a known temperature.

refractory metal thermocouple, *n.*—a thermocouple whose thermoelements have melting points above that of 60 percent platinum-40 percent rhodium, 1935°C (3515°F).

resistance, insulation (sheathed thermocouple wire), *n.*—the measured resistance between wires or between wires and sheath multiplied by the length of the wire expressed in megohms (or ohms) per foot (or meter) of length. (NOTE: The resistance varies inversely with the length.)

secondary reference points, *n.*—reproducible temperatures (other than the *defining fixed points*) listed in the text establishing the International Practical Temperature Scale as being useful for calibration purposes.

secondary standard thermocouple, *n.*—a thermocouple that has had its temperature-emf relationship determined by reference to a primary standard of temperature.

Seebeck coefficient, *n.*—the rate of change of thermal emf with temperature at a given temperature. Normally expressed as emf per unit of temperature. Synonymous with *thermoelectric power.*

Seebeck emf, *n.*—the net emf set up in a thermocouple under condition of zero current. It represents the algebraic sum of the Peltier and Thomson emf. Synonymous with *thermal emf.*

sheathed thermocouple, *n.*—a thermocouple having its thermoelements, and sometimes its measuring junction, embedded in ceramic insulation compacted within a metal protecting tube.

sheathed thermocouple wire, *n.*—one or more pairs of thermoelements (without measuring junctions(s)) embedded in ceramic insulation compacted within a metal protecting tube.

sheathed thermoelement, *n.*—a thermoelement embedded in ceramic insulation compacted within a metal protecting tube.

span, *n.*—the algebraic difference between the upper and lower range-values.

standard platinum resistance thermometer (SPRT), *n.*—a thermometer which meets all the requirements described in the text establishing the International Practical Temperature Scale.

standard thermoelement, *n.*—a thermoelement that has been calibrated with reference to platinum 67 (Pt-67).

test thermocouple, *n.*—a thermocouple that is to have its temperature-emf relationship determined by reference to a temperature standard.

test thermoelement, *n.*—a thermoelement that is to be calibrated with reference to platinum 67 (Pt-67) by comparing its thermal emf with that of a standard thermoelement.

thermal electromotive force (thermal emf), *n.*—the net emf set up in a thermocouple under conditions of zero current. Synonymous with *Seebeck emf.*

thermocouple, *n.*—two dissimilar thermoelements so joined as to produce a thermal emf when the junctions are at different temperatures.

thermocouple assembly, *n.*—an assembly consisting of a thermocouple element and one or more associated parts such as terminal block, connection head, and protecting tube.

thermocouple element, *n.*—a pair of bare or insulated thermoelements joined at one end to form a measuring junction and intended for use as a thermocouple or as part of a thermocouple assembly.

thermocouple, Type E, B, J, K, R, S, or T, *n.*—a thermocouple having an emf-temperature relationship corresponding to the appropriate letter-designated table in ASTM Standard E 230, Temperature-Electromotive Force (EMF) Tables for Thermocouples, within the limits of error specified in that Standard.

thermoelectric power, *n.*—the rate of change of thermal emf with temperature at a given temperature. Synonymous with *Seebeck coefficient*. Normally expressed as emf per unit of temperature.

thermoelectric pyrometer, *n.*—an instrument that senses the output of a thermocouple and converts it to equivalent temperature units.

thermoelement, *n.*—one of the two dissimilar electrical conductors comprising a thermocouple.

thermopile, *n.*—a number of thermocouples connected in series, arranged so that alternate junctions are at the reference temperature and at the measured temperature, to increase the output for a given temperature difference between reference and measuring junctions.

thermowell, *n.*—a closed end reentrant tube designed for the insertion of a temperature-sensing element, and provided with means for pressure-tight attachment to a vessel.

Thomson coefficient, *n.*—the rate at which heat is absorbed or evolved reversibly in a thermoelement, per unit temperature difference per unit current.

Thomson emf, *n.*—the product of the Thomson coefficient and the temperature difference across a thermoelement.

triple point (water), *n.*—the temperature of equilibrium between ice, water, and water vapor. This temperature is +0.01°C on the International Practical Temperature Scale of 1948.

upper range-value, *n.*—the highest quantity that an instrument is adjusted to measure.

working standard thermocouple, *n.*—a thermocouple that has had its temperature-emf relationship determined by reference to a secondary standard of temperature.

Index

L

M

N

O

P

R